Food Colloids and Polymers:
Stability and Mechanical Properties

Food Colloids and Polymers: Stability and Mechanical Properties

Edited by

E. Dickinson

Procter Department of Food Science, University of Leeds, UK

P. Walstra

Department of Food Science, Wageningen Agricultural University, The Netherlands

ROYAL
SOCIETY OF
CHEMISTRY

CHEM
seplae

The Proceedings of a Conference organized jointly by the Food Chemistry Group of the Royal Society of Chemistry and the Netherlands Society for Nutrition and Food Science, held at Lunteren, The Netherlands on 8–10 April 1992

Special Publication No. 113

ISBN 0-85186-325-6

A catalogue record of this book is available from the British Library.

Published by The Royal Society of Chemistry.
Thomas Graham House, Science Park, Cambridge CB4 4WF
Typeset by Keytec Typesetting Ltd, Bridport, Dorset
Printed by Redwood Press Ltd, Melksham, Wiltshire

Preface

Many important food products come under the category of food colloids and most of these contain polymers. This book describes recent developments in the experimental and theoretical investigation of model food colloids in terms of the properties of surface layers and the interactions between dispersed species (emulsion droplets, gas bubbles, or polymer particles). Emphasis is placed on understanding the microstructure and mechanical properties of solid and semi-solid materials, the stability and rheology of deformable particle systems, the dynamics of dispersions and fluid interfaces, and the adsorption and phase behaviour of mixed biopolymer systems.

This volume records the proceedings of an International Symposium on 'Food Colloids and Polymers' held at the De Blije Werelt Conference Centre, Lunteren, The Netherlands, on 8–10 April 1992. The conference was jointly organized by the Food Chemistry Group of the Royal Society of Chemistry (UK) and the Netherlands Society for Nutrition and Food Science, a subsidiary of the Royal Netherlands Chemical Society. The conference was the fourth in a series of biennial Spring Symposia on the subject of food colloids to be organized by the Food Chemistry Group of the Royal Society of Chemistry. The proceedings of each of the previous meetings—in Leeds (1986), Bedford (1988), and Norwich (1990)—have been published by the Society. The Lunteren conference programme was built around a set of invited overview lectures on the key topics of aggregation phenomena, polymer adsorption, fracture mechanics, protein–polysaccharide interactions, dynamic interfacial properties, and the physical properties of starch. These lectures were complemented by contributed oral presentations and a poster exhibition. The meeting was attended by 152 people from fourteen different countries. This book collects together most of the invited and contributed papers, together with short papers based on half of the poster presentations.

How does fundamental research on food colloids relate to current trends in the European food manufacturing and distribution business? This rather challenging question was addressed in the closing remarks to the meeting by Professor Peter Richmond, until recently the Head of the Norwich Laboratory at the AFRC Institute of Food Research. After identifying those talks he had found most exciting over the previous two days, Professor Richmond speculated about some of the broader issues that those concerned with food quality would have to address over the next few years—microbiological, nutritional, environmental, psychological, and so on. It was the speaker's view that—under pressure from consumers, retailers, and legislators—it will become necessary for those investigating food colloids to give greater consideration than at present to factors other than just the physical (or physico-chemical) ones represented in this volume. Great strides have certainly been made in applying the techniques and principles of surface and colloid science to food systems. According to Professor Richmond, it is now time to explore whether this basic knowledge has any role in the tackling of wider problems, such as the inhibition of microbial growth, the digestion of food components, the controlled release of flavour, or the development of healthier functional ingredients. It is too far-fetched to imagine that such topics could form part of the next food colloids symposium?

We are very pleased to acknowledge the important contributions from the other members of the organizing committee, *viz.* Dr R. D. Bee, Professor B. H. Bijsterbosch, Dr A. C. Juriaanse, Professor A. Prins, Professor P. Richmond, and Ms Y. Smolders.

E. Dickinson (Leeds)
P. Walstra (Wageningen)
June 1992

Contents

Structure, Rheology, and Fracture Properties

Interfacial Phenomena

Aggregation Phenomena

Introduction to Aggregation Phenomena in Food Colloids

By Pieter Walstra

DEPARTMENT OF FOOD SCIENCE, WAGENINGEN AGRICULTURAL
UNIVERSITY, PO BOX 8129, 6700 EV WAGENINGEN, THE NETHERLANDS

1 Introduction

Many foods are colloidal systems, containing particles of various kinds. The particles may aggregate, *i.e.* stay very close to each other for a much longer time than would be the case in the absence of attractive forces between them. Such aggregation may determine the rheological properties and the appearance of the product, as well as its physical instability, as reflected in a change in consistency or a loss of homogeneity. Many foods also contain macromolecules (polymers) which may affect aggregation and its rate in various ways.

Whether aggregation occurs depends primarily on the interaction forces between the particles, the classical subject of colloid science. One should, however, be cautious in applying colloid theory to foods. Colloid science typically considers the interaction between two identical, homogeneous, hard spheres, whereas most foods contain many particles, varying in size, shape, heterogeneity, and deformability, and making up a considerable volume fraction. In most studies on the effects of polymers, fairly simple molecules are considered, *e.g.* uncharged block copolymers, whereas foods typically contain proteins. Moreover, hydrodynamic forces are often prominent during food processing, and many phenomena may occur simultaneously. Finally, the food may show chemical changes, for instance due to enzyme action.

Nevertheless, the application of colloid science maybe very useful. Ever more complicated situations are being studied and the increasing possibilities for the application of intricate model calculations is tending to bridge the gap between simple model systems and actual foods. It is essential, however, to first know what system the food is, what particles it contains, and in what manner they aggregate, before applying colloid theory.

2 Interaction Forces

DLVO Theory

The Deryagin–Landau–Verwey–Overbeek (DLVO) theory considers the free energy G needed to bring two particles from infinite distance apart to a close distance between their surfaces h.[1-3] There are two terms which are additive. One describes the electrostatic repulsion between the particles, perhaps better explained as being due to the local increase in osmotic pressure where the ion clouds around the charged particles overlap. The attractive term is due to the ubiquitous van der Waals attractions between identical molecules, and hence particles. For two identical homogeneous spheres of radius a, we have for conditions as are common in food colloids (*e.g.* in water at room temperature)

$$G_R \approx 4.3 \times 10^{-9} a\ \psi_0^2 \ln (1 + e^{-\kappa h}) \tag{1}$$

and

$$G_A \approx - Aa/12h, \tag{2}$$

if expressed in SI units. Here ψ_0 is the surface potential of the particles, often taken as the more experimentally accessible, electrokinetic potential. κ is the Debye–Hückel parameter, or the inverse of the thickness of the electric double layer; in the present case it equals about $3.3 \sqrt{I}$ in nm, where I is the total ionic strength (molar). A is the Hamaker constant, which depends on the material of the particles and that of the interstitial liquid; tabulated values are available.[3]

One may thus calculate the interaction free energy from known or determinable parameters. An example is given in Figure 1a. If the minimum near C is deep compared to the average kinetic energy involved in the encounter of two particles, kT, the particles tend to aggregate, *i.e.* stay together at the corresponding value of h (aggregation in the secondary minimum). If the maximum near B is not large compared to kT, two particles may occasionally move over this energy barrier and become aggregated near A (in the primary minimum). Lowering the surface potential—*e.g.* by altering the pH—or increasing the ionic strength diminishes the electrostatic repulsion and thereby promotes aggregation. The Hamaker constant can usually not be manipulated.

The DLVO theory has been fairly successful in predicting the aggregation stability of inorganic colloids, except for the prediction of a considerable effect of particle size, which is mostly not observed. Moreover, the theory rarely holds at very small distances, say $h < 3$ nm, because of surface unevenness and because the presence of adsorbed material cannot easily be accounted for: it may cause additional repulsion (see below) and it interferes with the determination of the surface potential. Consequently, the DLVO theory is rarely exact for food colloids, although it often predicts trends fairly well.

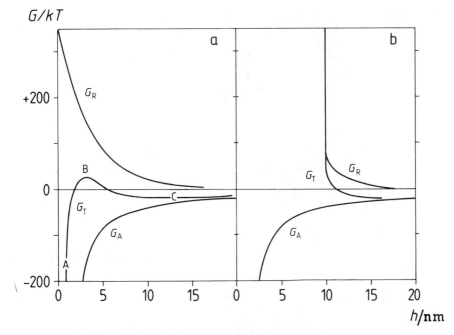

Figure 1 *Calculated examples of the repulsive (G_R), attractive (G_A), and total interaction free energy (G_T) as a function of surface separation distance* h *of two identical spheres: (a) electrostatic repulsion and van der Waals attraction (DLVO); (b) steric repulsion and van der Waals attraction*

Roles of Macromolecules

Polymers present in the continuous (usually aqueous) phase may adsorb onto the particles. If not, the polymers mostly cause the viscosity of the liquid to be higher, thereby slowing down any aggregation, or even causing a weak gel to be formed, thus preventing aggregation. Dissolved polymers may also, on the other hand, cause aggregation by depletion flocculation; see below. Adsorbed polymers may either prevent aggregation: steric stabilization; or cause it: bridging flocculation. The theory has now been fairly well developed and has been proved to be useful. We refer to a recent review by Fleer and Scheutjens.[4]

Figure 2 schematically shows how macromolecules may adsorb (or be grafted) onto surfaces. Whether adsorption occurs greatly depends on the solubility of the polymer. In this respect, it is enlightening to consider the osmotic pressure Π of a polymer solution. It can be expressed by

$$\Pi = RT \left(c_m + B \, c_m^2 + \ldots \right) \tag{3}$$

where R is the gas constant, c_m is the molar concentration, and B is the second virial coefficient, which can be fairly easily determined. Schematic

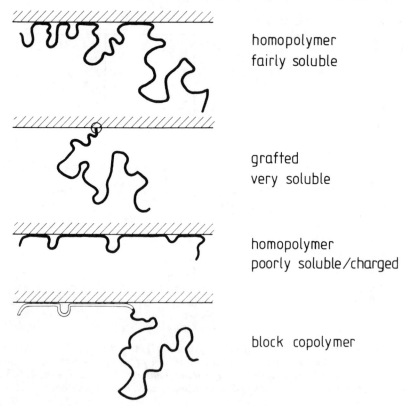

homopolymer
fairly soluble

grafted
very soluble

homopolymer
poorly soluble/charged

block copolymer

Figure 2 *Types of protruding macromolecules (highly schematic)*

examples are shown in Figure 3. For a good solvent (the upper curve refers
to xanthan in water), adsorption is very unlikely; for so called theta
conditions, it is likely; and for poor solvents, it is all but certain. To assure
adsorption, on the one hand, and a considerable protrusion of the
macromolecule into the solvent, on the other hand, (block) copolymers are
often used: part of the macromolecule is poorly soluble and adsorbs, and
another part is highly soluble and sticks out. For charged macromolecules
the relations are more complicated.[5] Proteins may adsorb in many ways:
almost unchanged, with some change in conformation, or almost fully
unfolded.

Steric Stabilization

Figure 4 attempts to explain the two mechanisms involved in steric
stabilization. When a surface approaches protruding macromolecules, the
latter become restricted in their freedom of motion; hence, a decrease
in entropy; hence, a repulsive free energy. This volume restriction free
energy term is always positive, *i.e.* it causes strong repulsion (unless the

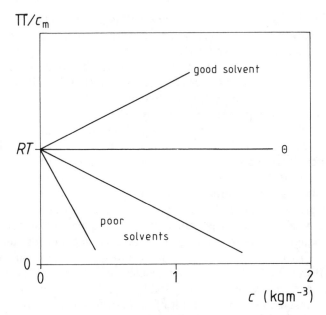

Figure 3 *Osmotic pressure divided by molar concentration* Π/c_m *versus mass concentration* c *of polymer solutions for various kinds of solvent quality. (θ denotes a theta solvent.)*

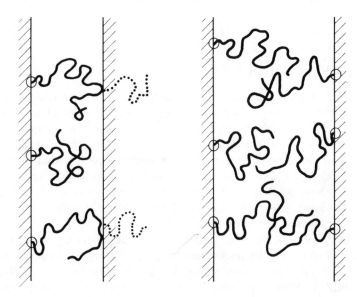

Figure 4 *Mechanisms of steric repulsion by protruding macromolecules (highly schematic): volume restriction (left) and mixing (right)*

macromolecule adsorbs onto the other surface; see below. Mere volume restriction, however, will rarely occur, since the other surface mostly bears macromolecules as well. Now the concentration of macromolecular matter increases as the polymer layers overlap and this may induce a local increase in osmotic pressure, causing repulsion (this is called the mixing term). Whether or not repulsion occurs primarily depends on the second virial coefficient in equation (3); see also Figure 3. If Π/c increases with concentration, there will be strong repulsion; if it markedly decreases, attraction may occur. (NB This treatment of the effect of the solubility of the polymer is to some extent an oversimplification, but it serves to illustrate the salient points.)

An example of the interaction free energy as a function of separation distance is given in Figure 1b. Mostly, repulsion increases very steeply with decreasing h, almost as a step function. Whether the secondary minimum is deep enough to allow aggregation primarily depends on the ratio of particle size to protrusion distance: large particles (strong van der Waals attraction) and fairly small macromolecules are associated with a deeper minimum, and hence with lower stability. Also a decrease in solvent quality, for instance by the addition of ethanol to an aqueous system, may lead to aggregation.

Steric stabilization is often of considerable importance in food colloids. Calculation of the interaction free energy is rarely possible, but a reasonable estimate of the protrusion distance of the macromolecules can often be made, thereby allowing one to roughly predict whether steric stabilization may be possible. Adsorbed proteins mostly provide stability against aggregation by the combined effect of both electrostatic and steric repulsion.

Bridging Flocculation

Adsorbed homopolymers between closely approaching particles will always make bridges (*i.e.* single molecules become adsorbed simultaneously onto two surfaces) if equilibrium is attained.[4] Equilibrium is, however, mostly not attained, because the time needed for reaching that conformation is much longer than the time during which particles in Brownian motion are close to each other. Nevertheless, one would theoretically expect bridging to occur when particles are close together for a long time, as in a sedimented layer. Even this is not widely observed in practice, presumably because copolymers are often used.

Bridging flocculation does occur if particles covered with adsorbed polymer are mixed with uncovered particles.[6] A prerequisite is that the concentration of the polymer in the solvent is very low, a condition that can be fulfilled because of the very high surface activity of many polymers. Consequently, the method of processing may determine, along with composition, whether bridging flocculation will actually occur.

Depletion Flocculation

As depicted in Figure 5, non-adsorbing macromolecules with a radius of gyration R_g leave a layer of approximate thickness R_g around any particle depleted of macromolecules. This causes the osmotic pressure of the system to be higher than it would be in the absence of particles. If now the particles come close to each other, the volume of solvent depleted is decreased; hence, a lower osmotic pressure; hence, a decrease in free energy; hence, a driving mechanism for flocculating the particles. For two hard spheres of radius a, the interaction free energy is given by[7]

$$G_{dep} = -\pi R_g^2 (2a + 4R_g/3) \, \Pi \, f(h) \qquad (4)$$

where Π is given by equation (3) and the function $f(h)$ decreases from 1 at zero separation distance to 0 for $h > 2R_g$. Depletion flocculation may thus occur, especially if R_g is large (high molar mass, good solvent) and if Π is high; the latter implies that, because of the high molar mass, the second virial coefficient B must be high—*i.e.* there is, again, a good solvent. The free energy minimum may become much deeper if the particles slightly flatten on close contact.[8]

Xanthan can in fairly low concentrations cause depletion flocculation,[9,10] for instance inducing rapid creaming in emulsions, but at somewhat higher concentration it may slow down creaming (and possibly aggregation), because of the increased apparent viscosity, especially at very low velocity gradients.[10] For several polysaccharides added to foods, it may be very difficult to determine whether they cause aggregation by (weak) bridging or by depletion.[11]

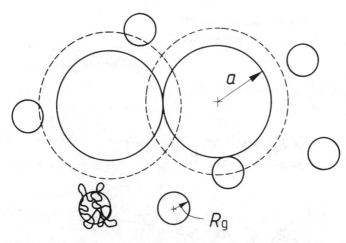

Figure 5 *Schematic explanation of the depletion of macromolecules, radius of gyration* R_g, *from the solution near spherical particles, radius* a, *and of the resulting depletion flocculation*

Other Aggregation Mechanisms

Figure 6 summarizes the interparticle region in various kinds of aggregates. Besides those already mentioned, one may distinguish the following mechanisms.

    (i) *Bridging by particles*, e.g. casein micelles between emulsion droplets ('homogenization cluster'),[12] fat crystals between emulsion droplets ('partial coalescence'),[12,13] and milk fat globules between air bubbles (whipped cream).[12]

    (ii) *Liquid necks*, e.g. tiny aqueous droplets between cocoa particles in cocoa mass (melted chocolate), which considerably enhance its consistency. Partial coalescence can also be envisaged as aggregation due to liquid necks, at least in some systems.

    (iii) *Cross-linking of adsorbed macromolecules*. Steric repulsion caused by the latter does not imply that the macromolecules cannot touch each other. If protein-covered particles are touching under conditions favouring cross-linking reactions (for instance during intensive heating), aggregation of the particles may readily occur. This happens, for instance, during the heat coagulation of homogenized cream.[14]

The diversity of this list conveys a warning: when aggregation phenomena occur in a food system, one should always try to find out first what the mechanism is before applying the theory. It also follows that the simple classical division of aggregation phenomena into flocculation (caused by polymers) and coagulation (caused by salt) makes little sense for food colloids.

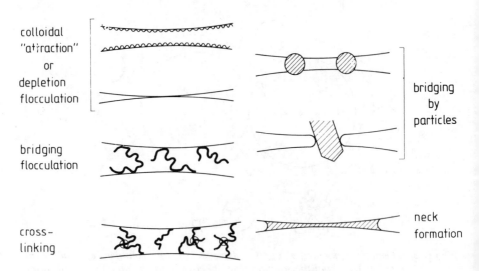

Figure 6 *Various kinds of particle aggregation (highly schematic)*

3 Consequences of Aggregation

The unhindered aggregation of particles in Brownian motion leads to the formation of fractal flocs. Such a floc is characterized by a relation between the number of particles n in the floc and its radius R,

$$n = (R/a)^D \qquad (5)$$

where a is the radius of the primary particles. The fractal dimensionality D is always < 3 (mostly $1.7-2.4$), implying that the flocs become ever more tenuous as they grow. Consequently, at a certain stage, the flocs take up the whole system volume and a gel forms. This has been discussed before.[15] The aggregation leads to a gel (even if weak) rather than to a coagulate, making the system more stable to sedimentation, rather than less so. This has also been discussed before.[16]

If a rearrangement of aggregated particles occurs as depicted in Figure 7 (top), a coagulate is formed. Very subtle differences may determine which structure is the result of aggregation in a quiescent system, and, although some determinant conditions have been identified,[17] this aspect needs more study, especially because of its great practical importance. If the liquid is stirred during aggregation, as is often the case when aggregation proceeds very fast, a coagulate is commonly formed; the fractal dimensionality then is almost 3. If the particles coalesce directly after aggregation, we have $D = 3$.

Figure 7 depicts some other forms of rearrangement that may occur after aggregation. Local repositioning is common for irregularly shaped particles,

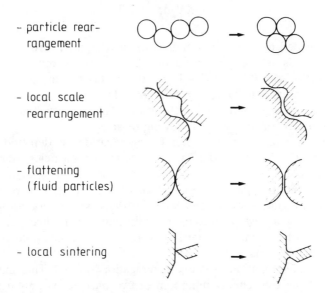

- particle rear-
  rangement

- local scale
  rearrangement

- flattening
  (fluid particles)

- local sintering

Figure 7 *Examples of bond strengthening mechanisms of aggregated particles (highly schematic)*

local flattening occurs with deformable ones, and sintering especially occurs with particles consisting of a material having a finite solubility in the continuous phase, *i.e.* with most crystals. All of these rearrangements lead to bond strengthening, implying that it becomes more difficult to redisperse the aggregates, or that the gel formed becomes stiffer and, mostly, stronger and shorter. To the examples shown in Figure 7, we may add slow polymer bridging and chemical cross-linking (Figure 6). All of these changes may also occur with particles residing in a sedimented layer, irrespective of any previous aggregation.

4 Aggregation Kinetics

Smoluchowski Theory

According to Smoluchowski the encounter frequency J ($m^{-3} s^{-1}$) of equal sized spherical particles due to Brownian motion is given by[18]

$$J = 4 \, kT \, N^2/3 \, \eta \tag{6}$$

where N is the particle concentration and η is the solvent viscosity. (Note that the particle radius is not involved in the equation: its effects on collision radius and diffusion rate cancel.) Assuming that any aggregate of particles is counted as one particle (of the same original radius), the so-called perikinetic aggregation rate is given

$$- \, dN/dt = J/W \tag{7}$$

The time needed to halve the number of particles then turns out to be *ca.* $d^3/10 \, \phi$ seconds if the particle diameter d is measured in μm. For a volume fraction $\phi = 0.1$ and $d = 1 \, \mu$m, this yields a value of 1 s, provided that all encounters lead to aggregation ($W = 1$).

In many cases aggregation is much slower. The stability factor W may be larger than unity for any of the following reasons.

  (i) There is a free energy barrier for aggregation, as depicted in Figure 1a, near separation B. This is typical for particles stabilized by electrostatic repulsion.

 (ii) Only a limited part of the surface of the particles is reactive (*i.e.* there are 'hot spots'). This may be the case for emulsion droplets containing a few protruding crystals that can induce partial coalescence, or for particles partly covered with protein, and partly with other surfactants.

(iii) Besides aggregation, some disaggregation occurs. This may be the case if the free energy minimum of two particles is only a few times kT, *e.g.* for flocculation in the secondary minimum (see Figure 1) or for depletion flocculation. This aspect has been insufficiently studied.

Complications

Smoluchowski's theory for perikinetic aggregation, equations (6) and (7), has been observed to be reasonably valid in several fairly simple model systems, albeit merely for the first few aggregation steps. In practice, large deviations are often observed. Possible causes for deviations have been extensively studied by Bremer,[17] and some highlights of his results will be mentioned here.

The first question to be asked is to what stage aggregation has to proceed to induce a perceptible change, since that is the crucial point in practice. It is often (tacitly) assumed that the time needed for such a change is a few times that needed to halve the number of particles, *i.e.*

$$t_{0.5} = 3 \, \eta \, W / 4 \, kT \, N \qquad (8)$$

for perikinetic aggregation. Often, $t_{0.5}$ is even designated as the coagulation time. This is very misleading. Let us assume that the first perceptible change is the emergence of visible particles, as may occur when the aggregated particles immediately coalesce or rearrange into compact flocs. We define a radius R_{vis} at which these flocs can be seen, say 0.2 mm, and obtain the corresponding time

$$t_{vis} = \pi \, R_{vis}^3 \, \eta \, W / 8 \, kT \, \phi \qquad (9)$$

If fractal aggregation occurs, the first perceptible change is usually the formation of a gel. Now we obtain for perikinetic aggregation a gelation time

$$t_{gel} = \pi \, a^3 \, \eta \, W \, \phi^{3/(D-3)} / kT \qquad (10)$$

which may be smaller than t_{vis} by several orders of magnitude (see Figure 8). Under some conditions (large primary particles, large density difference, small ϕ) the first perceptible change may be a visible separation into layers due to sedimentation of the growing aggregates.

There are more complications in the case of fractal aggregates. (i) The particles, and certainly the aggregates, are polydisperse, implying that aggregation proceeds faster than predicted. (ii) The collision radius of the aggregates may become markedly larger than the hydrodynamic (or diffusion) radius, thereby, again, causing faster aggregation. (iii) The volume fraction of the effective particles (*i.e.* fractal aggregates) may not any more be neglected (as was done by Smoluchowski) since it always becomes large; this may reduce the gel time by at least an order of magnitude.

Last but not least, the effects of velocity gradients in the system must be taken into account. Smoluchowski derived for the encounter frequency at a shear rate S:

$$J_S = 2 \, d^3 \, N^2 \, S \qquad (11)$$

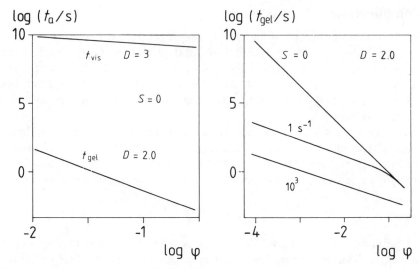

Figure 8 *Calculated examples of aggregation times* t_a *(either* t_{vis} *or* t_{gel}*) of spherical particles, radius 0.1 μm, in water at various volume fractions* φ. *Other variables: fractal dimensionality* D = 3 *(particles coalesce) or* 2.0 *(fractal flocculation); velocity gradient S. Sedimentation is neglected. (See text for further details.)*

This implies that, in water at room temperature, the ratio of this orthokinetic aggregation rate over that for perikinetic aggregation is $0.2\, d^3 S$, where d is in μm. Consequently, for shear-rates as small as $0.1\,\text{s}^{-1}$, which easily develop due to slight temperature fluctuations, orthokinetic aggregation becomes the determining factor for aggregates larger than about 4 μm. This will tend to occur quite often in practice; some results are shown in Figure 8.

There is much more to be said about these various aspects, and a full discussion is given elsewhere.[17]

5 Some Conclusions

(i) Aggregation phenomena largely determine the consistency, the appearance, and the physical stability of several types of food colloids.

(ii) Most food systems are too complicated for the direct application of the theories of elementary colloid science, although trends are often predicted correctly. The use of advanced numerical methods may greatly extend the applicability of colloid theory. However, one should first establish what is the actual aggregation mechanism.

(iii) Changes occurring after the primary formation of aggregates such as deflocculation, rearrangement of aggregate structure, and 'sintering' need more study. These changes may determine whether either a coagulate or a gel is formed, and the nature of their mechanical properties.

(iv) The theory of fractal aggregation is a powerful tool for describing aggregation and its consequences.
(v) Aggregation kinetics is far more intricate than suggested by simple perikinetic Smoluchowski theory. In particular, the formation of fractal aggregates and the presence of small velocity gradients may affect (by some orders of magnitude) the time needed for aggregation to become perceptible.

References

1. 'Colloid Science. Vol. I. Irreversible Systems', ed. H. R. Kruyt, Elsevier, Amsterdam, 1952.
2. E. Dickinson and G. Stainsby, 'Colloids in Foods', Applied Science, London, 1982.
3. J. Lyklema, 'Fundamentals of Interface and Colloid Science. Vol. I. Fundamentals', Academic Press, London, 1991.
4. G. J. Fleer and J. M. H. M. Scheutjens, in 'Coagulation and Flocculation: Theory and Applications', ed. B. Dobias, Dekker, New York, 1992.
5. G. J. Fleer, in 'Food Polymers, Gels and Colloids', ed. E. Dickinson, Special Publication No. 82, Royal Society of Chemistry, Cambridge, 1991, p. 34.
6. G. J. Fleer and J. Lyklema, *J. Colloid Interface Sci.*, 1974, **46**, 1.
7. A. Vrij, *Pure Appl. Chem.*, 1976, **48**, 471.
8. I. D. Evans and A. Lips, this volume, p. 214.
9. A. Lips, I. J. Campbell, and E. G. Pelan, in 'Food Polymers, Gels and Colloids', ed. E. Dickinson, Special Publication No. 82, Royal Society of Chemistry, Cambridge, 1991, p. 1.
10. H. Luyten, M. Jonkman, W. Kloek, and T. van Vliet, this volume, p. 224.
11. B. Bergenståhl, in 'Gums and Stabilizers for the Food Industry', eds. G. O. Phillips. P. A. Williams, and D. J. Wedlock, IRL Press, Oxford, 1988, Vol. 4, p. 363.
12. H. Mulder and P. Walstra, 'The Milk Fat Globule. Emulsion Science as Applied to Milk Products and Comparable Foods', Pudoc, Wageningen, 1974.
13. K. Boode and P. Walstra, this volume, p. 23.
14. P. Walstra, *J. Dairy Sci.*, 1990, **73**, 1965.
15. P. Walstra, T. van Vliet, and L. G. B. Bremer, in 'Food Polymers, Gels and Colloids', ed. E. Dickinson, Special Publication No. 82, Royal Society of Chemistry, Cambridge, 1991, p. 369.
16. T. van Vliet and P. Walstra, in 'Food Colloids', eds. R. D. Bee, P. Richmond, and J. Mingins, Special Publication No. 75, Royal Society of Chemistry, Cambridge, 1989, p. 206.
17. L. G. B. Bremer, 'Fractal Aggregation in Relation to Formation and Properties of Particle Gels', Ph.D. Thesis, Wageningen Agricultural University, 1992.
18. J. T. G. Overbeek, in 'Colloid Science, Vol. I. Irreversible Systems', ed. H. R. Kruyt, Elsevier, Amsterdam, 1952, p. 278.

The Fractal Nature of Fat Crystal Networks

By R. Vreeker, L. L. Hoekstra, D. C. den Boer, and W. G. M. Agterof

UNILEVER RESEARCH LABORATORIUM, OLIVIER VAN NOORTLAAN 120, 3133 AT VLAARDINGEN, THE NETHERLANDS

1 Introduction

In many foods, the desired product texture is provided by a network of aggregated fat particles. As a consequence, the rheology of aggregated fat dispersions has received much attention in the past. Attempts to describe the rheological behaviour quantitatively, however, have only been partially successful, as a result of the complex and random structure of the fat crystal aggregates.

Recently, considerable progress in describing the structure of aggregated particles has been made using the concept of fractals.[1] In this approach, the complex structure of aggregates is characterized by a single fractal dimension D which describes a relation between the number N of particles in the aggregate and its typical radius R, i.e. $N \sim R^D$. The higher the value of D, the more compact is the aggregate structure. Fractal growth models have been successfully studied for two limiting regimes of fractal aggregation, namely, diffusion-limited (or fast) aggregation, and reaction-limited (or slow) aggregation. In the diffusion-limited regime, aggregates are characterized by $D = 1.7-1.8$. Aggregates grown in the reaction-limited regime are characterized by a higher value of D, i.e. $D = 2.0-2.1$.

The fractal nature of aggregates has important consequences for the rheology of dispersions. For example, when a three-dimensional aggregate network is formed, fractal theory predicts $G' \sim \phi^\mu$, where G' is the elastic modulus, ϕ is the particle volume fraction, and μ is a constant depending on D.[2-7] A similar power-law relation is predicted for the variation of the yield stress with particle concentration. Also, the flow properties (i.e. the viscosity versus shear-rate behaviour) of dispersions are influenced by the fractal structure of the aggregates.[8,9]

In this paper the fractal nature of fat crystal aggregates is demonstrated by means of light scattering. Light scattering is a very powerful technique for studying fractal structures in dilute suspension.[10] It can be shown that, for not too large aggregates, the scattering intensity is related to the scattering vector q through the simple relation $I(q) \sim q^{-D}$, where $q = (4\pi/\lambda)\sin(\theta/2)$, and λ is the wavelength of light in solution, and θ is the scattering angle. This power-law relation provides a simple and accurate experimental method of determining D. An interpretation of rheological data for aggregated fat crystal networks can be given in the framework of fractal theory. It is shown here that the elastic modulus G' and the yield stress σ_y both vary with particle concentration according to a power-law. In addition, an analysis is given of the flow properties of aggregated fat crystal dispersions in a simple shear field.

2 Experimental

Dispersions containing 0.5 wt % glycerol tristearate (BDH, 95% pure) in olive or paraffin oil were prepared by crystallization from a melt. Crystallization was induced by rapid cooling of the melt from 90 °C to 2 °C. At 2 °C, glycerol tristearate crystals are formed in the crystallographic α-modification. This modification is thermodynamically unstable and changes rapidly into the stable β-modification. After completion of the crystallization process, the temperature of the dispersion was raised to 25 °C. At this temperature rapid aggregation of fat crystals could be observed. To obtain information concerning the internal structure of the aggregates, the dispersion was transferred to an optical cell for light-scattering experiments. The light source in the experiments was a 10 mW He–Ne laser operating at a wavelength of 633 nm. Viscosity *versus* shear-rate measurements were made using an Instron Rheometer with concentric cylinders (low shear-rates), a Haake viscometer with concentric cylinders (medium shear-rates), and a Ferranti viscometer with cone-and-plate geometry (high shear-rates).

3 Results and Discussion

Figure 1 shows the scattered light intensity for glycerol tristearate aggregates in olive oil as a function of scattering vector. A logarithmic plot of $I(q)$ *versus* q is seen to yield a straight line, indicating the fractal nature of the fat aggregates.[10] The fractal dimension as determined from the slope of the plot is $D = 1.7$. The same value for D was found for glycerol tristearate aggregates in paraffin oil. The low value of D is characteristic of aggregates with a very open structure. Aggregates with $D = 1.7-1.8$ have been observed previously for other colloidal systems and the structure is consistent with results from computer simulations of diffusion-limited aggregation.[1] Diffusion-limited aggregation is anticipated when the aggregation process is dominated by attractive forces. In the case that repulsive forces are also important, reaction-limited aggregation results, with $D = 2.1$. Repulsive forces can frequently result from electric charges at the surface of the aggregating particles. For fat particles in a non-polar

scattering intensity

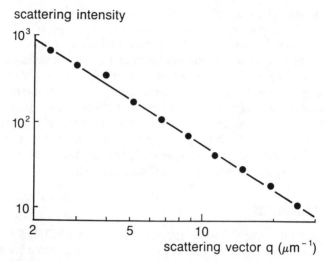

Figure 1 *Scattered light intensity as a function of the scattering vector q for a dilute dispersion of glycerol tristearate in olive oil. Drawn line is a best fit to the data according to* $I \sim q^{-D}$ *with* $D = 1.7$

medium, however, electric surface charges are expected to be small, and thus diffusion-limited aggregation is anticipated, in line with the experimental results.

Aggregates with a low fractal dimension are sensitive to spontaneous restructuring or ageing effects. For fat crystal dispersions, ageing occurs as a result of slow recrystallization processes, in which large crystals grow at the expense of smaller ones.[11] To see whether these recrystallization processes also influence the structure of fat aggregates, fractal dimensions were measured as a function of time. As is seen from Figure 2, *D* increases from 1.7 to 2.0 over several days of storage, indicating that the fat aggregates gradually become more compact. It is worth noting that the fractal nature of the aggregates is still conserved during the recrystallization process.

The experimental results described above have been obtained for dilute dispersions. When the volume fraction of fat particles is increased well above a certain level, formation of a volume filling network can be observed. The network can be considered as a collection of fractal aggregates closely packed throughout the sample. The question of the elastic properties of a network of aggregates has been addressed recently.[2–7] In the approach followed in reference 6, based on the work of Brown[3] and Kantor and Webman,[2] it is assumed that the network properties depend on the elasticity of the individual aggregates, which is dominated by the effective backbone of the aggregate. By approximating the backbone as a linear chain of springs, it can be shown that the elastic constant decreases with increasing aggregate size. Since the aggregate size is related to the particle volume fraction as $R \sim \phi^{1/(D-3)}$,[7] it follows that

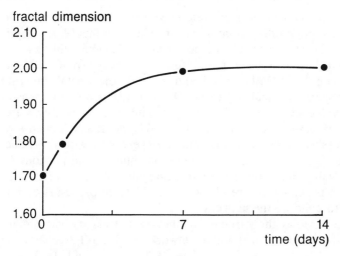

Figure 2 *Fractal dimension of glycerol tristearate aggregates as a function of time after initiation of the aggregation process. Drawn line is simply a guide to the eye*

the elastic constant of the flocs increases with increasing particle concentration. For the elastic modulus of the aggregate network, we obtain $G' \sim \phi^{\mu}$, with $\mu = (3 + x)/(3 - D)$ and x equal to the backbone fractal dimension, which varies between 1 and 1.3.[6]

Figure 3 shows elastic modulus as a function of weight fraction (at constant crystal size) for a network of glycerol tristearate crystals in

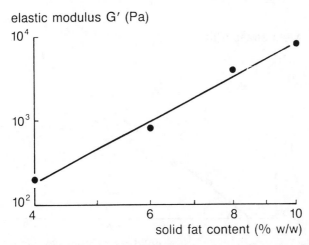

Figure 3 *Elastic modulus G' for glycerol tristearate in paraffin oil at different weight fractions as determined from small-amplitude oscillatory shear measurements (amplitude of deformation = 10^{-4}; frequency of modulation = 7 Hz). (Data are taken from reference 12.) Drawn line is a best fit to the data according to $G' \sim \phi^{\mu}$ with $\mu = 4.1$*

paraffin oil (data taken from reference 12). The moduli were determined from small-amplitude oscillatory shear measurements. A log–log plot of G' *versus* ϕ clearly demonstrates power-law behaviour with $\mu = 4.1$. Similar behaviour has also been found for silica, polystyrene, clay, and whey protein gels.[4-6,13] With $\mu = 4.1$ and $x = 1.3$, the fractal dimension of the fat aggregates is calculated to be $D = 2.0$. (D is not very sensitive to the precise value of x, which is unknown, and it changes only a few percent when x is varied between 1.0 and 1.3). The fractal dimension deduced from the shear measurements compares very well with the value deduced from the light-scattering experiments on dilute fat dispersions (*cf.* Figure 2). Its value indicates that the aggregates have a relatively compact structure, probably as a result of recrystallization processes during the time needed to build up the network.

Figure 4 shows the variation of the yield stress σ_y with weight fraction (at constant crystal size) for a network of glycerol tristearate crystals in paraffin oil (data taken from reference 14). The yield stress was taken as being the maximum of a stress–strain plot when subjecting the dispersion to a slow deformation in simple shear. From the log–log plot of Figure 4 it can be seen that σ_y conforms well to a power law, $\sigma_y \sim \phi^\nu$, with $\nu = 3.6$, over a range of approximately 4 decades. An interesting feature of the experimental results is that this power-law behaviour persists up to high volume fractions. The power-law index for the yield stress is somewhat lower than that for the elastic modulus, which is consistent with previous experimental findings.[4]

Figure 5 shows the flow behaviour (shear viscosity η *versus* shear-rate) of a 5 wt % glycerol tristearate dispersion in a simple shear field. The

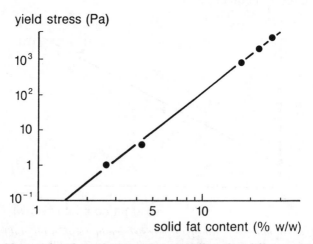

Figure 4 *Yield stress for glycerol tristearate in paraffin oil at different weight fractions as determined from stress–strain curves recorded at low shear rate (0.000553 s^{-1}). (Data are taken from reference 14.) Drawn line is a best fit to the data according to $\sigma_y \sim \phi^\nu$ with $\nu = 3.6$*

viscosity (Pa s)

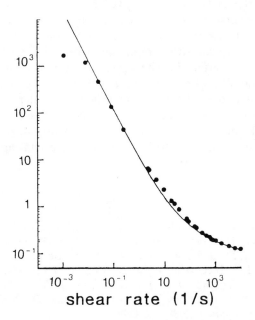

shear   rate   (1/s)

Figure 5 *Viscosity* versus *shear-rate dependence for a* 5 wt % *glycerol tristearate dispersion in paraffin oil. Drawn line is calculated as explained in the text*

viscosity decreases with increasing shear-rate due to breakup of the fat crystal aggregates into smaller fragments. For fractal aggregates it can be shown that the radius decreases with shear-rate according to a power law, $R \sim \gamma^{-m}$, where m depends on D and μ.[15] To calculate the flow properties of a dispersion, the effective volume fraction of aggregates, ϕ_{eff}, has to be considered instead of the real volume fraction ϕ. The quantity ϕ_{eff} is related to the radius of the aggregates by $\phi_{eff} = \phi(R/a)^{(3-D)}$. The aggregate volume fraction thus decreases with increasing shear-rate, which leads to a decrease in viscosity. When it is assumed that viscosity and volume fraction are related by the well-known Krieger–Dougherty relation, the following result is obtained:[8,9]

$$\eta/\eta_0 = [(\sqrt{((k/3)^3 + 1/4)} + 1/2)^{1/3} - (\sqrt{((k/3)^3 + 1/4)} - 1/2)^{1/3}]^{-\mu}$$

$$(1)$$

where $k = (\phi/\phi_m)(2.5\eta_0\gamma/\sigma_m)^{-1/\mu}$, ϕ_m and σ_m represent the volume fraction and yield stress of a system of closely packed particles, and η_0 is the solvent viscosity. Equation 1 is obtained under the condition $\mu = (15/2)\phi_m$. The drawn line in Figure 5 is a calculated viscosity *versus* shear-rate curve with $D = 2.0$, $\mu = 4.1$ (*cf.* Figure 3), $\phi = 0.05$, $\phi_m = 0.56$, $\eta_0 = 70$ mPa s, and $\sigma_m = 7.1 \times 10^5$ Pa. The value for σ_m is comparable to that found for

$Mg(OH)_2$ and $Fe(OH)_2$ dispersions $(3.2 \times 10^4$ and 1.0×10^4 Pa, respectively[9]). The theory is seen to give a very good description of the experimental results over a large range of shear-rates.

In summary, light-scattering measurements on dilute dispersions have demonstrated the fractal structure of aggregated glycerol tristearate crystals in olive oil and paraffin oil. The aggregates retain their fractal nature in concentrated fat dispersions, where a volume filling network is formed. Elastic modulus and yield stress show a power-law dependence on particle volume fraction, which can be explained within the framework of fractal network theory. The viscosity *versus* shear-rate dependence can be explained when the fractal structure of the fat crystal aggregates is taken into account.

References

1. For a review see, *e.g.*: P. Meakin, *Adv. Colloid Interface Sci.*, 1988, **28**, 249.
2. Y. Kantor and I. Webman, *Phys. Rev. Lett.*, 1984, **52**, 1891.
3. W. D. Brown, Ph.D. Thesis, University of Cambridge, 1987.
4. R. Buscall, I. J. McGowan, P. D. A. Mills, R. F. Stewart, D. Sutton, L. R. White, and G. E. Yates, *J. Non-Newtonian Fluid Mech.*, 1987, **24**, 183.
5. R. Buscall, P. D. A. Mills, J. W. Goodwin, and D. W. Lawson, *J. Chem. Soc., Faraday Trans. 1*, 1988, **84**, 4249.
6. W. H. Shih, W. Y. Shih, S. I. Kim, J. Liu, and I. A. Aksay, *Phys. Rev. A*, 1990, **42**, 4772.
7. L. G. B. Bremer, T. van Vliet, and P. Walstra, *J. Chem. Soc., Faraday Trans. 1*, 1989, **85**, 3359.
8. A. A. Potanin and N. B. Uriev, *J. Colloid Interface Sci.*, 1991, **142**, 385.
9. A. A. Potanin, *J. Colloid Interface Sci.*, 1991, **145**, 140.
10. See, *e.g.* J. Teixeira, in 'On Growth and Form', ed. H. E. Stanley and N. Ostrowsky, Nijhoff, Dordrecht, 1986, p. 145.
11. M. Knoester, P. De Bruyne, and M. van den Tempel, *J. Cryst. Growth*, 1968, **3,4**, 776.
12. J. M. P. Papenhuijzen, *Rheol. Acta*, 1971, **10**, 493.
13. R. Vreeker, L. L. Hoekstra, D. C. den Boer, and W. G. M. Agterof, *Food Hydrocolloids*, 1992, in press.
14. L. L. Hoekstra and M. van den Tempel, unpublished results.
15. R. C. Sonntag and W. B. Russel, *J. Colloid Interface Sci.*, 1987, **115**, 378.

Kinetics of Partial Coalescence in Oil-in-Water Emulsions

By K. Boode* and P. Walstra

DEPARTMENT OF FOOD SCIENCE, WAGENINGEN AGRICULTURAL
UNIVERSITY, PO BOX 8129, 6703 HD WAGENINGEN, THE NETHERLANDS

1 Introduction

In the absence of fat crystals, most oil-in-water emulsions containing droplets of not more than a few microns in diameter are fairly stable to coalescence at rest and in a flow field.[1] In many emulsions of practical interest, however, part of the oil may crystallize at a certain temperature. Strictly speaking, these emulsions have changed into three-phase systems; but as they originate from true oil-in-water (O/W) emulsions, we still call them emulsions. The presence of crystals in the fat globules of an O/W emulsion can enhance tremendously the instability of the emulsion.[2-4] The presence of a crystal structure can prevent complete coalescence of oil globules if the solid network is strong enough to resist the capillary forces involved. Instead of the energetically favoured spherical shape, clumps are formed of irregular size and shape.[5] The process is called partial coalescence.

Van Boekel and Walstra hypothesized[1] that coalescence may be caused by fat crystals at the surface of an oil globule (Figure 1). When two globules approach each other, a thin aqueous film usually remains between them due to the action of repulsion forces. The stronger the repulsion, the thicker the aqueous film. Crystals in the interface stick somewhat out of the globule, the distance depending on crystal size and shape, and also on the wetting properties (*i.e.* the contact angle). On approach of a second globule, such a crystal if protruding far enough may pierce the thin film between globules. If the globules orbit around each other, as frequently occurs in a shear flow, the probability that a protruding crystal will breach the film between two globules is greatly enhanced. As soon as a crystal touches the oil phase of another globule, partial coalescence is inevitable because the crystal is better wetted by oil than by water.

In the systems described by van Boekel,[2] the general kinetics of partial

* Present affiliation: Unilever Research Laboratorium, Vlaardingen, The Netherlands.

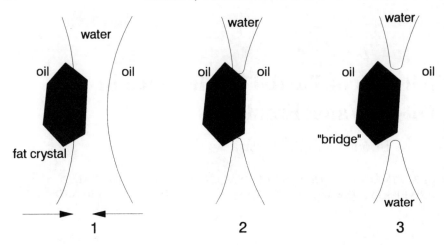

Figure 1 *Sketch of the hypothetical coalescence mechanism caused by a crystal*

coalescence was similar to that of normal coalescence, namely a gradual increase in average globule size with time until a critical size is reached above which the coalescence rate becomes accelerated and the emulsion breaks. Yet, there are situations where partial coalescence proceeds in a rather different manner.

Figure 2 briefly displays the effects of some key variables on emulsion stability. At first sight, the instability seems to be affected by all the variables in a different way. From these results, it seems that few conclusions can be drawn as to the effect of these variables on the emulsion properties. In order to provide a satisfactory means to characterize the emulsion stability, the factors that influence the reactivity of the emulsion globules (*i.e.* the probability that two globules closely encountering each other will partially coalesce) must be distinguished from factors affecting the frequency of encounters. To this end, we developed the simulation model described in the next section.

2 Simulation Model

The model attempts to describe the course of the partial coalescence process in simple shear. It is based on the Smoluchowski equation for orthokinetic aggregation, describing the encounter frequency of two-body encounters in polydisperse systems:

$$b_{ij} = \frac{G}{6} (d_i + d_j)^3 N_i N_j \qquad (1)$$

where G is the velocity gradient and N_i the number of globules of diameter d_i. For the purposes of this study, equation (1) is modified to

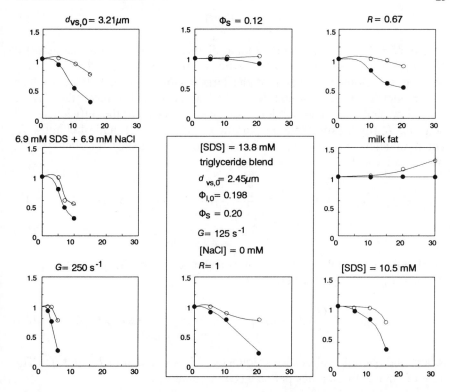

Figure 2 *Effect of several variables on emulsion stability in Couette flow. In the large box in the middle the standard emulsion is represented. The seven other boxes represent the changes found for emulsions that deviate in one respect (as given at the top of each box) from the standard emulsion. Abscissa: time (min); ordinate: relative normalized globule size* $(d_{vs}/d_{vs,0})$ (○), *and normalized residual fat content* $(\phi_l/\phi_{l,0})$ (●), *i.e. fat not immediately creaming*

account for the colloidal interactions and for the influence of the presence of fat crystals in the oil globules. (We denote all emulsion droplets and aggregates, whether or not containing crystals, as 'globules'.) The model distinguishes between singlets containing crystals, S_i, singlets without crystals, s_i, and clumps, C_i (by definition containing crystals, since crystals are necessary for partial coalescence to occur). Globules with crystals ('reactive' globules, denoted with a capital) can cause partial coalescence (although this will not necessarily occur for all encounters), whereas singlets without crystals can only become part of a clump by collision with a reactive globule.

The partial coalescence efficiency, J, of collisions in which (at least) one reactive globule is involved, is affected by some factors that depend on globule size (the fit parameter is m) and some that do not (the fit parameter is P):

$$J = P(d_i^*)^m \qquad (2)$$

where $d_i^* = d_i/d_{min}$ [d_{min} is the globule diameter of the smallest globules] if the smaller globules are the more efficient ones ($m < 0$), and $d_i^* = d_i/d_{max}$ [d_{max} is the globule diameter of the largest globules (*i.e.* 9.75 μm) which the model takes into account] if the larger globules have the higher efficiency ($m > 0$). The model distinguishes between the efficiencies of singlets, J_S (by means of the fit parameters P_S and m_S), and of clumps, J_C (fit parameters P_C and m_C). The coalescence frequency for singlets, b_{ij}^S, now becomes:

$$b_{ij}^S = J_S \frac{G}{6} (d_i + d_j)^3 [N_{S,i}(N_j - N_{C,j}) + N_{S,j}(N_i - N_{C,i}) - N_{S,i}N_{S,j}]$$

$$(3a)$$

If clumps are involved in partial coalescence then the coalescence frequency, b_{ij}^C, can be described by:

$$b_{ij}^C = J_C \frac{G}{6} (d_i + d_j)^3 [N_{C,i}N_j + N_{C,j}N_i - N_{C,i}N_{C,j}] \qquad (3b)$$

where the overall coalescence frequency, b_{ij}, is the sum of b_{ij}^S and b_{ij}^C. The rate of change of globules per size class i equals:

$$\frac{dN_i}{dt} = \frac{1}{2} \sum_{j=1}^{j=i-1} b_{ik} - \sum_{j=1}^{\infty} b_{ij} \qquad (4)$$

Analogously to equation (4), the rates of change of reactive singlets and of clumps are estimated, so that for all kind of globules the changes in the size distribution can be followed as a function of time.

This model can be used to evaluate the course of the coalescence process, *i.e.* the rate at which singlets and clumps partially coalesce. Instability of an emulsion can only occur if singlets are able to aggregate, and this ability is expressed by the 'initial coalescence efficiency' (α_{init}), the fraction of encounters leading to partial coalescence at the beginning of the process, *i.e.*

$$\alpha_{init} = t^*/t \qquad (5)$$

where t^* is the time needed to reach the same change in globule-size distribution as would be observed in time t if *all* collisions would lead to partial coalescence.

The progress of the coalescence process is represented by the dependence on globule size (represented by m) and by the ratio P_C/P_S, provided that the dependence on globule size is the same for singlets and clumps (*i.e.* $m_S = m_C$, which appears to be the case for all experiments discussed here). The larger the m value, the greater is the tendency for larger globules to aggregate as compared to small globules, leading to the formation of a bimodal size distribution. The larger is P_C/P_S, the sooner newly formed doublets will coalesce further, with the resulting formation of big aggregates that cream during the heating step.

3 Results and Discussion

In Table 1 the values of the three fit parameters that result from the model calculations are listed for three types of fat globules. It was found that reactivity to partial coalescence is low in globules with a single (large) crystal. In globules with numerous crystals presumably formed by secondary nucleation reactivity increases with crystal size. It was further shown that there is a dependence on reactivity on globule size. Since m is positive, a large globule implies a more reactive globule. These results are obtained under the constant controlled conditions denoted in the heading of Table 1. In the following sections the consequences of changes to these conditions are discussed.

Total Fat Content

The total fat content, $\phi_{1,0}$, is expected not to affect the fraction of collisions leading to partial coalescence. The encounter frequency will be proportional to $\phi_{1,0}$ and so will the coalescence rate. We observe, with our model, that the coalescence efficiencies are indeed similar, irrespective of the fat volume fraction. It is concluded that the Smoluchowski equation (1) can be used in our situation, implying that multiple-globule interactions and disturbance of the flow field can be left out of consideration.

Initial Average Globule Size

From model calculations (Table 1), it appears that the coalescence efficiency of encounters increases with globule size; m values calculated here were above zero. From this result come two important points. Firstly, any simplification of the model calculations to monodisperse systems may lead

Table 1 *Typical values of the fit parameters in the simulation model as a function of the crystallization habit. Shear rate = 125 s^{-1}; total fat content = 20%; fraction of solid fat = 20%; initial average globule diameter = 2.5 μm*

Fit Parameters			
Relative initial coalescence efficiency	0.5	1	35
Reactivity clumps/singlets	0.001	0.5	3
Dependence of reactivity on d	2	2	4

to erroneous results. Secondly, larger values of $d_{vs,0}$ result in higher initial coalescence rates (α_{init} values), since the fraction of globules with a higher reactivity is higher.

Shear-rate

The shear-rate mainly influences the initial coalescence efficiency (Figure 3). A minimum shear rate, G_{min}, is necessary to obtain a film thickness that is smaller than the protrusion distance of a crystal into the aqueous phase. As expected, G_{min} is lower for globules with larger crystals. Above G_{min} a sharp transition from stable to unstable emulsions occurs. Furthermore, at some higher G values, α_{init} increases only slightly. We conclude that freshly formed aggregates can be disrupted again in some cases. Such disruption may even occur in some conditions at a shear-rate as low as 125 s^{-1}.

Solid Fat Content

The fraction of fat solid, ϕ_s, has a great impact on emulsion stability (Figures 4 and 5). At the lowest solid fat fractions examined, the emulsions remained stable in shear. There appear to be insufficient crystals present to form a continuous network and therefore partial coalescence does not occur.[5] Beyond the minimum content and irrespective of the crystal size, α_{init} increases with increasing solid fat content in the oil globules. Furthermore we see that the parameter m increases, *i.e.* larger globules become relatively more sensitive to partial coalescence with increasing ϕ_s. This

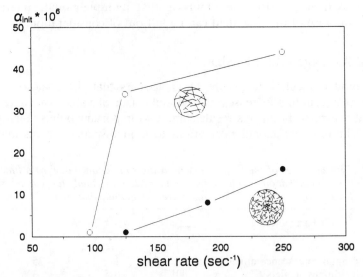

Figure 3 *The initial coalescence efficiency, α_{init}, as a function of shear-rate for globules with large (O) and with small (●) crystals*

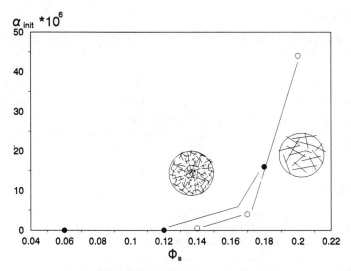

Figure 4 *The initial coalescence efficiency α_{init} as a function of solid fat content ϕ_s for globules with large (○) and small (●) crystals*

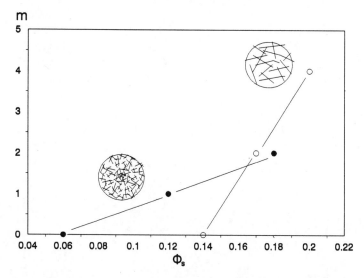

Figure 5 *The parameter* m *denoting dependence of reactivity on globule size as a function of solid fat content ϕ_s for globules with large (○) and small (●) crystals*

behaviour is observed for singlets as well as for clumps. The most likely explanation is that more crystals will tend to protrude from the O/W interface if more solid fat is present. The increase in α_{init} with increasing ϕ is in agreement with this.

4 Conclusions

The model presented here appears to be a helpful tool in unravelling the factors affecting partial coalescence of emulsion globules in a shear field, whether they be emulsion properties or other effects like the encounter frequency or the distance of approach of the globules. The model predicts that lowest reactivity of emulsion globules for partial coalescence is obtained if the following qualifications are met: small globules, small crystals, and as little solid fat as possible (sometimes achievable by arranging for supercooling to persist).

References

1. M. A. J. S. van Boekel and P. Walstra, *Colloids Surf.*, 1981, **3**, 109.
2. M. A. J. S. van Boekel, Ph.D. Thesis, Wageningen, 1980.
3. D. F. Darling, *J. Dairy Res.*, 1982, **49**, 695.
4. J. P. Melsen, Ph.D. Thesis, Wageningen, 1988.
5. K. Boode, P. Walstra, and A. de Groot-Mostert, to be published.

On the Physics of Shear-induced Aggregation in Concentrated Food Emulsions

By A. Lips, T. Westbury, P. M. Hart, I. D. Evans, and I. J. Campbell

UNILEVER RESEARCH, COLWORTH LABORATORY, COLWORTH HOUSE, SHARNBROOK, BEDFORD MK44 1LQ, UK

1 Introduction

The sensitivity to shear of oil-in-water food emulsions, such as dairy creams, is well known. This contribution to an already extensive literature[1-4] seeks to review and adapt recent theories in the physics of aggregation processes and introduce new experimental approaches. A particular interest here is whether these processes display criticality in the classical sense of gelation.

2 Theoretical

By **critical** kinetic behaviour we mean processes of aggregation which result in gelation in finite time (for mathematical description see later). Criteria for such types of aggregation behaviour have been widely discussed within the framework of the Smoluchowski coagulation equation:[5]

$$\frac{\mathrm{d}c_k}{\mathrm{d}t} = \frac{1}{2} \sum_{i+j=k} K_{ij} c_i c_j - c_k \sum_{j=1}^{\infty} K_{kj} c_j \qquad (1)$$

where $c_k(t)$ represents the number concentration of k-fold aggregates at time t, and the coagulation kernel K_{ij} describes the coalescence mechanism of an i- and a j-cluster, *i.e.* the kernel represents the dependence of coalescence rates of i and j. Models for K_{ij} have, in general, to address factors relating to both collision efficiency and collision probability (velocity). Our concern here is mainly with the latter. Implicitly it is assumed, therefore, that collision efficiencies are not dependent on i and j, or at least much less so than collision probabilities. This assertion is believed to be reasonable for modelling the physics of perikinetic (Brownian) coagulation and for a wide range of conditions of orthokinetic (shear-induced) coagulation. The kernel for diffusion-limited aggregation is of the form:

$$K_{ij} = (D_i + D_j)(R_i + R_j) \sim (i^{-1/D} + j^{-1/D})(i^{1/D} + j^{1/D}) \qquad (2)$$

where D_j ($\propto 1/R_j^h$) is the diffusion coefficient of the j-cluster, R_j is its collision radius, and it is assumed that the hydrodynamic diffusion radius $R_j^h \approx R_j$. Self similarity arguments suggest the scaling indicated in equation (2) with the fractal dimension $D \simeq 2$ in three dimensions. Note that K_{ij} is homogeneous in the mathematical sense of being comprised of a sum of terms $i^\mu j^\nu$ with variables μ and ν but which all share the same degree λ defined by $\lambda = \mu + \nu$. In this Brownian case the degree λ is close to zero; to obtain analytic solutions, Smoluchowski in fact assumed $K_{ij} = 1$ (*i.e.* $\lambda = 0$) which leads to the moment equations[5]

$$dM_0(t)/dt = -M_0^2/2,$$

$$dM_2(t)/dt = M_1^2 \text{ [with the } p\text{th moment } M_p = \sum k^p c_k(t)] \qquad (3)$$

The zeroth moment M_0 of the distribution represents the total number of clusters, the first moment M_1 the sol mass, and the second moment M_2 the average degree of polymerization (accessible by low angle light scattering).

The kernel expected for shear-induced coagulation primarily reflects mutual collision volumes between i- and j-clusters with

$$K_{ij} \sim (R_i + R_j)^3 \qquad (4)$$

Again to obtain analytic solutions, Smoluchowski[5] considered j-clusters to be fully coalesced, so that $R_i \propto i^{1/3}$ ($D = 3$), and he approximated $(R_i + R_j)^3$ by i^3 ($i \gg j$). It is especially important to appreciate that K_{ij} is then homogeneous of degree $\lambda = 1$ and conforms to the general moment equations for the sum kernel $i + j$, *viz.*:[6]

$$dM_0/dt = -M_0 M_1, \quad dM_2/dt = 2M_1 M_2 \qquad (5)$$

Both sets of equations (4) and (5) are for general initial distributions. The higher degree λ for shear aggregation implies a more rapid rate of decrease in cluster number M_0 (exponential with time rather than inverse linear as is the case with Brownian coagulation).[6]

Much understanding has been generated in recent years concerning the behaviour and physical relevance of a wider range of kernels than the classical cases delineated above. In part this has been to establish scaling rules for dynamic cluster size distributions within the framework of the mean field Smoluchowski equation (and so to link with computer experiments of aggregation). A second objective has been to define more closely the criteria which determine **critical** kinetic behaviour. We can distinguish between gelling and non-gelling aggregation mechanisms. In the former, a **divergence** is observed in the mean cluster size as t approaches the gel point t_c. Similarly M_2 and higher moments of the distribution diverge at the gel point and beyond the gel point the sol mass M_1 ceases to be conserved (it decreases as sol is rapidly transformed into gel). The gel point is analogous to a critical point. Gelling mechanisms display divergence in the above properties in finite time t_c; non-gelling mechanisms display divergence at infinite time.

A well known example of a kinetic mechanism for gelation is provided by the kernel $K_{ij} = ij$ ($\lambda = 2$) which yields solutions (mathematically) isomorphous with those of the statistical Flory–Stockmayer RA_f model.[7,8] By contrast, the Brownian diffusion-limited model cannot display gelation in finite time, either in its simplest form with $K_{ij} = 1$, or in its elaboration in terms of fractals as above. Therefore cluster size and molecular weight increase continuously with time without divergence. Also the model originally proposed by Smoluchowski for shear induced aggregation (based on the sum kernel $i + j$) does not give critical behaviour in finite time.

Van Dongen and Ernst[9] have shown that, for homogeneous kernels $i^\mu j^\nu$ of degree $\lambda = \mu + \nu$, gelation will occur only if $\lambda > 1$. One implication of this striking result is that **perikinetic** formation of particle gels requires reaction control as opposed to strict diffusion control, since it is only the former for which kernels of the required degree can be expected (*e.g.* $\lambda = 1$ has been suggested for reaction-limited fractal flocs[10] and $\lambda = 2$ as an upper bound consistent with Flory–Stockmayer gelation). On the other hand, it is clearly not necessary to invoke reaction control for **orthokinetic** gelling mechanisms since the classical Smoluchowski model based on an extreme assumption of full coalescence ($D = 3$) already defines $\lambda = 1$ as a low bound; more realistic orthokinetic kernels such as

$$K_{ij} \simeq (i^{1/D} + j^{1/D})^3 \tag{6}$$

where D is an effective fractal parameter with values expected in the range $2 < D < 3$, yield $\lambda > 1$. We may conclude then that, on the basis of mutual collision volumes alone, partially coalescing particles should show gelation in orthokinetic regimes since we have $\lambda > 1$.

Next we consider whether a solution is available for the kernel defined by equation (6). Here we note the inequality $K_{ij} < C(i^{3/D} + j^{3/D})$ for all positive i, j where C is a constant. As an approximate upper bound for the kernel defined in (6), we can therefore consider the sum kernel $(i^\alpha + j^\alpha)$ ($\alpha \geqslant 1$). This has been studied in some detail by Ernst and Hendriks[11] who provide a prescription for a sequential solution of (1) in the pre-gel regime with the help of the appropriate moment equation

$$dM_0/dt = -M_0 M_\alpha \tag{7}$$

and derive an upper bound for the gel time when $\alpha(= 3/D) > 1$, *viz.*:

$$\tau_c \leq \left[2 \left(\frac{3}{D} - 1 \right) M_2{}^{(3/D-1)} (0) \right]^{-1} \tag{8}$$

Studies for related kernels, *e.g.* $(ij^l + i^l j)$, indicate similar dependence of the critical time on initial polydispersity $M_2(0)$ and the power index $l(< 0)$, and it is important to appreciate that models such as equation (8) in general prescribe a quantitative insight into the dependence of the critical time both on the degree of floc ramification (as characterized by D) and the initial polydispersity of the colloid. Those with absolute faith in the self-similar nature of flocs will identify D as the fractal dimension and expect it to be time invariant. While this is probably not correct we will imply that D is an effective floc structure parameter, in the range $2 < D < 3$, and probably closer to 3 in concentrated dispersions. Note that, in the limit $D = 3$ (full coalescence), we have $\tau_c \to \infty$ and so gelation does not occur. In general, therefore, the gel time can be strongly dependent on the type of floc structure.

The variables τ_c and $M_2(0)$ (with the convention $M_1(0) = 1$) in equation (8) are reduced variables, respectively, of time and concentration. In terms of real time t_c the scaling is $\tau_c = t_c A_{11} \phi_0$, where ϕ_0 is the phase volume of the colloid and A_{11} is a binary reaction rate constant for successful monomer–monomer encounters. For diffusion-limited coagulation of monodisperse spheres, we have $A_{11} = 4kT/\eta\pi a^3$ with k the Boltzmann constant, η the viscosity of the continuous phase, and a the radius of the monomer particles. The equivalent rate constant A_{11}, for orthokinetic conditions and unit collision efficiency, is $A_{11} = 4G/\pi$ where G is the shear rate. An important point is that the critical time t_c scales inversely with sol concentration (ϕ_0^{-1}). This is expected for any model based on the Smoluchowski equation (1) which is uni-directional, *i.e.* does not incorporate fragmentation (*e.g.* shear break-up) processes.

It is of relevance to reference recent work by Walstra and co-workers[12–14] in which the formation of particulate gels is viewed simply as the consequence of ramified fractal growth of clusters to a stage where these fully occupy the available volume. On the assumption that the kinetics of cluster growth follow the simplest Smoluchowski kernel $K_{ij} = 1$, expressions have been derived for an apparent gel time here defined as the time for the effective disperse phase volume, increased over the initial phase volume ϕ_0 due to fractal growth with $D < 3$, to become of the order of unity. The proposed simple incorporation of a spatial argument—in effect an allowance for excluded volume—is appealing. Both the classical statistical theories of gelation (*e.g.* Flory–Stockmayer) and the mean field kinetic models referenced above neglect excluded volume. While very close to the sol–gel transition this renders their use inappropriate, it is often held[15] that such models can nevertheless be successful below and sufficiently close to the gel point, the latter implying that gel points can be represented by these models to an accuracy better than is typically achievable by measurement. It needs to be appreciated that the characteristic time defined by the fractal aggregation model of Walstra is not a **critical** time (*i.e.* a gel time or percolation time in the classical sense), it having been derived within the constraint of coagulation kernels of low degree ($\lambda \ll 1$) which do not meet the requirement for critical behaviour, *i.e.* **divergence** in cluster size and mass in finite time. The classical theories consider the rheological transition at the gel point to be almost coincident with the onset of divergence in cluster mass. The characteristic time suggested by Walstra relates solely to a criterion of excluded volume and does not incorporate the concept of *differential* increase in aggregate reactivity with size. The latter is the major feature of the classical statistical and kinetic theories and is also integral to percolation models. The sol–gel transition is treated then as a critical phenomenon within the framework of phase transition theory implying that the nature of the gel (degree of conversion of potential reactive groups) at the gel points is invariant with initial sol concentration. Such an invariance is not implicit in the fractal aggregation model and we note that the characteristic time suggested by Walstra scales with $\phi_0^{3/(D-3)}$ as opposed to ϕ_0^{-1}. While not incorporating excluded volume, the model of equation (8) has the merit of self consistency and of predicting critical divergence, which, as we will show, can be observed experimentally with concentrated aggregating food emulsions.

3 Experimental

All the studies were carried out on a model food emulsion stabilized with UHT treated buttermilk protein and an additional emulsifier (Tween 60). The emulsion was prepared by homogenization at 100 bar. The concentration of fat (coconut/hardened palm kernel) was 35 wt %. On the basis of the $d(3, 2)$ size (= 0.7 μm) and typical adsorbed layer thickness measurements for caseinate coats, we estimate an effective disperse phase volume ϕ of *ca.* 0.45. Studies of concentration dependence involved dilution of the cream with its own serum, with dairy serum, or with distilled water. Temperature studies spanned the melting transition of the fat globules (fully molten at *ca.* 30 °C).

Orthokinetic aggregation was carried out in a specially constructed apparatus of Couette geometry (strain-rate controlled) and in the standard parallel-plate and cone-and-plate geometries of a strain-rate controlled Weissenberg rheogoniometer.

Measurements of aggregation extent and structure were based on small-angle laser light scattering using the Malvern Mastersizer with non-standard interpretation as explained in the text. The scattering measurements involved *ca.* 1000-fold dilution of the concentrated emulsion which was sampled at regular time intervals. Direct observation of floc structure using a light microscope was also carried out.

4 Results and Discussion

The technique of small-angle laser light scattering (SALLS) is particularly suited to detailed studies of the physics of aggregation processes. A striking illustration is shown in Figure 1 for the time evolution of the volume-weighted size distribution as measured by the Malvern Mastersizer. In this sequence our concentrated model emulsion was continuously sheared at 20 °C in a Couette geometry at 397 s^{-1} and sampled at regular intervals.

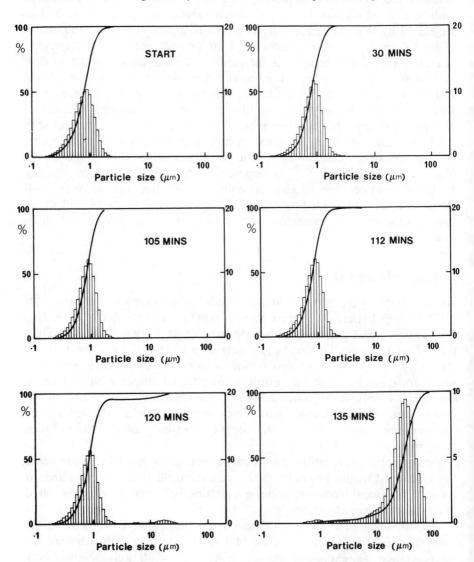

Figure 1 *Orthokinetic aggregation of a concentrated model food emulsion. The six diagrams show the time evolution of the volume-weighted size distribution inferred by small-angle laser light scattering (SALLS). (20 °C, Couette geometry, G = 397 s^{-1})*

Up to a time of 105 min we were unable to detect any obvious change in the state of aggregation of the emulsion. Beginning at 112 min, however, we see the **sudden** development of a 'gel' phase with apparent size $> 10 \ \mu m$. At this stage, and at 120 min, the distribution within the main 'sol' peak has changed very little. At 135 min virtually complete conversion from 'sol' to 'gel' phase has taken place. These changes are characteristic of a critical kinetic process, and this is further amplified in Figure 2 by the divergent time evolution of the variance in mean cluster size at the critical time of *ca.* 100 min and the loss of 'sol' mass at that point. Note that the variance in mean cluster size V is given in terms of the moments of the distribution defined in equation (3), *viz*.:[6]

$$V = [M_2 M_0 - M_1^2]/M_1^2 \equiv [(d(6,3)/d(3,0))^3 - 1] \qquad (9)$$

SALLS with Mie inversion can at best provide continuous as opposed to discrete distributions. Equation (9) provides the link between the two implicating the average diameter $d(6,3)$ and $d(3,0)$ as shown: divergence in V arises from divergence in M_2 and equivalently in $d(6,3)$.[3] Note also that, with the present Mie inversion algorithm, the Malvern Mastersizer interprets light scattering data on the basis of continuous distributions of polydisperse spheres. The size moments so derived are only approximate estimates in the present case where the aggregation does not always progress to full coalescence. However, we will show elsewhere that, for the purpose of estimating critical times, the $d(6,3)$ parameter currently derived (purporting to reflect M_2) is quite adequate. A more rigorous approach

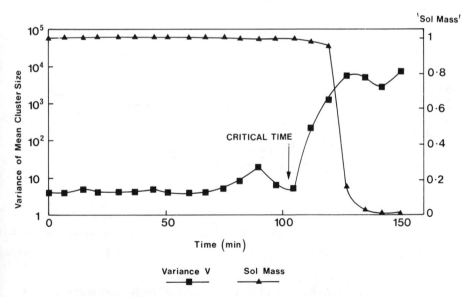

Figure 2 *Time dependence of variance in mean cluster size (■) and sol mass (▲) by SALLS. Conditions as in Figure 1*

would circumvent the Mie inversion algorithm and concentrate on changes in scattering at very low angles in the limit where this is strictly proportional to M_2.[16] Our measure of the sol mass M_1 is the fraction of the distribution in particles with $d < 10\ \mu$m.

Figure 3 shows that the shear-induced aggregation also produces a divergence in the viscosity of the concentrated emulsion at a 'rheological' critical point which has an inverse dependence on shear rate and a complicated dependence on rheometer geometry as shown in Figure 4. Not unreasonably, the rheological critical times are slightly longer than the critical times inferred from light scattering, and increase with the gap width. We believe that the rheological transition point reflects a more advanced state of the gelation corresponding to a mean cluster size comparable with the gap width. Note that up until this transition there are no complications from slip in the measurement of viscosity.

Figures 5 and 6 show the dependence of, respectively, the light scattering and the rheological critical times on the disperse phase volume of the emulsion. Our estimate of the phase volume of the starting emulsion is $\phi_0 = 0.45$ (see Section 3). Figure 6 shows that there is the same behaviour irrespective of whether the starting emulsion was diluted with its own serum, with a dairy cream serum, or with distilled water. Both the figures illustrate a high power law region $\phi^{-\beta}$ (with $\beta > 6$) in the range $0.3 < \phi < 0.45$. In this region the viscosity of the dispersion rises steeply with phase volume, resulting in increased energy dissipation and local shear

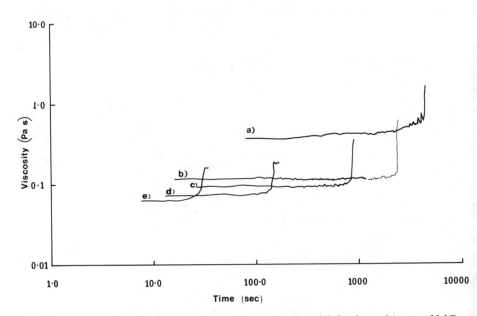

Figure 3 *Orthokinetic aggregation of a concentrated model food emulsion at 20 °C showing the critical divergence in viscosity. Weissenberg cone-and-plate geometry with shear-rates* (s⁻¹): *(a)* 10; *(b)* 100; *(c)* 200; *(d)* 500; *(e)* 1000

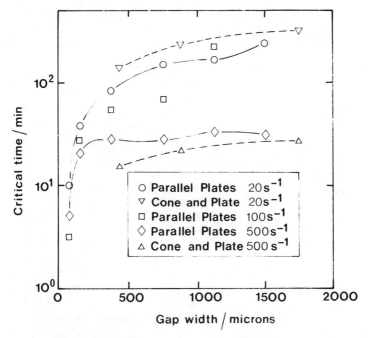

Figure 4 *Dependence of critical time of shear-induced gelation at* 20 °C *on Weissenberg rheometer geometry* (*strain rate controlled*)

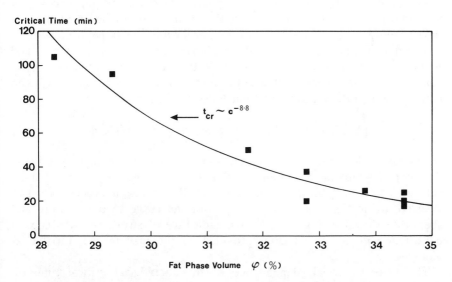

Figure 5 *Dependence of critical time on disperse phase volume of fat. Critical times at* 20 °C *are inferred by SALLS* (*Couette geometry*, 1057 s^{-1})

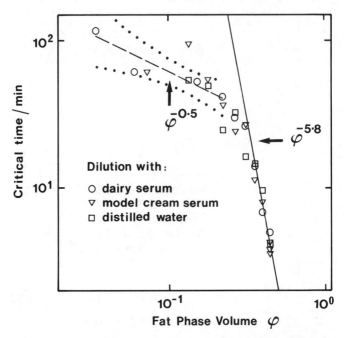

Figure 6 *Dependence of critical time of shear induced gelation on the disperse phase volume of fat globules. Critical time at 20 °C is inferred from the divergence of the viscosity (Weissenberg cone-and-plate geometry, 800 s^{-1})*

stresses between fat globules in strain-controlled shear application. We believe this to be the reason for the high power law behaviour. At moderate concentrations the disperse phase affects the starting viscosity of the emulsions in only a minor way. We then observe (Figure 6) the low power law ($\beta \sim 1$) expected for a second order uni-directional process.

The dependence of the rheological critical time t_{cr} on shear-rate G and temperature is illustrated in Figure 7. Within an order of magnitude, we can estimate the collision efficiency ε using equation (8) to infer A_{11} and comparing with the maximum Smoluchowski collision rate, *viz.*:

$$[2(3/D - 1)M^{(3/D-1)}] = t_{cr}A_{11}\phi_0 = t_{cr}(\varepsilon 4G/\pi)\phi_0 \qquad (10)$$

Note that, if the starting distribution is not excessively pre-aggregated (*i.e.* $M_2 \approx 1$) and the fractal dimension D is not too close to 3, the left hand side of equation (10) is of the order 1. The collision efficiencies then inferred are in the range 10^{-8} to 10^{-6} for the range of temperature and shear-rate shown in Figure 7. This is indicative of strong reaction control consistent with a steric barrier provided by the protein coat on the fat globules.

The observed weak dependence of the rheological critical time on shear-rate ($G^{-\delta}$ with $\delta < 1.8$) is surprising. The experimental conditions span the Péclet number range $100 < Pe < 10^6$ (where $Pe = 6\pi\eta Ga^3/kT$).

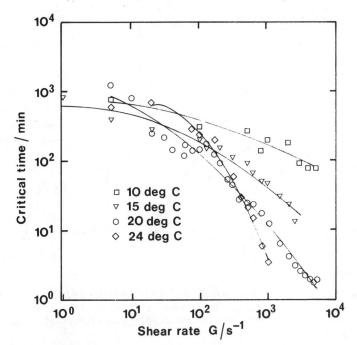

Figure 7 *Dependence of critical time on shear-rate and temperature. Critical times are inferred from the divergence of the viscosity (Weissenberg cone-and-plate and parallel-plate geometries)*

Predominantly orthokinetic behaviour is expected, for which theories based on trajectory models[17] require $\delta \approx 0.8$ when barriers are not rate limiting; for strong reaction control, δ should be substantially greater than unity, yet we can actually observe $\delta < 0.5$! At present we have no completely satisfactory explanation for this, and further experimental work is in progress. The possibility of a change in floc structure with increasing shear-rate could be relevant. This would follow from equation (8) if D increased with shear-rate. It is important to note that the critical time is expected to be not just a function of the primary reactivity, A_{11}, but also of the type of floc structure. Polydispersity in the starting emulsion could also be a controlling variable in that the larger droplets are expected to interact with high collision efficiencies and may 'seed' the coagulation process. Certainly, we can demonstrate that pre-aggregation of the emulsion (by quiescent storage at 40 °C for 4 hours) leads to a reduction in critical time (Figure 8) as would be expected on the basis of equation (8). An additional relevant factor is shear break-up of large aggregates. Fragmentation processes may be more important at the higher shear-rates.[18] In general their effect is to counteract the capacity for critical tendency inherent in the growth kernels.

Figure 7, and the inset to Figure 8, implies a decrease in stability of the emulsion with increase in temperature. In a previous model study[19] we

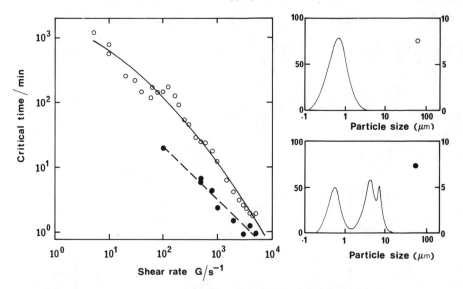

Figure 8 *Role of initial polydispersity on orthokinetic aggregation of concentrated model food emulsion. Critical times at 20 °C are inferred from the divergence of the viscosity (Weissenberg cone-and-plate and parallel-plate geometries)*

have attributed such an effect to a loss in stabilizing power of the protein coat. Extensive studies to be presented elsewhere of the **perikinetic** Brownian forward aggregation rate A_{11} of our model emulsion clearly show a decrease in stability with increasing temperature. This is **continuous** in the range of temperature 10–60 °C, *i.e.* we do not observe a discontinuity in the Brownian rate in the temperature range 10–30 °C where fat crystal content in the globules changes drastically, and the intrinsic perikinetic reactivity between particles increases well beyond the temperature, 35 °C, at which all the globules are fully molten. By contrast **orthokinetic** behaviour is more critical in nature and it displays a discontinuity with temperature. This is illustrated in Figure 9 for a shear-rate 397 s^{-1} where critical times have been measured by the light-scattering method. In view of the perikinetic studies suggesting a continuous increase in particle reactivity with temperature (due to a progressive weakening in the protein coat), we suggest that the **primary** reason for the observed minimum in orthokinetic critical time with temperature is the progressive transition from ramified partially coalesced aggregate structures to full coalescence as would be consistent with equation (8) with D approaching 3. Microscopic investigations of aggregation extent and structure are in accord with this view. We do not therefore believe that the proposed **direct** kinetic role of fat crystals in rupturing films[2] is the major controlling factor with our model emulsions. The crucial function of fat crystals is rather to ensure and to mediate partial coalescence.

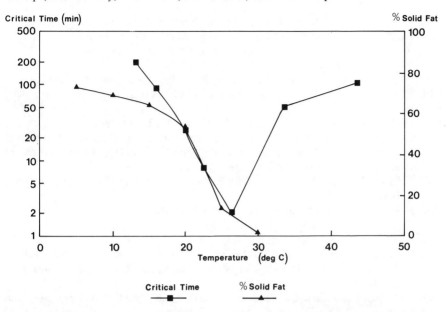

Figure 9 *Dependence of critical time on temperature through the melting transition of the fat. Critical times (■) are inferred by SALLS (Couette geometry, 397 s⁻¹). Solid fat content (▲) is determined by pulsed low-resolution proton NMR*

In conclusion, we have shown that protein-stabilized oil-in-water emulsions display kinetic behaviour with a tendency to criticality under ortho-kinetic conditions. Theoretical reasoning, based on the uni-directional Smoluchowski equation with physically realistic kernels allowing for mutual collision volumes, shows that this is to be expected for orthokinetic aggregation with partial coalescence. A qualitatively successful model [equation (8)] has been proposed which combines the variables of primary particle reactivity, initial polydispersity, and floc structure represented by an effective fractal parameter. Critical kinetic behaviour can arise merely from the increase in mutual collision volumes with aggregate size and the consequent increase in *differential* reactivity of large aggregates. Additional reaction control, due to steric or electrostatic barriers, can reinforce the critical nature of this process whereas shear break-up can counteract it.

References

1. J. H. Labuschagne, Ph. D. Thesis, University of Wageningen, 1963.
2. M. A. J. S. van Boekel, Ph. D. Thesis, University of Wageningen, 1980.
3. D. J. Darling, *J. Dairy Res.*, 1982, **49**, 695.
4. K. Boode, Ph. D. Thesis, University of Wageningen, 1992.
5. M. Smoluchowski, *Z. Phys. Chem.*, 1917, **92**, 129.

6. R. M. Ziff, in 'Fractal Aggregation and Gelation', ed. F. Family and D. P. Landau, Elsevier Science Publishers, Amsterdam, 1984, p. 191.
7. R. J. Cohen and G. B. Benedek, *J. Phys. Chem.,* 1982, **86**, 3696.
8. W. H. Stockmayer, *J. Chem. Phys.*, 1973, **11**, 45.
9. P. G. J. van Dongen and M. H. Ernst, *Phys. Rev. Lett.,* 1985, **54**, 1396.
10. R. C. Ball, D. A. Weitz, T. A. Witten, and F. Leyvraz, *Phys. Rev. Lett.*, 1987, **58**, 274.
11. M. H. Ernst and E. M. Hendriks, *Phys. Lett.*, 1982, **92A**, 267.
12. P. Walstra, T. van Vliet, and L. G. B. Bremer, in 'Food Polymers, Gels and Colloids', ed. E. Dickinson, Special Publication No. 82, Royal Society of Chemistry, Cambridge, 1991, p. 369.
13. L. G. B. Bremer, T. van Vliet, and P. Walstra, *J. Chem. Soc., Faraday Trans. 1,* 1989, **85**, 3359.
14. L. G. B. Bremer, Ph. D. Thesis, University of Wageningen, 1992.
15. K. Dusek, *Brit. Polym. J.*, 1985, **17**, 185.
16. A. Lips, *J. Chem. Soc., Faraday Trans. 2,* 1987, **83**, 221.
17. T. G. M. van de Ven and S. G. Mason, *Colloid Polym. Sci.,* 1977, **225**, 468.
18. F. E. Torres, W. B. Russel, and W. R. Schowalter, *J. Colloid Interface Sci.,* 1991, **142**, 554.
19. A. Lips, I. J. Campbell and E. G. Pelan, in 'Food Polymers, Gels and Colloids', ed. E. Dickinson, Special Publication No. 82, Royal Society of Chemistry, Cambridge, 1991, p. 1.

Denaturation of Whey Proteins Studied by Thermal Analysis and Chromatography

By P. J. J. M. van Mil and S. P. F. M. Roefs

NETHERLANDS INSTITUTE FOR DAIRY RESEARCH (NIZO), KERNHEMSEWEG 2, 6718 ZB EDE, THE NETHERLANDS

1 Introduction

Whey proteins are on the verge of wide-spread application in food products; they are being used as water holding and gelling agents, foaming and emulsifying agents, and recently also as a fat substitute.[1,2] Part of the functional behaviour is related to the physico-chemical changes which are induced by the various treatments of the whey proteins before introduction in foods or during food manufacture. One of the key unit operations in food processing is heating, and it is this which causes the functional properties of the whey protein to be altered most drastically.[3-5]

Denaturation of whey protein is generally assumed to be a process consisting of at least two steps: a transformation of the native state into an activated state (often called 'unfolding'), and a subsequent aggregation of the activated molecules.[3] Both steps have been studied widely in model systems, but hardly ever in actual food systems. The changes in physico-chemical properties of the different whey proteins which occur during denaturation in a food system under various conditions are the result of a complicated set of reaction mechanisms, and they cannot be quantified by a simple summation of known reactions established for model systems (i.e. a single component in a buffer solution). Thus, the control of the functional properties implies the study of denaturation in more-or-less actual food systems.

We have studied different aspects of the unfolding and aggregation phenomena in concentrated mixtures of whey proteins. It is our aim to form a consistent picture of how the changes in physico-chemical properties relate to the functional properties. The preliminary results presented here are part of a more elaborate study.

2 Unfolding

Differential scanning calorimetry (DSC) is one of the established methods of studying the unfolding of globular proteins. Over many years, this technique has been used to elucidate the kinetics of unfolding of β-lacto-globulin and other whey proteins.[6-10]

In this work, we have used a 'heat-flux' calorimeter (DSC, Dupont model 990) in the temperature range 40–135 °C at a heating rate of 5–10 °C min^{-1}. The amount of sample was 20 mg; the protein concentration must be greater than 5% in order to get a large enough signal from the endothermal reaction. The α-lactalbumin (α-la) and the β-lactoglobulin (β-lg) were prepared by an ion-exchange process and freeze dried. The whey protein concentrate (WPC) was prepared from fresh whey by ultrafiltration and spray dried.[11]

In a simple buffer solution (pH 6.0), the transition temperature T_{tr} of pure β-lactoglobulin is 78 °C.[7] The whey proteins have a transition temperature range of 62–78 °C.[6] T_{tr} is affected by the heating rate and pH of the solution. Our experiments show that, for the unfolding of β-lg, there is an increase in the transition temperature when the pH is lowered. The reaction rate characteristics—the activation energy E_A between 70 and 75 °C and the reaction rate constant k at 75 °C—are changed also on lowering the pH (Table 1), indicating a more stable β-lg molecule. The transition enthalpy hardly changes, however, which leads to the conclusion that the unfolding could be to the same extent regardless of the pH (at least in this range). Our measurements demonstrate also that there is no influence of pH on the T_{tr} of α-la (in the same pH range). The different behaviour of β-lg *vis-à-vis* α-la is attributed to the thiol-group of β-lg which becomes protonated at low pH giving a more stable molecule (against unfolding). The α-la is without a thiol-group and less susceptible to changes of pH. However, in the literature, a different influence of pH on unfolding is reported which is in conflict with our results, and also in contradiction to other results.[7,8]

In more complicated systems like milk or milk products, other effects have to be taken into account also. For instance, not only the lowering of

Table 1 *Thermodynamic and Kinetic Characteristics of β-lg at Various pH Values*

pH	T_{tr} (°C)	ΔH (J g^{-1})	E_A (kJ mol^{-1})	k (75 °C) (ms^{-1})
6.75	75.1	15.1	485	21.3
6.22	77.5	14.0	562	7.0
6.06	78.5	14.5	644	3.1
5.92	79.1	15.3	740	1.6
5.78	79.6	15.6	847	0.8

T_{tr}, temperature of transition; ΔH, reaction enthalpy; E_A, activation energy; k, reaction rate constant.

the pH, but also the increasing of the lactose concentration, has a stabilizing effect on the thermal unfolding of the whey proteins; this is a result of the reduced water activity.[7] We have combined an increased content of lactose and a reduced pH, and have shown that the influence on the denaturation of β-lg is not a summation of two effects, but is an interactive and more complicated process. As an illustration, some thermodynamic and kinetic characteristics (for a first-order mechanism of unfolding) are given in Table 2.

In whey protein concentrate (WPC) or whey protein isolate (WPI), the ratio of α-la to β-lg is not always constant. The unfolding reaction of a protein molecule will be influenced by the presence of the other whey proteins. We have changed the relative concentrations of α-la and β-lg and have observed that the temperature of the transition does not change, nor does the reaction enthalpy (ΔH); both are controlled by β-lg. However, the activation energy, in particular for β-lg, is decreased by 30% when the ratio of α-la to β-lg changes from 0.20 to 0.48 (Figure 1). A decrease of the concentration of β-lg (or an increase of the ratio of α-la to β-lg) does not lead to a change in the activation energy, which for pure β-lg is about 500 kJ mol^{-1}. The explanation is presumably that, since the transition temperature of α-la is lower than that for β-lg, this results (at the higher transition temperature of β-lg) in a destabilization of β-lg by α-la.[12]

By increasing the concentration of pure whey protein, the transition of α-la or β-lg is slowed down, and is perhaps not complete due to the reduced water activity (Figure 2 and Table 3). At concentrations of more than 60 wt % whey protein, the enthalpy ΔH decreases significantly for β-lg only, and the transition temperature, T_{tr} increases for both α-la and β-lg. At the transition temperature, the Gibbs free energy change of the transition is $\Delta G_{tr} = 0$; and thus, from the equation $\Delta G = \Delta H - T\Delta S$, at the transition temperature we have $\Delta S_{tr} = \Delta H / T_{tr}$. Calculations with the data from Figure 2 show that the transition entropy ΔS_{tr} at concentrations of 10 wt % pure whey protein (α-la or β-lg) is 20 J g^{-1} °C^{-1}, and at a concentration of 75 wt % ΔS_{tr} is 3.8 and 9.7 J g^{-1} °C^{-1} for β-lg and α-la, respectively. The reduced water content hampers the unfolding of β-lg more than that of α-la. The denaturation of whey proteins in concentrated

Table 2 *Thermodynamic and Kinetic Characteristics of β-lg at Reduced pH and Raised Lactose Content*

Lactose (wt %)	pH	T_{tr} (°C)	ΔH (J g^{-1})	E_A (kJ mol^{-1})	k (75 °C) (ms^{-1})
0.0	6.75	75.1	15.1	485	21.3
0.0	6.22	77.5	14.0	562	7.0
10.0	6.75	77.6	15.8	340	9.9
10.0	6.22	82.5	18.6	615	0.3

Abbreviations: see Table 1

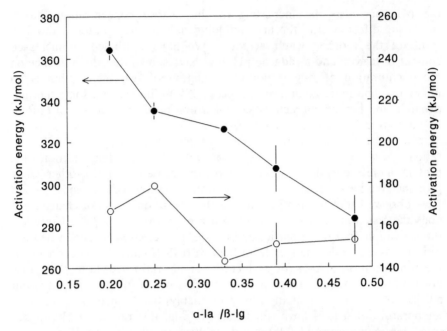

Figure 1 *Activation energy of the kinetics of the thermal denaturation of mixtures of*
α-lactalbumin (○) and β-lactoglobulin (●) as a function of the weight ratio
α-la:β-lg

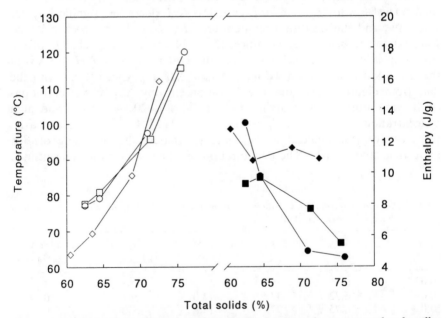

Figure 2 *Transition temperature and transition enthalpy of concentrated α-lactalbu-*
min (◇), β-lactoglobulin (○), and whey protein concentrate (□). Tem-
perature: open symbols; enthalpy: filled symbols

Table 3 *Kinetic Characteristics of Unfolding of Concentrated Samples of α-la and β-lg*

	α-lactalbumin			β-lactoglobulin	
TS (wt %)	E_A (kJ mol^{-1})	k (75 °C) (ms^{-1})	*TS* (wt %)	E_A (kJ mol^{-1})	k (75 °C) (ms^{-1})
10	–	–	10	461	25
60.5	461	11900	62.5	438	23
63.5	450	750	64.5	351	11
68.9	440	0.65	71.0	331	0.05
72.6	351	0.00042	76.0	294	0.0004

TS, total solids; other abbreviations: see Table 1.

WPC (up to a concentration of 75 wt %; molecular ratio α-la:β-lg = 12.5:53.2) is dominated by β-lg (Figure 2), but the influence is somewhat less at high concentration.

We conclude that the mechanism and kinetics of unfolding of α-la and β-lg in buffer solutions is different from solutions with lactose added, mixed solutions with various ratios of α-la to β-lg, or solutions with a high total solids content (WPC). The mechanism of unfolding in mixtures is more complicated than that given by the sum of the mechanisms of α-la and β-lg unfolding in single-component solutions.

3 Aggregation

Aggregation experiments were carried out by heating glass tubes filled with whey protein isolate solutions (5 ml of a commercial sample) at 75 °C in a waterbath. The total solids concentration varied from 0.2 to 10.0 wt %. At regular time intervals tubes were taken out from the water thermostat and cooled in ice water; subsequently, the protein solutions were acidified to pH 4.6 and centrifuged at 20 000 g for 10 min; the concentrations of non-aggregated α-la and β-lg were determined by means of an HPLC technique.[11]

The time-dependent decrease of concentration, $C(t)$, of non-aggregated α-la or non-aggregated β-lg is expressed for each component using the general equation for reaction kinetics

$$- dC(t)/dt = k_n C^n, \tag{1}$$

where t is the time, k_n is the reaction rate constant (g^{1-n} l^{n-1} s^{-1}), and n is the order of the reaction. For each concentration the reaction order n is determined for α-la (n_α) and for β-lg (n_β) according to equation (1) using the differential method.[13]

Values are found for n_α between 0.9 and 1.1 with an average of 1.10 and for n_β between 1.4 and 1.9 with an average of 1.57. For a low total solids concentration of 0.2 to 0.5 wt%, however, n_β is 2.2 or higher. Based on these results the concentration curves of α-la were fitted with a first-order equation and those of β-lg with an equation of order 1.5, *i.e.*

$$C(t)/C_0 = \exp(-k_{1,\alpha}t) \tag{2}$$

where $k_{1,\alpha}$ is the first-order reaction rate constant for aggregation of α-la, and

$$C(t)/C_0 = (1 + \tfrac{1}{2}k_{1.5,\beta}C_0^{1/2}t)^{-2} \tag{3}$$

where $k_{1.5,\beta}$ is the reaction rate constant of order 1.5 for aggregation of β-lg. For α-la, good fits are obtained (Figure 3), which indicates that at each concentration the decrease is described adequately by a first-order reaction. The ratio $C(t)/C_0$, however, does not coincide for all concentra-

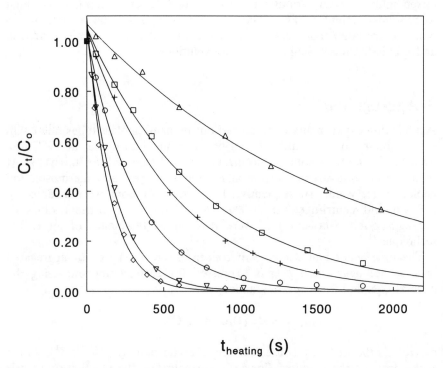

Figure 3 *Fraction of non-aggregated α-lactalbumin in whey protein isolate (WPI) after heating a solution at 75 °C. Concentration of WPI: 0.2% (△); 0.5% (□); 0.7% (+); 1.4% (○); 3.5% (▽); 10% (◇)*

tions that were expected for a real first-order process. The reaction rate clearly increases with concentration as illustrated in Figure 4, where the reaction rate $k_{1,\alpha}$ is plotted as a function of the initial α-la concentration. After an increase at low concentrations, $k_{1,\alpha}$ levels off at high concentrations. Although the concentration $C(t)$ can be described by a first-order decay curve, the aggregation of α-la in the presence of β-lg apparently does not occur via a simple first-order process.

Since the experiments were carried out with a whey protein isolate, the α-la:β-lg weight ratio was fixed at 0.24, and the initial β-lg concentration varied to the same extent as the initial α-la concentration. In a separate experiment, purified β-lg was added (12.54 g l^{-1}) to a whey protein isolate solution with an initial α-la concentration of 1.3 g l^{-1} (see filled symbol in Figure 4). For this sample, $k_{1,\alpha}$ is increased by a factor of two, which indicates that the rate of α-la aggregation is strongly affected by the β-lg concentration. If $k_{1,\alpha}$ is plotted as a function of the initial β-lg concentration, a smooth curve is found which passes through the value for the sample with extra β-lg. In other words $k_{1,\alpha}$ seems to depend on the β-lg concentration only.

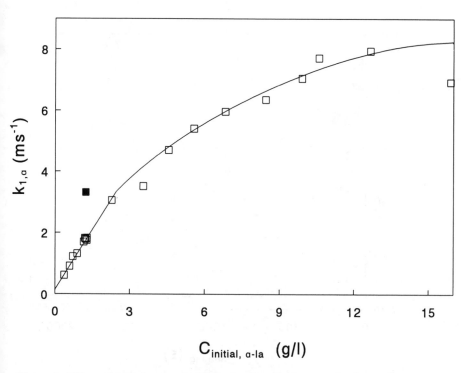

Figure 4 *Effect of initial concentration of α-lactalbumin on the first-order reaction rate constant* $k_{1,\alpha}$. *The filled symbol indicates the experiment where* 12.5 g l^{-1} *of β-lg was added to the solution of 0.7 wt % whey protein isolate*

For the concentration of non-aggregated β-lg as a function of heating time, good fits were obtained with equation (3), except for the lowest and highest isolate concentrations (Figure 5). At the highest concentrations an adequate detection of the residual β-lg concentration after prolonged heating times was disturbed by the formation of stiff gels in the glass tubes. These samples could not be acidified to pH 4.6. At the lowest concentration the order of the reaction was found to increase above 2.0. Therefore, only the first part of the concentration curve was fitted (see 0.2 wt % sample in Figure 5). The ratio $C(t)/C_0$ for a reaction with order 1.5 should decrease stronger in time with increasing initial concentration (equation (3)). This behaviour is observed in Figure 5 and the fits are quite good. However, $k_{1.5,\beta}$ plotted as a function of initial β-lg concentration shows a slight increase (Figure 6). A reaction order of 1.5 and a slight concentration dependence of $k_{1.5,\beta}$ suggest that the aggregation mechanism of β-lg may be very complicated. The $k_{1.5,\beta}$ value of the sample with extra β-lg added fits in with the other values, which suggests that the aggregation of β-lg is hardly dependent on the concentration of α-la.

It can be concluded that the decrease in concentration of α-la and β-lg during heating at 75 °C in a whey protein isolate can be well described by, respectively, a first-order reaction and a reaction of order 1.5. However,

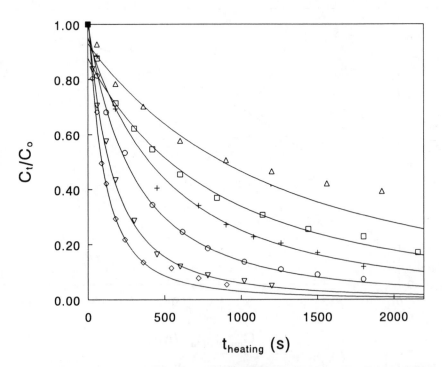

Figure 5 *Fraction of non-aggregated β-lactoglobulin in whey protein isolate (WPI) after heating a solution at 75 °C. Concentration of (WPI): 0.2% (△); 0.5% (□); 0.7% (+); 1.4% (○); 3.5% (▽); 7% (◇)*

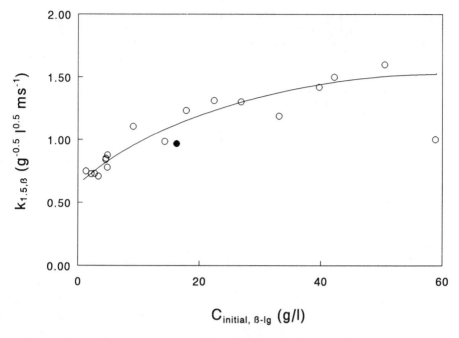

Figure 6 *Effect of initial concentration of α-lactalbumin on the reaction rate constant $k_{1.5,\beta}$. The filled symbol indicates the experiment where $12.5\,\mathrm{g\,l^{-1}}$ of β-lg was added to the solution of $0.7\,\mathrm{wt\%}$ whey protein isolate*

the real aggregation mechanisms may be far more complicated than suggested by these relatively simple reactions; the aggregation of α-la is, for instance, strongly dependent on the β-lg concentration.

Acknowledgement

The authors thank P. M. Dekker and J. Grandia for performing most of the experiments.

References

1. J. N. de Wit, in 'Developments in Dairy Chemistry 4, Functional Milk Proteins', ed. P. F. Fox, Elsevier Applied Science, London, 1989, p. 323.
2. N. S. Singer, S. Yamamoto, and J. Latella, Protein Product Base, US Patent 4, 734,287, 1988.
3. D. M. Mulvihill and M. Donovan, *Irish J. Food Sci. Technol.*, 1987, **11**, 43.
4. J. N. de Wit, in 'Developments in Dairy Chemistry 4, Functional Milk Proteins', ed. P. F. Fox, Elsevier Applied Science, London, 1989, p. 285.
5. J. E. Kinsella, *CRC Crit. Rev. Food Sci. Nutr.*, 1984, **21**, 197.

6. M. Ruegg, U. Moor, and B. Blanc, *Biochim. Biophys. Acta*, 1975, **400**, 334.
7. J. N. de Wit and G. A. M. Swinkels, *Biochim. Biophys. Acta*, 1980, **624**, 40.
8. K. H. Park and D. B. Lund, *J. Dairy Sci.*, 1984, **67**, 1699.
9. M. Paulsson, P.-O. Hegg, and H. B. Castberg, *Thermochim. Acta*, 1985, **95**, 435.
10. P. Relkin and B. Launay, *Food Hydrocolloids*, 1990, **4**, 19.
11. J. N. de Wit, *J. Dairy Sci.*, 1990, **73**, 3602.
12. T. Sienkiewicz, *Nahrung*, 1981, **25**, 329.
13. C. C. Hill and R. A. Grieger-Block, *Food Technol.*, 1980, **94**, 56.

Milk Clotting Time as a Function of Volume Fraction of Casein Micelles

By C. G. de Kruif

NETHERLANDS INSTITUTE FOR DAIRY RESEARCH (NIZO), KERNHEMSEWEG 2, 6718 ZB EDE, THE NETHERLANDS

1 Introduction

The coagulation of milk under the influence of rennet has been studied quantitatively for more than 100 years. It is well established that the casein micelles lose their colloidal stability under the action of chymosin.[1] This enzyme splits off the glycomacropeptide (GMP) part of the κ-casein at the (Phe_{105}–Met_{106}) peptide bond. The split-off caseino-macropeptide is highly lyophilic and remains in solution. The casein micelles progressively lose their colloidal stability, meaning that the initial repulsive interaction between them changes gradually into an increasingly attractive interaction. The repulsive interaction of intact casein micelles is predominantly of the so-called steric or polymeric stabilization type.[2] This follows from the fact that ionic strength hardly influences the stability of micelles. Changing the pH is a different situation, since it changes the solvent quality.

The flocculation of casein micelles is usually modelled as a classical flocculation process described by Smoluchowski kinetics.[3,4] The use of fractal geometry helps in describing the structural properties, but it does not provide a kinetic model. The resulting microstructure can, however, be related to certain kinetic growth models of the flocs by means of extensive computer simulations.

It is the purpose of this paper to develop an alternative model for the interaction between the casein micelles. In this model we emphasize the fact that the strength of the interactions between the micelles parallels the enzymatic splitting of the κ-casein. In other words, the weak but increasing attractive interactions are already present before flocculation occurs, where finally the attractions become strong enough to give permanent flocs. In order to describe the flocculation kinetics it must thus be realized (and this cannot be overemphasized) that the attraction grows gradually in relation to the amount of κ-casein split off. In the model presented here, the flocculation time (*i.e.* the time between addition of rennet and flocculation) is naturally related to the strength (the free energy) of the interactions.

In fact, the question arises as to how 'flocculation time' should be defined. The definition of flocculation time as originally given by Berridge[5,6] involves the visible observation of flocs in the milk. Ruettiman and Ladish[7] define flocculation also as the observation of flocs in a milk film. They implicitly assume that this is equivalent to definitions based on turbidity, viscosity, and gel-firmness. Dalgleish[8] also defines coagulation time as the point of appearance of visible clots in the milk. Guthy, Auerswald, and Buchheim[9] define coagulation time as the time when the viscosity reaches its original value after the initial drop on addition of rennet. Often flocculation time is defined as the amount of time, after adding chymosin, until an observable quantity asymptotically reaches some limiting value. In itself this is already ambiguous, since it is not defined to what degree the asymptotic value must be reached. Plotting the reciprocal of the observable quantity and extrapolation to zero may give a better definition. Many papers—for example Dalgleish and Horne[10] and Dulac *et al.*[11]—discuss in great detail the correlation between clotting times observed with different techniques.

Despite all this, it is realized that no matter what measurement technique is used the physics of the underlying process is not changed. This implies that it should be possible to scale the results of the different techniques properly. This does not necessarily mean that there is a one-to-one relation between differently observed flocculation times, since different properties may be assessed.

It is the purpose of this paper to show that 'clotting' and 'clotting time' can be defined unambiguously, and that the definition can be used to scale and interpret experimental observations. For this we will use a model in which the attractive interactions between the casein micelles gradually increase in relation to the amount of split-off casein. This proteolysis time-scale is rate determining since it is long compared to the diffusive time-scale of the casein micelles. We therefore suppose that the changing equilibrium microstructure of the micellar dispersion is attained at all moments in time. Furthermore, the quasi-equilibrium properties are assumed to be adequately described by an appropriate statistical mechanics theory.

2 Theoretical Background

System Description

Skim milk is considered here as being a dispersion of casein micelles of average size about 200 nm in a continuous solvent. The solvent, being water, includes ions and whey proteins that are much smaller than the casein micelles. The interaction between the casein micelles can be decribed as being hard-sphere repulsive. The viscosity of a skim milk dispersion containing 10 vol % casein micelles (volume fraction $\phi = 0.1$) is 1.31 times the viscosity of the continuous phase, which is obtained by ultrafiltration of the skim milk.[12] Results of experiments by Griffin et al.[13] at high volume fractions have led to the same conclusion that casein micelle dispersions follow hard-sphere viscosity behaviour.

Proteolysis

The enzyme chymosin splits off the GMP from the surface of the micelles. The enzymatic reaction kinetics are adequately described by[14]

$$\ln(1 - [P]/[P_\infty]) = -k \frac{[E]}{[E_0]} t \qquad (1)$$

where $[P]/[P_\infty]$ is the fraction of the GMP split-off, $[E]$ $[E_0]$ is the normalized amount of enzyme, k is the reaction constant, and t is time. Rearranging this equation to

$$t = \frac{-1}{k} \frac{[E_0]}{[E]} \ln (1 - [P]/[P_\infty]) \qquad (2)$$

shows the basis of the so-called Holter plot[15] in which it is assumed that clotting time t_c is inversely proportional to enzyme concentration. This assumption is usually valid at a given volume fraction, in which case the clotting time will occur at the same value of $(1 - [P]/[P_\infty])$. A change in volume fraction will change the ratio $[P]/[P_\infty]$.

The action of chymosin is two-fold: firstly it decreases the effective micellar size, and secondly it decreases the steric repulsive barrier to the mutual approach of two micelles, allowing the van der Waals attraction forces to become operative.

Change in Volume Fraction

The change in micellar radius is accounted for by a change in volume fraction

$$\phi = \phi_0 (1 - \alpha) \qquad (3)$$

where ϕ_0 is the original volume fraction, and

$$\alpha = 3 \frac{f}{a_0} \cdot \frac{[P]}{[P_\infty]}$$

Here f is the 'hydrodynamic' length of the GMP, which is much smaller than a_0, the initial (i.e. hard-sphere) radius of the casein micelles. The use of a linear relationship in equation (3) is justified by the experiments of Griffin[13] and de Kruif et al.[16]

Van der Waals Attraction

Concomitant with the loss of stabilizing surface chains is the increase in van der Waals attraction between the casein micelles. In view of the stability of casein micelles at 4 °C, there may be hydrophobic interactions as well. Also, specific hydrogen bonding or ion bonding may be present. The precise nature of the attraction is irrelevant, but it must be short ranged.

The pair potential is modelled as a combination of a hard-sphere repulsion and a short-range attraction:

$$V(r) = \begin{cases} \infty & (0 < r < \sigma) \\ -\varepsilon & (\sigma < r < \sigma + \Delta) \\ 0 & (\sigma + \Delta < r) \end{cases} \qquad (4)$$

The width of the well, Δ, is small with respect to particle radius. The depth of the well, ε, is a function of time since the casein micelles gradually lose their steric stabilization. In a previous paper,[16] we related the potential well depth to the degree of splitting of the κ-casein from the surface of the micelles. The well depth is solely determined by the progression of the enzymatic reaction if all other conditions are kept constant. This gives

$$\frac{\varepsilon}{kT} = hk \frac{[E]}{[E_0]} t \qquad (5)$$

where $[E]$ is the enzyme concentration relative to a standard concentration $[E_0]$, and h is an experimentally determined proportionality constant (≈ 1.6).

The Second Virial Coefficient

For short-ranged attractions the precise shape of the potential well does not influence the macroscopic equilibrium properties.[17] In the dilute or semi-dilute region, the macroscopic equilibrium properties are described by the value of the second osmotic virial coefficient. For colloidal systems, the pair potential $V(r)$ of the particles can be influenced by changing ionic strength, pH, or the nature of the polymeric stabilization. The pair potential and the number density of the particles are sufficient to characterize the system.

The second osmotic virial coefficient is given by[18]

$$B_2' = 2\pi \int_0^\infty (1 - \exp(-V(r)/kT)r^2 \mathrm{d}r \qquad (6)$$

At a given moment in time, or for a certain degree of splitting of the

GMP, we may substitute the pair potential given by equation (4) and perform the integration which leads to

$$B_2 = 4 + 12 \frac{\Delta}{\sigma} (1 - \exp(+\,\varepsilon/kT)) \tag{7}$$

with

$$B_2 = B_2' / \frac{1}{6} \pi \sigma^3 = B_2'/V_{HS}$$

The macroscopic properties of hard-sphere (HS) colloidal systems can be calculated from statistical mechanics.[18] Similarly, Baxter[19] has made calculations for a so-called 'adhesive' hard-sphere system (AHS) which we use as a model description for the casein micelles in solution. Baxter used a mathematically convenient potential to solve the statistical mechanics equations. These calculations led to a dimensionless parameter τ which is a measure of the strength of the particle attraction. It can be considered as a dimensionless temperature. This so-called Baxter τ parameter is related to the second osmotic virial coefficient by

$$1/\tau = 4 - B_2 \tag{8}$$

Both equilibrium and transport properties of colloidal dispersions can be understood and described consistently using this formalism provided that pair interactions are predominant and that they are of the short-ranged attractive type.

Percolation

With increasingly strong attractions, the particles form temporal clusters whose sizes are such as to form an infinite network. That is, the system eventually forms percolated clusters. In statistical mechanics the percolation threshold is determined from the value of the pair-connectedness function, which diverges at percolation. The analytical results of Chiew and Glandt[20] for percolation describe the results of the computer simulations of Kranendonk and Frenkel[21] quite well at medium and higher volume fractions. The analytical approximation relating the attraction parameter τ to volume fraction at the boundary between percolating and non-percolating systems is

$$\tau_{perc} = \frac{19\phi^2 - 2\phi + 1}{12(1 - \phi)^2} \tag{9}$$

The percolation transition has been related to gelation phenomena in several investigations.[16,20,22,23] It appears, however, that equation (9) is not adequate for volume fractions below $\phi \approx 0.1$. In these dilute suspensions the percolation line seems to follow the spinodal decomposition curve. The spinodal line is the line where the second derivative of the Gibbs free energy of the system with respect to composition is zero. Equivalently, it is the line where the osmotic compressibility diverges. In light-scattering measurements the system shows critical opalescence and the turbidity diverges. The reciprocal osmotic compressibility is expressed in a virial series by

$$\frac{V_{HS}}{kT}\frac{d\pi}{d\phi} = 1 + 2B_2\phi \tag{10}$$

Along the spinodal line for $\phi < 0.1$, we have

$$1 + 2B_2\phi = 0$$

or

$$1/\tau_{perc} = 1/2\phi + 4 \quad (0.01 \leqslant \phi \leqslant 0.1) \tag{11}$$

The combination of equations (9) and (11) presumably describes the percolation line of sticky hard-sphere dispersions as a function of volume fraction. At this percolation line, various observable quantities, such as scattered light intensity (at infinite wavelength) and high-frequency elastic modulus, are seen to diverge.[24] Also viscosity will diverge, but since the system becomes extremely shear thinning[25] the applied shear-rate (stress) must become infinitely small, which is not the case when visible flocculation tests are made. Here it will be demonstrated that the AHS model predicts observable quantities quite well and that it can be used to calculate the flocculation time using the concept of percolation. With these equations the flocculation time of rennet clotting, as observed after adding chymosin to milk, can be calculated.

3 Results

Using equations (9) and (11) the percolation line has been calculated as a function of volume fraction. The result is shown in Figure 1, together with the computer simulation data of Kranendonk and Frenkel.[21] In addition, the osmotic equivalent of the gas–liquid co-existence region is shown. As a result of the (relatively) weak attractions the colloidal system separates into a dilute (gas) phase and a concentrated (liquid) phase at volume fractions below the critical point ($\phi < 0.13$). The fact that, at higher concentrations, the percolation line lies well above the region of demixing may explain why the transition is not readily observed: when the system is in the percolation phase, the kinetics are too slow to allow a physical phase separation even when attractions are of such strength that on thermodynamic grounds this would be expected.

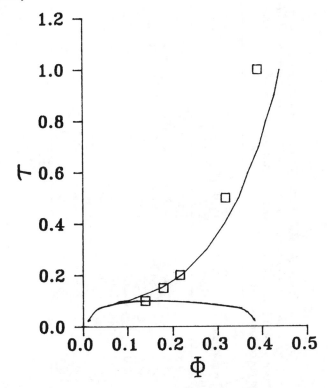

Figure 1 *Phase behaviour of a dispersion of adhesive hard spheres. The parameter τ is a dimensionless temperature. The points represent the percolation transition as derived from Monte-Carlo computer simulations.*

From Figure 1 it is clearly seen that gelation/percolation depends strongly on volume fraction. The volume fraction of casein micelles in normal skim milk is $\phi \approx 0.1$. The value of τ_{perc} is thus 0.1, and the second virial coefficient is -5.82. At each different volume fraction the percolation boundary is reached at a different value of τ_{perc}. These data are collected in Table 1. If we now use the equation for the enzyme kinetics in combination with the equation for the well depth and then calculate B_2 (which is equivalent to the τ-parameter), we can calculate the clotting time. Here clotting time is defined as the time when the system passes through the percolation line. The calculations are made keeping the ratio $[E]/[S]$ constant, corresponding to 'true' dilutions. For comparison we also performed the calculations with $[E]$ as a constant. The calculated clotting times are given in columns 6 and 7 of Table 1 and plotted in Figure 2 (column 6) as the dashed curve. The strong increase in clotting time at low volume fractions is mainly due to the decrease in enzyme concentration, but it occurs also because the attractions must be stronger.

To check these results experimentally would require light-scattering intensities in the limit of zero scattering angle, which are not available. We

Table 1 *Calculated milk clotting times at various volume fractions of casein micelles in the skim milk. The symbols have the following meanings:* ϕ = *volume fraction of micelles,* τ_{perc} = *value of Baxter* τ *parameter at the percolation line,* B_2 = *second osmotic virial coefficient,* ε = *depth of the attractive well,* t_{clott} = *clotting time under various conditions,* [E] = *enzyme concentration,* [S] = *substrate concentration, and* η_{rel} = *dispersion viscosity relative to the continuous phase*

ϕ	τ_{perc}	B_2 percolation	B_2 spinodal	ε/kT	t_{clott1} [E]/[S] = constant (min)	t_{clott2} [E] = constant (min)	η_{rel} t = 0 [E]/[S] = constant	t_x [E]/[S] = constant (min)	t_{clott3} ε = 5 kT (min)
0.4	0.75	2.67	–	3.05	9.5	38.0	2.94*	11.5*	16
0.3	0.36	1.21	–	3.80	15.5	46.5	2.28*	15.5*	21
0.2	0.18	-1.65	–	4.44	27.0	54.0	1.74	24	31
0.15	0.13	-3.69	–	4.75	38.5	57.8	1.51	34	41
0.10	0.10	-5.82	-5.0	5.00	62.0	62.0	1.31	54	62
0.075	0.0937	-6.73	-6.7	5.10	84.0	63.0	1.22	75	82.5
0.05	0.0714	–	-10	5.35	133	66.5	1.140	122	124
0.025	0.0417	–	-20	5.90	290	73.0	1.066	274	246
0.01	0.0185	–	-50	6.70	830	83	1.026	785	620
0.0075	0.0127	–	-75	7.10	1167	116	1.019	1100	820

* Extrapolated values.

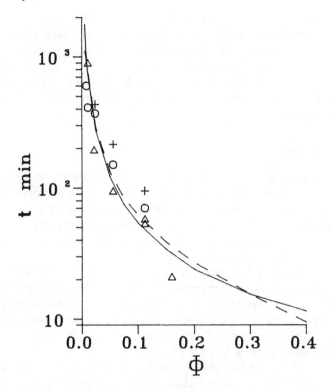

Figure 2 *Clotting time* t *as a function of volume fraction* φ *at a constant casein/ rennet ratio. Experimental points are from transmission measurements at* λ = 650 nm (○) *and at* λ = 1200 nm (+) *and from viscosity measurements* (△). *The dashed line denotes the calculated time for the system to reach percolation; the solid line denotes the time after which the viscosity again reaches the initial value*

have, however, made transmission measurements at finite wavelengths (λ from 600 to 1200 nm) as a function of volume fraction.[26] Extrapolation to zero transmission gives an apparent clotting time. This time depends on the wavelength used. A full discussion of these transmission measurements is given elsewhere.[26] The data are plotted in Figure 2. Since we have shown that viscosity curves as a function of time can be predicted quite accurately, we can also use the time after which the initial drop in viscosity due to the decreased volume fraction is compensated by the attractions between the casein micelles. This point is unambiguously defined and is amenable to experimental determination. Column 8 of Table 1 gives the viscosity (calculated) at the start of the experiment, and column 9 the time when this original value was reached again. It should be mentioned that the theory which calculates the dispersion viscosity is probably not correct for the two highest volume fractions, but nevertheless it gives the trend. In Figure 2 we have plotted these data as a full line together with the experimentally determined 'crossing times' t_x. The agreement between theory and experiment is quite good. Furthermore the ratio of clotting time to t_x is nearly constant. The last column in Table 1 gives the predicted time after which the pair potential reaches a depth of 5 kT. Note that this time for the high volume fractions is considerably longer than the time to reach the percolation line. At low volume fractions $(0.1 > \phi > 0.01)$ the difference is small.

4 Discussion

Figure 2 shows that the concept of percolation allows an unambiguous definition of the onset of flocculation. During the experimental determination, non-invasive methods should be used, such as optical transmission measurements. It should be realized that measurements must be extrapolated to $1/\lambda \to 0$, or a correction should be applied. With the theoretical and experimental results currently available,[26,27] this is entirely possible. Since the proteolysis is a very slow process, the system is always in a quasi-equilibrium state and therefore statistical mechanics can be used to describe the actual properties. Casein micelles are somewhat polydisperse. Dynamic light scattering gives a standard deviation from the mean (second cumulant) of 0.3–0.4. It is our experience that this degree of polydispersity only slightly affects the properties discussed here.

A method often used to define clotting time is the Berridge test,[5,6] in which the appearance of visible clots on the wall of a rotating glass vessel marks the clotting time. This method applies a shear stress to the system—just as a 'hot wire' gauge or a 'gelograph' does. If we estimate the shear-rate to be *ca.* $1\,s^{-1}$, it follows from a simple calculation that the interaction potential between the casein micelles must be greater than 5 kT in order for the flocs not to be disrupted by the shear stress. So, the Berridge test actually relates to a certain threshold value of the pair potential.

The last column of Table 1 gives the time required to reach the 5 kT value. If the enzyme concentration were kept constant, this time would be essentially constant if the enzyme kinetics remained strictly first order. This aspect is reflected in the last column of Table 1, since the predicted time varies directly in proportion to the enzyme concentration, which changes with substrate concentration.

It is concluded that the adhesive hard-sphere model can be used to describe and understand the enzymatic clotting of casein micelles. The model allows a consistent description of the different properties probed by viscosity and turbidity measurements.

References

1. P. Walstra and R. Jenness, 'Dairy Chemistry and Physics', Wiley, New York, 1984.

2. D. H. Napper, in 'Polymeric Stabilization of Colloidal Dispersions', Academic Press, London, 1983.

3. T. A. J. Payens, *Adv. Colloid Interface Sci.*, 1989, **30**, 31.

4. D. G. Dalgleish and E. W. Robson, *J. Dairy Res.*, 1984, **51**, 417.

5. N. J. Berridge, *J. Dairy Res.*, 1952, **19**, 328.

6. N. J. Berridge, *Analyst (London)*, 1952, **77**, 57.

7. K. W. Ruettiman and M. R. Ladish, *Enzyme Microb. Technol.*, 1987, **9**, 578.

8. D. G. Dalgleish, *J. Dairy Res.*, 1980, **47**, 231.

9. K. Guthy, D. Auerswald, and W. Buchheim, *Milchwissenschaft*, 1989, **44**, 560.

10. D. G. Dalgleish and D. S. Horne, *Milchwissenschaft*, 1991, **46**, 417.

11. A. Dulac, C. Durier, J. L. Bellow, J. P. Quiblier, and O. Cerf, *Sci. Aliments*, 1990, **10**, 749.

12. Th. J. M. Jeurnink and C. G. de Kruif, *J. Dairy Res.*, in press.

13. M. C. A. Griffin, J. C. Price, and W. G. Griffin, *J. Colloid Interface Sci.*, 1989, **128**, 223.

14. A. C. M. van Hooydonk, C. Olieman, and H. G. Hagedoorn, *Neth. Milk Dairy J.*, 1984, **38**, 207.

15. H. Holter, *Biochem. Z.*, 1932, **255**, 160.

16. C. G. de Kruif, Th. J. M. Jeurnink, and P. Zoon, *Neth. Milk Dairy J.*, 1992, **46**, 123.

17. C. Regnaut and J. C. Ravey, *J. Chem. Phys.*, 1989, **91**, 1211.

18. D. A. McQuarrie, 'Statistical Mechanics', Harper and Row, New York, 1976.

19. R. J. Baxter, *J. Chem. Phys.*, 1968, **49**, 2770.

20. Y. C. Chiew and E. D. Glandt, *J. Phys. A., Math. Gen.*, 1983, **16**, 2599.

21. W. Kranendonk and D. Frenkel, *Molecular Physics*, 1988, **64**, 403.

22. S. A. Safran, I. Wehman, and G. S. Crest, *Phys. Rev., A, Math. Gen.*, 1985, **32**, 506.

23. M. Tokita, R. Niki, and K. Hikichi, *J. Phys. Soc. Japan*, 1984, **53**, 480.

24. A. J. T. M. Woutersen, C. Blom, J. E. Mellema, and C. G. de Kruif, submitted.

25. A. J. T. M. Woutersen and C. G. de Kruif, *J. Chem. Phys.*, 1991, **94**, 5739.

26. C. G. de Kruif, submitted.

27. C. G. de Kruif, submitted.

Calcium Induced Flocculation of Emulsions Containing Adsorbed Phosvitin or β-Casein

By Josephine A. Hunt, Eric Dickinson, and David S. Horne[1]

PROCTER DEPARTMENT OF FOOD SCIENCE, UNIVERSITY OF LEEDS, LEEDS LS2 9JT, UK
[1]HANNAH RESEARCH INSTITUTE, AYR KA6 5HL, UK

1 Introduction

The state of aggregation of protein-coated oil droplets is an important factor affecting the structure, rheology, and stability of many food colloids. For casein-stabilized emulsions the degree of droplet flocculation is sensitive to the calcium concentration.[1] An interesting issue in such systems is the reversibility of the calcium-induced flocculation.[1,2]

It is the presence of several phosphoserines on the caseins (α_{s1}- and β-) which is the cause of the strong calcium binding. Another protein which strongly binds calcium is the egg-yolk protein phosvitin. This protein is exceptional in having over half of its constituent amino acids phosphorylated.[3] Phosvitin is a good emulsifying agent,[4] but we would expect its emulsion stabilizing properties to be extremely sensitive to calcium ions—possibly more so than the milk protein β-casein. This paper examines the flocculation behaviour of phosvitin-stabilized emulsions in the presence of calcium ions in comparison to β-casein-stabilized emulsions. Particular emphasis is placed on the reversibility aspects.

2 Materials and Methods

Freeze-dried β-casein[5,6] and phosvitin[7,8] were prepared as described previously. AnalaR-grade n-tetradecane was obtained from Sigma Chemicals. Buffer solutions were prepared from BDH AnalaR-grade reagents using double-distilled water. Calcium chloride ($CaCl_2.H_2O$) of AnalaR-grade was obtained from BDH.

Oil-in-water emulsions (0.5 wt% protein, 20 wt% n-tetradecane, pH 7) were prepared as described previously.[9] Calcium addition was achieved by adding a solution of twice the required strength to an equal volume of emulsion.

The extent and reversibility of flocculation was inferred from changes in apparent droplet size distribution measured by the Malvern Mastersizer.[4] Reversibility both towards dilution with buffer solution and towards dilution with calcium-free emulsion was investigated.

3 Results and Discussion

Figures 1 and 2 show how the droplet size distributions of phosvitin- and β-casein-stabilized emulsions, respectively, are affected by making the aqueous phase of the emulsions up to 5 mM in calcium. We observe that, compared to the emulsion prior to calcium addition, there is no significant alteration in the size-distribution for the β-casein emulsion after calcium addition, but that in the case of phosvitin a bi-modal distribution is formed, indicating flocculation. From binding isotherms,[10] we know that phosvitin possesses a much greater affinity for calcium than does β-casein. This explanation for the greater calcium sensitivity of phosvitin emulsions with respect to flocculation may be related to the emulsions being stabilized primarily by charge repulsion.[4] β-Casein-stabilized emulsions, which become flocculated at a calcium concentration of 10 mM, are stabilized both sterically and electrostatically.

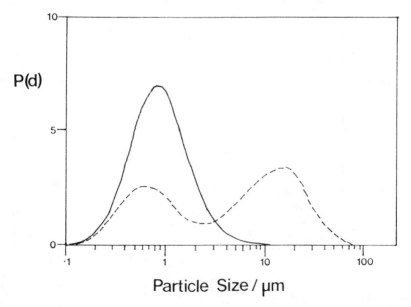

Figure 1 *The effect on the droplet size distribution of adding calcium (5 mM) to a phosvitin-stabilized emulsion (0.5 wt % protein, pH 7): ——, before calcium addition; - - - - -, after calcium addition. The number frequency of particles of diameter* d, P(d), *is plotted against the particle diameter*

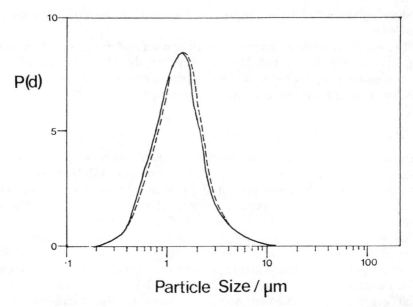

Figure 2 *The effect on the droplet size distribution of adding calcium* (5 mM) *to a β-casein-stabilized emulsion* (0.5 wt % *protein*, pH 7): ——, *before calcium addition*; – – – –, *after calcium addition*

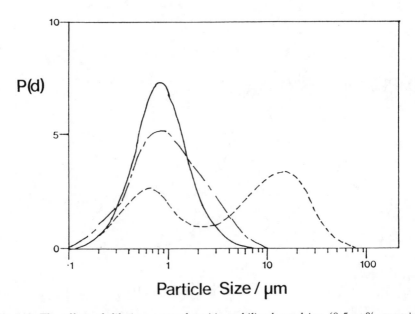

Figure 3 *The effect of dilution on a phosvitin-stabilized emulsion* (0.5 wt % *protein*, pH 7), *flocculated by the addition of calcium* (5 mM): ——, *control* (*no calcium*); – – – –, *flocculated emulsion*; – – –, *flocculated emulsion measured after 10 minutes in the Mastersizer*

This calcium-induced flocculation is reversible upon dilution with buffer, as demonstrated in Figure 3 for a phosvitin-stabilized emulsion. The flocculated emulsion was left circulating in the Mastersizer sample tank and, as the time increases, we see that the emulsion size-distribution approaches that of the original calcium-free emulsion. Similar deflocculation on dilution for β-casein-stabilized emulsions has previously been reported.[2]

By mixing various proportions of two phosvitin-stabilized emulsions— one made with no calcium and the other with 15 mM calcium—deflocculation is observed (see Figure 4). The emulsion made with 15 mM calcium is highly flocculated. However, on mixing it with various proportions of the calcium-free emulsion, there is a redistribution of calcium ions resulting in a certain degree of flocculation.

Clearly calcium ions have considerable influence on the flocculation behaviour of both β-casein- and phosvitin-stabilized emulsions. We can also confirm that this influence extends to other interfacial properties such as surface shear viscosity and the competitive displacement behaviour.

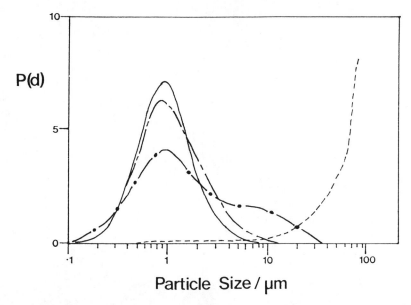

Figure 4 *Droplet-size distributions of mixed phosvitin-stabilized emulsions. Emulsion A was calcium-free and emulsion B was made with* 15 mM *calcium:* —, *emulsion* A; ----, *emulsion* B; – – –, A:B = 6:1; – • –, A:B = 3:1

References

1. E. Dickinson, *Colloids Surf.*, 1989, **42**, 191.
2. E. Dickinson, R. H. Whyman, and D. G. Dalgleish, 'Food Emulsions and Foams', ed. E. Dickinson, Special Publication No. 58, Royal Society of Chemistry, London, 1987, p. 40.

3. G. E. Perlman, *Isr. J. Chem.*, 1973, **11**, 393.
4. E. Dickinson, J. A. Hunt, and D. G. Dalgleish, *Food Hydrocolloids*, 1991, **4**, 403.
5. C. A. Zittle and J. H. Custer, *J. Dairy Sci.*, 1963, **46**, 1183.
6. W. D. Annan and W. Manson, *J. Dairy Res.*, 1969, **36**, 259.
7. F. J. Joubert and W. H. Cook, *Can. J. Biochem.*, 1958, **36**, 399.
8. A. Tziboula and D. G. Dalgleish, *Food Hydrocolloids*, 1990, **4**, 149.
9. E. Dickinson, A. Murray, B. S. Murray, and G. Stainsby, in 'Food Emulsions and Foams', ed. E. Dickinson, Special Publication No. 58, Royal Society of Chemistry, London, 1987, p. 86.
10. J. A. Hunt, E. Dickinson, and D. S. Horne, *Food Hydrocolloids*, 1992, **6**, 359.

Interactions and Contributions of Stabilizers and Emulsifiers to Development of Structure in Ice-cream

By H. Douglas Goff

DEPARTMENT OF FOOD SCIENCE, UNIVERSITY OF GUELPH, GUELPH, ONTARIO N1G 2W1, CANADA

1 Introduction

Ice-cream is a complex food colloid that consists of air bubbles, fat globules, ice crystals, and an unfrozen serum phase. The air bubbles are usually lined with fat globules[1] and the fat globules are coated with a protein–emulsifier layer[2]. The serum phase consists of the sugars and high-molecular-weight polysaccharides in a freeze-concentrated solution. Various steps in the manufacturing process, including pasteurization, homogenization, ageing, freezing, and hardening, contribute to the development of this structure. Two categories of additives are usually incorporated into ice-cream mix formulations, the emulsifiers and the stabilizers. Each of these contributes to the final structure and texture of the ice-cream, but their mode of action is very different. Emulsifiers (mainly small-molecule surfactants) are used to promote dryness during extrusion and a smooth texture with slow meltdown. Stabilizers (mainly polysaccharides) are used in frozen foods such as ice-cream to protect the product from the development of a coarse texture during temperature fluctuations which may occur during storage and distribution. The objectives of this research were to examine the mode of action of these two additives.

2 Materials and Methods

The action of the emulsifiers was explored using various combinations of proteins and surfactants in both mixes and model systems by measuring the fat–serum interfacial tension with a du Nuoy ring surface tensiometer; measuring the surface excess on the emulsion droplets by centrifugal removal of the fat globules and determining adsorbed protein by difference (Kjeldahl); examining interfacial layers by transmission electron microscopy (TEM) and image analysis; and measuring emulsion destabilization by

spectrometry during freezing of ice-cream mixes.[2-4] The action of the
stabilizers was studied using various concentrations and types of polysaccharides, in both mixes (11% fat, 11% milk solids, non-fat, 12% sucrose,
4% 42DE corn syrup solids) and model systems, by measuring the glass
transition temperature T_g, the amount of frozen water, and the onset of
melting by differential scanning calorimetry (DSC) and thermomechanical
analysis (TMA); measuring the viscosity at subzero temperatures by TMA
parallel-plate rheometry; and examining the ice-crystal size distributions as
a function of storage times and temperatures by low-temperature (cryo-)
scanning electron microscopy (LT-SEM) and image analysis.[5,6]

3 Results and Discussion

The Behaviour of the Surfactants

The adsorption of milk proteins to fat globules lowered the interfacial
tension of the fat–serum interface from $8.26\,\mathrm{mN\,m^{-1}}$ to $5.5\,\mathrm{mN\,m^{-1}}$.
However, the addition of a surfactant (polysorbate 80) lowered the
interfacial tension further than was accomplished by the proteins alone, to
$2.24\,\mathrm{mN\,m^{-1}}$ in the presence of the milk protein, thus favouring its
preferential adsorption to the fat globule surface. This led to a reduction in
the amount of protein which was adsorbed, from 15.9 wt % of the total
protein in the ice-cream mix to 7.8 wt % in the presence of the polysorbate
80. TEM techniques demonstrated that there were significantly more casein
micelles adsorbed to the fat globules in the absence of the surfactant
(3.82 ± 0.25 per fat globule section) than in the presence (2.20 ± 0.18 per
fat globule section). The emulsifier had no effect on the size distribution of
the globules in the mix. The lowering of the surface excess causes the
ice-cream-mix emulsion to be less stable to the shear forces encountered in
the barrel freezer which result from the formation of ice crystals and the
mechanical action of the knives and dashers. This results in a controlled
amount of fat globule aggregation (emulsion destabilization) around the air
bubbles. Partial coalescence imparts desirable structure and texture to the
ice-cream.

The Behaviour of the Polysaccharides

The controlling effect of the stabilizers on ice crystal growth was demonstrated by LT-SEM. Ice-cream mixes without stabilizer and with 0.15 wt %
locust bean gum and 0.02 wt % carrageenan added, which were continuously frozen and blast hardened ($-25\,°C$), were compared after preparation and after 24 weeks of storage at temperatures fluctuating daily
between $-25\,°C$ and $-10\,°C$. Image analysis of the micrographs from this
study illustrated the growth in size of the ice crystals that occurred as a

function of time during the temperature-abusive storage, and also demonstrated the protective effect that the stabilizers provided. Both the initial size of the crystals and the rate of growth of the crystals were reduced in the presence of the stabilizers.

The effect of polysaccharides on the phase behaviour of sucrose solutions was examined to determine whether differences existed in low temperature thermal events occurring in DSC or TMA scans, *i.e.* the glass transition temperature T_g of the maximally freeze-concentrated solution (a glass being defined as an amorphous metastable solid with a high viscosity of $>10^{13}$ Pa s), the onset temperature of the melting endotherm, the peak maximum temperature, or the enthalpy of the melting endotherm (hence the amount of frozen water). The polysaccharide was found to have no effect on the thermal behaviour of the sucrose solution. The viscosity of the ice-cream mixes with and without added stabilizer at sub-ambient temperatures was calculated from sample deformation measurements in the parallel plate rheometer attachment of the TMA. A large divergence in the two samples was seen at temperatures below $-14\,°C$. At a temperature of $-18\,°C$, the viscosity of the stabilized sample was 2.7×10^6 Pa s as compared with 1×10^6 Pa s in the unstabilized sample at the same temperature.

Although the polysaccharides have been shown to provide beneficial sensory attributes in frozen systems, it has been difficult to demonstrate their mode of action.[7] Recent research on the subject of cryostabilization[8,9] suggests that kinetic mobility is the overriding mechanism in frozen food stability. The results presented here show that polysaccharides increase viscosity at temperatures above T_g thus reducing molecular mobility in the freeze-concentrated visco-elastic serum phase and providing resistance to recrystallization and structural collapse.

Acknowledgements

The author wishes to acknowledge the contributions of M. Liboff (Cornell University), J. Kinsella (University of California—Davis), K. Caldwell (Ault Foods), and M. Sahagian (University of Guelph) for their contributions to this research, and to thank the Wisconsin Milk Marketing Board, the Natural Sciences and Engineering Research Council of Canada, and the Ontario Ministry of Agriculture and Food for support of this project.

References

1. K. B. Caldwell, H. D. Goff, and D. W. Stanley. *Food Struct.*, 1992, **11**, 1.
2. H. D. Goff, M. Liboff, W. K. Jordan, and J. E. Kinsella. *Food Microstruct.*, 1987, **6**, 193.
3. H. D. Goff and W. K. Jordan. *J. Dairy Sci.*, 1989, **72**, 18.
4. H. D. Goff, J. E. Kinsella, and W. K. Jordan. *J. Dairy Sci.*, 1989, **72**, 385.
5. K. B. Caldwell, H. D. Goff, and D. W. Stanley. *Food Struct.*, 1992, **11**, 11.

6. H. D. Goff, K. B. Caldwell, D. W. Stanley, and T. J. Maurice. *J. Dairy Sci.*, 1992, submitted.
7. A. H. Muhr and J. M. V. Blanshard. *J. Food Technol.*, 1986, **21**, 683.
8. H. Levine and L. Slade, in 'Thermal Analysis of Foods', ed. V. R. Harwalker and C. Y. Ma, Elsevier Applied Science, New York, 1990, p. 221.
9. H. D. Goff. *Food Res. Int.*, 1992, accepted.

Polymer–Polymer Interactions

Protein–Polysaccharide Interactions in Food Colloids

By Eric Dickinson

PROCTER DEPARTMENT OF FOOD SCIENCE, UNIVERSITY OF LEEDS, LEEDS
LS2 9JT, UK

1 Introduction

Proteins and polysaccharides are present together in nearly all food
colloids.[1,2] In the formulation of food colloids, proteins are best known for
their emulsifying and foaming properties, and polysaccharides for their
water-holding and thickening properties. In addition, both proteins and
polysaccharides contribute to the structural and textural characteristics of
many food colloids through their aggregation and gelation behaviour.[3-5]
Ideally, one would like to be able to predict the stability and rheology of a
system from the list of ingredients which go into making it. This is,
however, no simple task for a complex multiphase system containing a
mixture of biopolymers, since the colloidal properties are dependent not
only on the functionality of the individual protein and polysaccharide
components present, but also on the nature and strength of the protein–
polysaccharide interactions. While much is now known about the functional
properties of individual food proteins and polysaccharides in model systems, it is fair to say that our knowledge of the role of protein–polysaccharide interactions is still rather limited. Against this background, the
present article describes some recent research at Leeds on the influence of
protein–polysaccharide interactions in the stabilization of emulsions. To
put our work into context, we first briefly recall some of the relevant
concepts and terminology.

Macromolecular interactions are multifarious: they may be weak or
strong, specific or non-specific, attractive or repulsive. Repulsive interactions are always non-specific and of transient duration. They usually arise
from excluded volume effects and/or electrostatic interactions, and they
tend to be relatively weak, except at very close range or very low ionic
strength. Net repulsive protein–polysaccharide interactions are most likely
to be found in mixtures of proteins with non-ionic polysaccharides, or with
anionic polysaccharides at a pH above the protein isoelectric point pI.
Attractive biopolymer interactions may be weak or strong, and either

specific or non-specific. A covalent linkage between protein and poly-saccharide represents an attractive interaction which is specific, strong, and permanent. Non-specific net attractive protein–polysaccharide interactions arise as a result of averaging over a multitude of individual specific chemical interactions between groups on the biopolymers: ionic, dipolar, van der Waals, hydrogen bonding, hydrophobic, *etc.* Strong attractive interactions may occur between positively charged proteins (pH < pI) and anionic polysaccharides, especially at low ionic strength, and weak attractive interactions may occur between uncharged or negatively charged proteins (pH > pI) and polysaccharides. In any particular system, the protein–polysaccharide interaction may change from net repulsive to net attractive, or *vice versa*, on changing the temperature or the solvent conditions (pH, ionic strength). Any changes in protein–protein or poly-saccharide–polysaccharide interactions will also indirectly affect the strength of the unlike interaction.

Aqueous solutions of protein plus polysaccharide may exhibit phase separation at finite concentrations. It is well recognized that there are two alternative kinds of behaviour: coacervation[6] and incompatibility.[7,8] Coacervation (or complex coacervation, as it is sometimes called) involves the spontaneous separation into solvent-rich and solvent-depleted phases, the latter containing both protein and polysaccharide; it is caused by co-precipitation of protein–polysaccharide complexes under the influence of net attractive (non-specific) protein–polysaccharide interactions. Thermodynamic incompatibility, on the other hand, involves the spontaneous separation into two solvent-rich phases, one composed predominantly of protein, and the other of polysaccharide; it is caused by demixing of non-dilute protein and polysaccharide solutions under the influence of net repulsive protein–polysaccharide interactions. Whether a particular binary combination of protein and polysaccharide exhibits coacervation or incompatibility (or neither) depends on the concentration, the temperature, and the solvent conditions.

The way in which macroscopic behaviour is affected by the nature of the protein–polysaccharide interactions is perhaps seen most directly in the gelation behaviour.[9,10] Depending on the interactions, a gel formed from a mixture of two biopolymers may contain a coupled network, an inter-penetrating network, or a phase-separated network.[11,12] In food colloids, the two most important proteinaceous gelling systems are gelatin and casein micelles. An example of a covalent protein–polysaccharide inter-action is that produced[13] when gelatin reacts with propylene glycol alginate under mildly alkaline conditions. Non-covalent non-specific attractive interactions occur in mixed gels of gelatin with sodium alginate or low-methoxy pectin.[14,15] In systems containing mixtures of gelatin with the non-gelling polysaccharide dextran, the rate of gelatin triple-helix forma-tion is greatly increased by the presence of net repulsive protein–poly-saccharide interactions.[10] Under conditions of thermodynamic incompatib-ility, gelatin forms phase-separated gels with agar.[16] Under acidic conditions, casein micelles aggregate to produce particle gels,[17] and the presence of attractive protein–polysaccharide interactions in mixed milk gelling systems (*e.g.* casein and carrageenan[9,18]) can produce composite particle-polymer gels. With systems containing casein and some polysaccharides (carrageenan, alginate, *etc.*), the presence of small concentrations of calcium ions induces gelation via specific attractive protein–polysaccharide linkages.

In a food emulsion containing protein and polysaccharide, any of these kinds of interactions may take place in the aqueous phase of the system with consequences for structure, rheology, and stability. In addition, it is necessary to consider how the nature of the protein–polysaccharide interaction affects the surface behaviour of the biopolymers and the aggregation properties of the dispersed droplets. To illustrate the various types of phenomena involved, we separately consider the cases of (i) weak reversible protein–polysaccharide interaction and (ii) permanent protein–polysaccharide covalent linkage.

2 Weak Protein–Polysaccharide Interactions

Let us begin by considering a mixture of milk protein and hydrocolloid: sodium caseinate and xanthan. Sodium caseinate is a very important proteinaceous emulsifying agent used in the formulation of food colloids; it is itself a multicomponent interacting mixture of proteins, the main surface-active components being β-casein and, to a lesser extent, α_{s1}-casein.[1] Xanthan is an extracellular anionic polysaccharide ($\sim 2 \times 10^6$ daltons) widely used in food as a thickening agent and as a synergistic gelling agent (with locust bean gum); a notable feature of xanthan in solution is its extreme pseudoplasticity, which is maintained over a wide range of temperature, pH, and ionic strength.[19]

The influence of xanthan, added after emulsification, on the creaming of micron-sized oil-in-water emulsions (10 wt% mineral oil, 0.5 wt% caseinate, pH 7) is illustrated in Figure 1 as a plot of serum layer thickness *versus* storage time. Visual inspection of the reference emulsion (no added polysaccharide) shows no serum separation discernible by eye over the observation period of 72 hours.[20] (Ultrasonic velocity scanning of the emulsion does reveal[21] the existence of an oil-depleted serum layer a few millimetres thick after 72 hours; this is not readily apparent to the unaided eye, though it does become so after 1 week.) The presence of 0.125 wt% xanthan in the aqueous phase also gives an emulsion that exhibits no discernible serum separation over the observational time-scale. On the other hand, in emulsions with lower polysaccharide content, distinct creaming in quite evident, and at xanthan concentrations of 0.025 and 0.05 wt% the rate of serum separation at the bottom of the sample is very rapid.

Emulsion stability is determined by the relative contributions of protein–polysaccharide and polysaccharide–polysaccharide interactions.

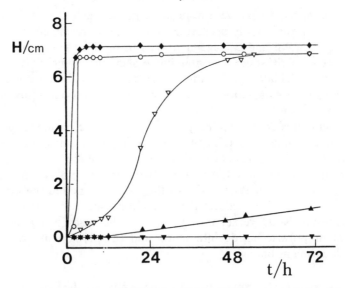

Figure 1 *Creaming of oil-in-water emulsions* (10 wt % *mineral oil,* 0.5 wt % *sodium caseinate,* 0.005 M *phosphate buffer* pH 7.0) *containing various concentrations of xanthan* (*Shell "Maxaflo"*) *in samples of total height* 10 cm. *The height* H *of the serum layer at the bottom of the sample is plotted against the time* t *of storage at* 25 °C *for various concentrations of added polysaccharide:* ▼, 0 wt % *and* ≥ 0.125 wt %; ◆, 0.025 wt %; ○, 0.05 wt %; ▽, 0.0625 wt %; ▲, 0.1 wt %

The inhibition of creaming at xanthan concentrations ≥ 0.125 wt % is due to immobilization of dispersed oil droplets in a weak gel-like network with a high low-stress shear viscosity.[19] Above 0.25 wt % xanthan, there is no serum or cream layer detectable ultrasonically after storage for several weeks.[21] At low xanthan concentrations (below the weak gelation threshold), enhancement of creaming is attributed[20–22] to depletion flocculation of protein-coated emulsion droplets by non-adsorbing polysaccharide. Recent surface viscosity experiments have suggested[21] that xanthan in solution does not become complexed or associated in any way with casein adsorbed at the oil–water interface. This is consistent with a protein–polysaccharide interaction which is net repulsive. It implies thermodynamic incompatibility between xanthan in solution and protein around the emulsion droplets leading to separation of a dense polymer-depleted emulsion phase from a polymer-rich serum phase.[1,2]

The destabilization of emulsions by water-soluble polymers appears to be a general phenomenon. Figure 2 compares the creaming behaviour for 0.1 wt % xanthan with that for the same concentration of three other water-soluble polymers: a 'medium viscosity' carboxymethylcellulose (CMC7MF), a 'high viscosity' carboxymethylcellulose (CMC7HOF), and a microbial polysaccharide succinoglycan. From these results and others presented elsewhere,[20,21] we note that the relative abilities of the four

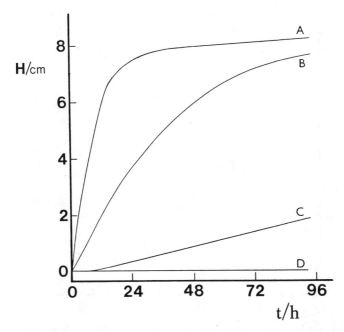

Figure 2 *Effect of four different polysaccharides* (0.1 wt %) *on the creaming of oil-in-water emulsions* (10 wt % *mineral oil, 0.5 wt % caseinate, pH 7.0) in samples of total height 10 cm. The serum layer height* H *is plotted against storage time* t *at 25 °C: A, CMC7MF; B, CMC7HOF; C, xanthan; D, succinoglycan*

polymers to confer emulsion stability under neutral pH conditions are: succinoglycan > xanthan > CMC7HOF > CMC7MF. The creaming stabilities in the polysaccharide concentration range 0.1–0.3 wt % are found to be qualitatively consistent with the dynamic viscosities of the polymer solutions ($\sim 10^{-2}$ Hz). For complete inhibition of creaming in the presence of added polysaccharide, the whole emulsion should possess gel-like behaviour with a significant yield stress. This is illustrated by the rheological data presented in Figure 3. In the emulsions containing xanthan or succinoglycan (0.25 wt %), which show no creaming over several weeks, we find[21] a substantial yield stress,* as well as appreciable shear hysteresis over a period of a few minutes for shear-rates below *ca.* 2 s^{-1}. By way of contrast, the emulsion containing 0.25 wt % CMC7MF, which exhibits extensive serum separation in just a few hours, is essentially Newtonian

---

*This is an *apparent* yield stress as opposed to a genuine Bingham-like yield value (see ref. 2). Measurements with a more sensitive constant-stress rheometer operating at shear stresses in the range $10^{-2}-1\,\mathrm{N\,m^{-2}}$ indicate a viscosity plateau in the limit of very low shear-rates ($< 10^{-4}\,\mathrm{s^{-1}}$). That is, in reality, the emulsions containing 0.25 wt % xanthan or succinoglycan are extremely pseudoplastic, with apparent viscosities some 10^5 times larger than that for the (Newtonian) emulsion containing no polysaccharide, but there is apparently no shear stress below which flow is entirely inhibited.

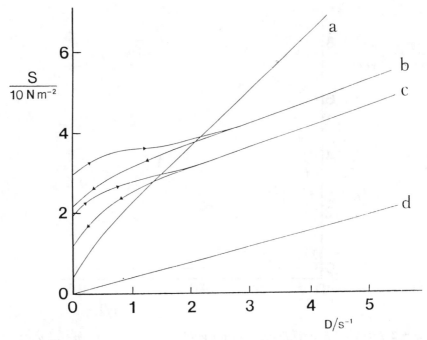

Figure 3 *Effect of four different polysaccharides* (0.25 wt %) *on the rheology of oil-in-water emulsions* (10 wt % *mineral oil,* 0.5 wt % *caseinate,* pH 7.0, 25 °C). *The shear stress* S *is plotted against the shear-rate,* D: (a) *CMC7HOF;* (b) *succinoglycan;* (c) *xanthan;* (d) *CMC7MF. Curves* (b) *and* (c) *have apparent yield stresses; 'the arrows indicate differences in measured stress for increasing and decreasing shear-rates*

down to low shear-rates, with no discernible yield stress or hysteresis. Though the origin of rapid creaming in emulsions containing carboxymethylcellulose is undoubtedly a reflection of polymer-induced flocculation, it is not certain that this is attributable entirely to a depletion mechanism (as postulated for xanthan). A net attractive protein–polysaccharide interaction could destabilize by a bridging mechanism.[23] Surface shear viscosity measurements provide evidence[21] for an interfacial complex between adsorbed casein and CMC7MF, as do measurements of electrophoretic mobility. This implies that some bridging flocculation of casein-coated droplets by CMC is a distinct possibility.

We turn next to emulsions containing the neutral non-gelling polysaccharide dextran. Previous work has demonstrated[24,25] that this polysaccharide induces serum separation in emulsions made with casein or gum arabic at moderate ionic strength. Here, we report recent data[26] on emulsions made with 11S globulin (broad bean) and dextran, where both polysaccharide and protein are present during emulsification. The pair interactions in mixed aqueous solutions of 11S globulin and dextran are known[27] from static light scattering. The cross second virial coefficient is positive corresponding to a considerable net repulsive protein–polysaccharide of entropic origin.[28] This means that, when the protein and polysaccharide components are mixed, there is an increase in the chemical potentials of the biopolymers which leads to phase separation (i.e. thermodynamic incompatibility) at high concentrations. At the critical point of 11S globulin and dextran, the biopolymer concentrations are 3.1 wt % and 1.3 wt %, respectively.

Figure 4 presents stability results[26] for oil-in-water emulsions (10 vol % hydrocarbon oil, 0.5 wt % protein, pH 8, 0.1 M) made with 11S globulin $(3.2 \times 10^5$ daltons) and low-molecular-weight dextran T40 $(4 \times 10^4$ daltons) at a polysaccharide/protein molar ratio, $R = 5$. The emulsion containing dextran T40 is more stable with respect to creaming than that made with protein alone [Figure 4(a)]. This is consistent with the average emulsion droplet diameter d_{43} being smaller after emulsification with dextran present, and it remaining so over the storage period [Figure 4(b)]. Hence, the dextran increases the emulsifying capacity (i.e. the area of fresh oil–water interface created during emulsification) of the protein, and improves the stability of the emulsion with respect to creaming/coalescence. The emulsifying capacity is presumably increased because the added dextran T40 raises the thermodynamic activity of the protein in solution (whilst maintaining a one-phase system), thereby enhancing the rate and extent of protein adsorption during emulsification, and enabling the stabilization of smaller droplets which are less prone to creaming. The tendency of the unadsorbed polysaccharide to repel more protein towards the droplet surface also results in a thicker proteinaceous steric stabilizing layer, which favours improved stability with respect to coalescence.

Replacement of dextran T40 by the high-molecular-weight dextran T500 $(5 \times 10^5$ daltons) leads to droplet sizes which are much more sensitive to polysaccharide/protein molar ratio R (see Figure 5). With dextran T500 present at $R = 0.2$, there is a similar improvement in emulsifying capacity of 11S globulin to that found with dextran T40 for $1 \leqslant R \leqslant 10$. On the other hand, the presence of a large quantity of dextran T500 present during emulsification $(R \geqslant 3)$ leads to much coarser emulsions being produced than with the protein alone.* The larger droplet size, together with depletion flocculation induced by the high-molecular-weight polysaccharide, gives rapid emulsion creaming with dextran T500 at $R \geqslant 3$. The reduced emulsifying capacity of 11S globulin seems to be related to a greatly increased protein surface coverage in the presence of dextran T500, i.e. 7.1 mg m^{-2} as compared with 2.6 mg m^{-2} for the protein alone.[26] This

---

*Though droplet sizes produced by high-pressure homogenization are relatively insensitive to the viscosity of the continuous phase, there is probably some small increase in d_{43} which is attributable to the viscosity increase of the aqueous phase at high dextran concentrations (especially T500). At low polysaccharide/protein molar ratios $(R \leqslant 1)$, the effect of viscosity modification of the continuous phase on the emulsifying capacity is estimated to be negligible. Another effect of polymer on emulsification efficiency is 'turbulence depression'; this also tends to lead to an increase in average droplet size—for molecular dimensions exceeding the Kolmogorov eddy size (P. Walstra, Chem. Eng. Sci., 1974, 29, 882).

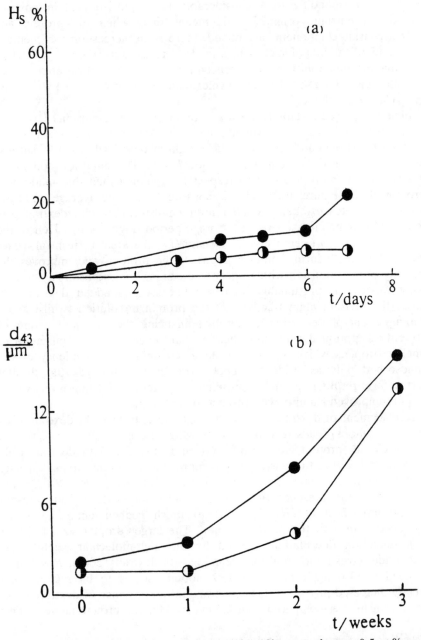

Figure 4 *Stability of oil-in-water emulsions (10 vol% n-tetradecane, 0.5 wt% pro-tein, 0.1 M phosphate buffer pH 8.0) made with 11S globulin Vicia faba and dextran T40 (1:5 molar composition). Filled and open symbols represent data for emulsions made with protein alone and protein and polysaccharide, respectively. (a) Serum layer thickness H is plotted against storage time t at 25 °C; (b) Average droplet size d_{43} is plotted against storage time t*

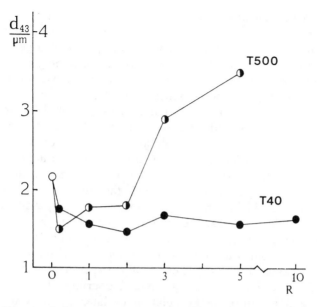

Figure 5 *Influence of dextran on the emulsifying capacity of 11S globulin. The average droplet diameter d_{43} of the emulsions (10 vol % n-tetradecane, 0.5 wt % protein, pH 8.0, ionic strength 0.1 M) is plotted against the polysaccharide/protein molar ratio R. Filled and open symbols represent dextrans T40 and T500, respectively*

high protein surface concentration is indicative of protein aggregation in bulk solution and multilayer formation at the interface under the influence of strong protein–polysaccharide incompatibility. So, while the moderate increase in thermodynamic activity of protein produces improved emulsifying capacity (and stability) with dextran T40, the opposite is the case for the high-molecular-weight dextran T500. It appears[26] that, under conditions of incipient phase separation during emulsification, it is difficult for protein to get quickly enough to the surface of new droplets to stabilize them against recoalescence, when it is already associated in the aqueous phase or adsorbed as thick layers on other droplets. The high concentration of dextran T500 increases the bulk protein chemical potential to such an extent that there is pre-wetting[29] of existing oil–water interface by the protein; this is the origin of the high surface coverage. Once the emulsion has been formed, stability is further impaired by depletion flocculation, which theory predicts[30] to be much more important for a high-molecular-weight polymer like dextran T500 than for dextran T40.

We have also studied the effect of dextran on the emulsifying properties of bovine serum albumin.[26,30] The general trends of emulsifying capacity as a function of polysaccharide/protein molar ratio and dextran molecular weight are similar to those with 11S globulin, but the change in average droplet size is much less significant in emulsions made with BSA and dextran.[26] A further difference is seen when we consider relative stabilities of emulsions with and without dextran T40. Even though the initial value of d_{43} is slightly lower in the presence of the polysaccharide ($R = 3$), the resulting emulsions are much less stable towards creaming and coalescence than those made with BSA alone. The relative insensitivity of the emulsifying capacity of BSA to the presence of dextran T40 or T500 may be a reflection of the lack of thermodynamic incompatibility in the BSA and dextran system. This would be consistent with the negative cross second virial coefficient found[28] in dilute solutions of mixtures of BSA with certain polysaccharides, corresponding to a protein–polysaccharide interaction which is weakly attractive. Such an interaction implies that there might be a weak association of dextran around the surface of protein-coated droplets, which could in turn lead to destabilization by bridging flocculation and hence loss of creaming stability. Analysis of the aqueous serum phase following centrifugation of emulsions made with BSA and dextran has suggested[26] a small adsorbed concentration ($< 0.1\ \mathrm{mg\,m^{-2}}$) of dextran on the BSA-coated droplets. There is also experimental evidence in the literature for adsorption of dextran onto the surface of BSA-coated latex particles[31] and for bridging flocculation of latex particles by dextran.[32] Some bridging flocculation in the emulsions made with BSA and dextran seems therefore to be quite plausible.

3 Covalent Protein–Polysaccharide Conjugates

The idea of using soluble protein–polysaccharide complexes to stabilize emulsions is an appealing notion.[22,23] Ideally, the two biopolymers should be linked together covalently to form a stable conjugate combining the favourable surface-active properties of the hydrophobic protein component with the potential steric stabilizing properties of the hydrophilic polysaccharide, thereby avoiding the undesirable complications of flocculation and phase separation encountered in systems with weak non-specific protein–polysaccharide interactions, as just described. It is, of course, important that conjugates intended for food use should be made without recourse to legally unacceptable chemicals. Recently, it has been shown[30,34–37] that conjugates having excellent emulsifying properties can be made by simple controlled heating of proteins and polysaccharides at 60 °C under conditions of low water activity.

Figure 6 shows droplet-size distributions of fresh oil-in-water emulsions (10 vol % n-tetradecane, 0.5 wt % protein, pH 8, ionic strength 0.1 M) made with Maillard complexes of low-molecular-weight dextran T40 with either 11S globulin or BSA at dextran/protein molar ratio $R = 3$. The complexes were prepared by dry heating at 60 °C for 3 weeks.[36] Also recorded in Figure 6 are data for emulsions made with simple (unreacted) mixtures of protein and dextran T40 ($R = 3$) and data for emulsions made with each protein alone ($R = 0$). We can see that emulsification with the globulin–dextran conjugate or the BSA–dextran conjugate generates

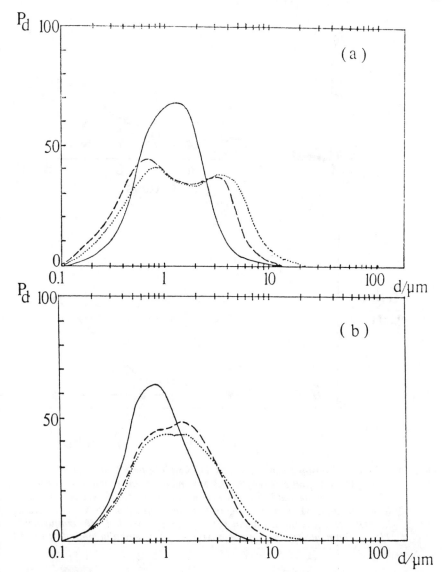

Figure 6 *Droplet-size distributions of fresh oil-in-water emulsions (10 vol% n-tetra-decane, 0.5 wt% protein, pH 8.0, ionic strength 0.1 M) containing (a) 11S globulin and (b) bovine serum albumin. Probability function P_d is plotted against droplet diameter, d: · · · ·, protein alone; - - -, simple mixture with dextran T40 (R = 3); ——, protein–dextran conjugate (R = 3)*

smaller droplets with a narrower size distribution than with 11S globulin or BSA alone or in simple admixture with the polysaccharide. The stability data recorded in Figure 7 indicate that emulsions made with the globulin–dextran conjugate ($R = 3$) have good stability with respect to creaming and

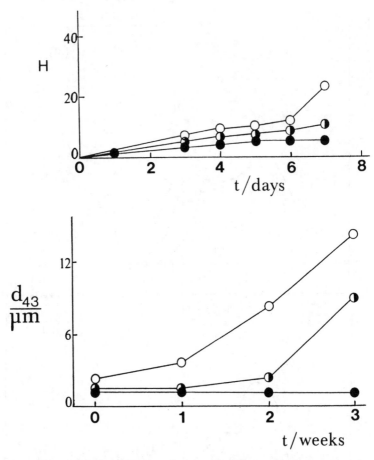

Figure 7 *Stability of oil-in-water emulsions* (10 vol% *n-tetradecane*, 0.5 wt% *protein*, pH 8.0, *ionic strength* 0.1 M) *made with globulin–dextran T40 (R = 3) conjugate made by dry-heating at 60 °C for 3 weeks. (a) Serum layer thickness H is plotted against storage time t at 25 °C. (b) Average droplet size d_{43} is plotted against storage time t. Symbols:* ●, *conjugate;* ○, *11S globulin alone;* ◑, *simple mixture of 11S globulin and dextran T40 (R = 3)*

coalescence.[37] We have also found that Maillard complexes having good emulsion stabilizing properties can be made from 11S globulin, β-lactoglobulin or BSA with high-molecular-weight dextran T500.[30,36,37]

To investigate the effect of the heat treatment on the protein itself (*i.e.* independent of any putative interaction with polysaccharide), emulsions were made with just 11S globulin or BSA which had been subjected to the same dry-heat treatment (60 °C for 3 weeks) as for the protein–polysaccharide conjugates. Table 1 lists values of the average droplet sizes, d_{32} and d_{43}, measured for the fresh emulsions.[37] It is clear that the improvement in emulsifying capacity (and also in emulsion stability[30,36]) on forming the conjugate is not due to the effect of dry-heating on the protein itself, but is a genuine reflection of an interaction with the polysaccharide induced by the dry-heating. Indeed, the dry-heated 11S globulin gives a substantially coarser emulsion than does the native (unheated) protein. With BSA, there is relatively little difference between native and dry-heated protein samples in terms of emulsion droplet-size distribution. The loss of functionality of pure 11S globulin on dry-heating is consistent with a 25% reduction in solubility.[37] The same protein may be dry-heated with dextran T40 at 60 °C for 3 weeks without any apparent loss of solubility. The heating time of 3 weeks appears to be a good compromise for this protein–polysaccharide system. Extending the heating time to 5 weeks does lead to visible precipitation of the globulin–dextran conjugate and a reduction in emulsifying capacity. Figure 8 shows that, whereas at high dextran contents ($R = 5$ or 20) the average droplet size d_{43} is little affected by heating time, at low contents ($R = 0.5$ or 1) there is a large increase in d_{43} with heating time. The more polysaccharide there is present, the more the heated protein appears to be protected against denaturation, coagulation, and loss of emulsifying capacity. This means that the amount of polysaccharide present during dry-heating should not be too low.

We now consider the effects of polysaccharide molecular weight and polysaccharide/protein molar ratio on the emulsifying capacity of the protein–dextran conjugates. Figure 9 indicates that the high-molecular-weight dextran T500 is more effective in producing small droplets at low molar ratios ($R \leqslant 3$ for 11S globulin,. $R \leqslant 0.5$ for BSA).[37] The emulsifying capacity with the low-molecular-weight dextran T40 is gradually improved with increasing proportions of polysaccharide (up to $R \approx 5$). This is consistent with the idea that more dextran T40 molecules can bind successfully to protein to make a surface-active complex than can dextran T500 molecules. While the 1:1 Maillard complex of 11S globulin with dextran T500 is a good emulsifying agent, the conjugate made from dry heating BSA and dextran T500 in a 1:1 ratio is rather poor. Certainly, the plot of d_{43} against R in Figure 9(b) closely resembles that found[26] for emulsions made with simple mixtures of BSA and dextran T500. This

Table 1 *Average droplet diameters in fresh oil-in-water emulsions* (10 vol % *n-tetra-decane, 0.5 wt % protein, pH 8.0, ionic strength 0.1 M) made with (i) pure (unheated) protein, (ii) pure protein dry-heated at 60 °C for 3 weeks, or (iii) protein and dextran T40 dry-heated at 60 °C for 3 weeks.*

Sample	d_{32}/μm	d_{43}/μm
11S globulin *Vicia faba*	0.86	2.2
bovine serum albumin (BSA)	0.86	2.0
dry-heated 11S globulin	1.38	3.0
dry-heated BSA	0.95	1.8
globulin–dextran conjugate ($R = 3$)	0.85	1.35
BSA–dextran conjugate ($R = 3$)	0.66	1.02

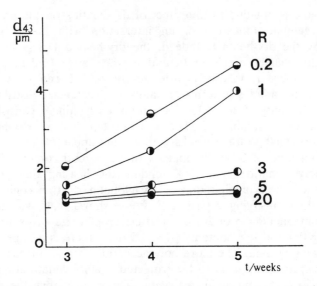

Figure 8 *Effect of dry-heating time* (60 °C) *on average droplet size* d_{43} *of oil-in-water emulsions* (10 wt% *n-tetradecane*, 0.5 wt% *protein*, pH 8.0, *ionic strength* 0.1 M) *made with globulin–dextran T40 conjugates of different polysaccharide/protein molar ratio:* ◒, $R = 0.2$; ◑, $R = 1$; ◐, $R = 3$; ○, $R = 5$; ●, $R = 20$

suggests that even one bulky hydrophilic dextran T500 molecule linked to BSA reduces its emulsifying and surface-active properties. On the other hand, one presumes that the high-molecular-weight protein moiety of the globulin–dextran T500 conjugate is still readily accessible to the oil–water interface during emulsion formation.

Under optimized conditions, globular protein–dextran conjugates give excellent emulsion stability with respect to coalescence and creaming. It is important, however, that there should be no free polymer present in the aqueous phase which might cause depletion flocculation.[38] Free unadsorbed dextran could be present in protein/polysaccharide samples subjected to dry-heating for too short a time or with too high a polysaccharide content. This puts a further restriction on the molar ratio R, especially in systems containing high-molecular-weight polysaccharides.

Improvement in the emulsification properties of globular proteins may be achieved by dry-heating with other polysaccharides such as propylene glycol alginate, dextran sulphate, or amylopectin.[30,36] Covalent complexation of the disordered protein β-casein with dextran, however, leads to a substantial loss of emulsifying capacity.[37] The reason for this difference between the β-casein–dextran conjugate and those formed with the globular proteins is not clear at present. One possibility is that the disordered casein molecule becomes partly entangled within the skeleton of the larger polysaccharide molecule(s), thereby shielding many of the hydrophobic

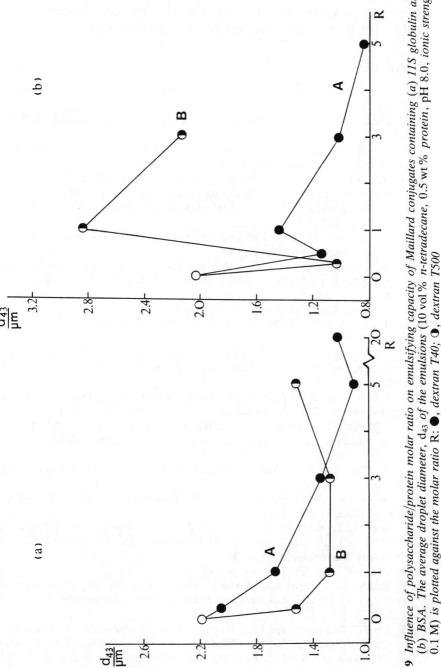

Figure 9 *Influence of polysaccharide/protein molar ratio on emulsifying capacity of Maillard conjugates containing (a) 11S globulin and (b) BSA. The average droplet diameter, d_{43} of the emulsions (10 vol % n-tetradecane, 0.5 wt % protein, pH 8.0, ionic strength 0.1 M) is plotted against the molar ratio R: ●, dextran T40; ◐, dextran T500*

protein residues from rapidly adsorbing at the oil–water interface during emulsification. Further experimental information on the molecular weight and surface properties of these conjugates is required, however, before anything like a thorough understanding can begin to emerge.

References

1. E. Dickinson and G. Stainsby, 'Colloids in Food', Applied Science, London, 1982.
2. E. Dickinson, 'An Introduction to Food Colloids', Oxford University Press, 1992.
3. J. R. Mitchell and D. A. Ledward, ed., 'Functional Properties of Food Macromolecules', Elsevier Applied Science, London, 1986.
4. A. H. Clark and S. B. Ross-Murphy, *Adv. Polym. Sci.*, 1987, **83**, 57.
5. P. Harris, ed., 'Food Gels', Elsevier Applied Science, London, 1990.
6. H. G. Bungenberg de Jong, in 'Colloid Science', ed. H. R. Kruyt, Elsevier, Amsterdam, 1949, Vol. 1, p. 232.
7. P.-A. Albertsson, 'Partition of Cell Particles and Macromolecules', 2nd Edn., Wiley Interscience, New York, 1971.
8. V. B. Tolstoguzov, in 'Functional Properties of Food Macromolecules', ed. J. R. Mitchell and D. A. Ledward, Elsevier Applied Science, London, 1986, p. 385.
9. G. Stainsby, *Food Chem.*, 1980, **6**, 3.
10. E. R. Morris, in 'Food Gels', ed. P. Harris, Elsevier Applied Science, London, 1990, p. 291.
11. G. J. Brownsey and V. J. Morris, in 'Food Structure—Its Creation and Evaluation', ed. J. M. V. Blanshard and J. R. Mitchell, Butterworths, London, 1988, p. 7.
12. V. J. Morris, in 'Food Polymers, Gels and Colloids', ed. E. Dickinson, Special Publication No. 82, Royal Society of Chemistry, Cambridge, 1991, p. 310.
13. J. E. McKay, G. Stainsby, and E. L. Wilson, *Carbohydr. Polym.*, 1985, **5**, 223.
14. M. A. Muchlin, E. S. Wajnermann, and V. B. Tolstoguzov, *Nahrung*, 1976, **20**, 313.
15. G. Ja. Tschmak, E. S. Wajnermann, and V. B. Tolstoguzov, *Nahrung*, 1976, **20**, 321.
16. A. H. Clark, R. K. Richardson, S. B. Ross-Murphy, and J. M. Stubbs, *Macromolecules*, 1983, **16**, 1367.
17. P. Walstra, T. van Vliet, and L. G. B. Bremer, in 'Food Polymers, Gels and Colloids', ed. E. Dickinson, Special Publication No. 82, Royal Society of Chemistry, Cambridge, 1991, p. 369.
18. P. M. T. Hansen, *Prog. Food Nutr. Sci.*, 1982, **6**, 127.
19. E. R. Morris, in 'Extracellular Microbial Polysaccharides', ed. P. A. Sandford and A. Laskin, ACS Advances in Chemistry Series, Washington, DC, 1977, Vol. 45, p. 81.
20. Y. Cao, E. Dickinson, and D. J. Wedlock, *Food Hydrocolloids*, 1990, **4**, 185.
21. Y. Cao, E. Dickinson, and D. J. Wedlock, *Food Hydrocolloids*, 1991, **5**, 443.
22. E. Dickinson and S. R. Euston, in 'Food Polymers, Gels and Colloids', ed. E. Dickinson, Special Publication No. 82, Royal Society of Chemistry, Cambridge, 1991, p. 132.

23. E. Dickinson and L. Erikkson, *Adv. Colloid Interface Sci.*, 1991, **34**, 1.
24. S. Bullin, E. Dickinson, S. J. Impey, S. K. Narhan, and G. Stainsby, in 'Gums and Stabilizers for the Food Industry', ed. G. O. Phillips, D. J. Wedlock, and P. A. Williams, IRL Press, Oxford, 1988, Vol. 4, p. 337.
25. E. Dickinson, D. J. Elverson, and B. S. Murray, *Food Hydrocolloids*, 1989, **3**, 101.
26. E. Dickinson and M. G. Semenova, *J. Chem. Soc., Faraday Trans.*, 1992, **88**, 849.
27. M. G. Semenova, G. E. Pavlovskaya, I. A. Popello, and V. B. Tolstoguzov, *Carbohydr. Polym.*, 1992, in press.
28. V. B. Tolstoguzov, *Food Hydrocolloids*, 1991, **4**, 429.
29. E. Dickinson, in 'Gums and Stabilisers for the Food Industry', ed. G. O. Phillips, D. J. Wedlock, and P. A. Williams, IRL Press, Oxford, 1988, Vol. 4, p. 249.
30. E. Dickinson and V. B. Galazka, in 'Gums and Stabilisers for the Food Industry', ed. G. O. Phillips, D. J. Wedlock, and P. A. Williams, Oxford University Press, 1992, Vol. 6, p. 351.
31. R. Edwards and P. R. Rutter, *J. Colloid Interface Sci.*, 1980, **78**, 304.
32. A. Lips, I. J. Campbell, and E. G. Pelan, in 'Food Polymers, Gels and Colloids', ed. E. Dickinson, Special Publication No. 82, Royal Society of Chemistry, Cambridge, 1991, p. 1.
33. V. B. Tolstoguzov and E. E. Braudo, *J. Dispersion Sci. Technol.*, 1985, **6**, 575.
34. A. Kato, Y. Sasaki, R. Furuta, and K. Kobayashi, *Agric. Biol. Chem.*, 1990, **54**, 107.
35. A. Kato and K. Kobayashi, *ACS Symp. Ser.*, 1991, **448**, 213.
36. E. Dickinson and V. B. Galazka, *Food Hydrocolloids*, 1991, **5**, 281.
37. E. Dickinson and M. G. Semenova, *Colloids Surf.*, 1992, **64**, 299.
38. M. R Shaw and D. Thirumalai, *Phys. Rev. A*, 1991, **44**, R4797.

Thermodynamic Incompatibility of Food Macromolecules

By Vladimir B. Tolstoguzov

INSTITUTE OF FOOD SUBSTANCES OF THE RUSSIAN ACADEMY OF SCIENCES,
VAVILOV STREET 28, MOSCOW 117813, GSP-1, RUSSIA

1 Introduction

This paper will discuss aspects of the thermodynamic incompatibility of proteins and polysaccharides. Why is this incompatibility so important for the functionality of food biopolymers? We will take up this question, as well as the reasons why the phenomenon has many technological implications.

2 Thermodynamic Incompatibility of Food Macromolecules

Figure 1 shows two phase diagrams for mixed solutions of the 11S broad bean globulin (called legumin) plus dextran and skimmed milk protein plus carboxymethylcellulose (CMC). In the composition region lying outside the binodal curve, aqueous biopolymer solutions are completely miscible. The region lying within the binodal curve represents compositions of two-phase systems. This means that, on mixing aqueous solutions of different biopolymers, the mixture breaks down into two liquid phases and a 'water-in-water' (W/W) emulsion can be formed.[1-5] The upper and lower branches of the binodals represent compositions of the co-existing phases. Binodal branches are joined together at the critical point. The thin lines are tie-lines. They connect the points corresponding to the compositions of the co-existing phases. Each phase mainly contains one of the biopolymers. Thermodynamic incompatibility of biopolymers is observed under conditions of weaker attractive forces between macromolecules of different types. As a result each macromolecule shows preference for being surrounded by its own type. The investigation of the affinity of biopolymers for each other and for the solvent (water) in dilute solution by the light-scattering technique enables the phase behaviour of their mixed concentrated solutions to be predicted.

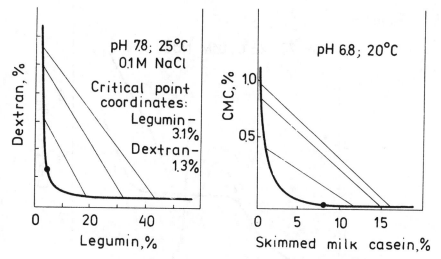

Figure 1 *Phase diagrams of two protein and polysaccharide systems. (a) The system 11S broad bean globulin (legumin), dextran (2.5 × 10⁵ daltons), and water (pH 7.8, 0.1 M NaCl, 20 °C). (b) The system skim milk protein, CMC (degree of polymerization = 500, degree of substitution = 0.85), and water (pH 6.8, 20 °C). The point (●) is the critical point. Thick curves are the binodals; thin lines are tie-lines*

We turn now to look at the matter in terms of the functional properties of mixtures of proteins and polysaccharides.[5-7]

3 Solubility

Phase diagrams can be regarded as the co-solubility profiles of a given biopolymer pair.[5,7] This approach is very informative, since it takes account of the multicomponent nature of food systems. Many functional properties of food biopolymers depend on their co-solubility in aqueous media.

4 Stabilization of Emulsions and Foams

Figure 2 shows protein adsorption on n-decane droplets dispersed in aqueous solutions of legumin (curve 1) or in mixed solutions of legumin and 1% dextran (curve 2). The protein adsorption isotherms are notable for an extended plateau. Its existence shows that a monolayer of adsorbed protein molecules depresses further adsorption of the same protein. This feature of protein adsorption seems to result from the thermodynamic incompatibility of the adsorbed 'surface oriented and denatured' protein with the same native protein dissolved in the aqueous dispersion medium. Presumably, two main interrelated reasons may be responsible for the fact that protein molecules dissolved in the dispersion medium can not 'identify'

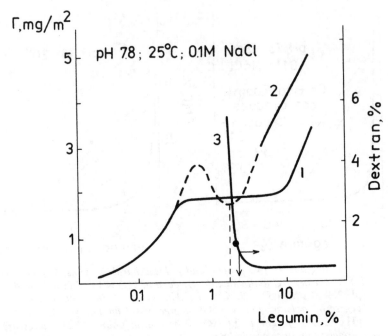

Figure 2 *Adsorption isotherm of legumin and its mixture with dextran* (1 wt%, 2.5×10^5 daltons) (pH 7.8, 0.1 M NaCl, 20 °C). *Surface concentration Γ is plotted against protein concentration*: (1) *without dextran*, (2) *with dextran. Curve* (3) *shows the phase diagram of the mixture*

the adsorbed molecules as being the same. The first is protein conformational change arising from its partial 'surface denaturation'. The second is a relative increase in net charge and hydrophilicity of adsorbed protein molecules resulting from exposure of the protein hydrophilic residues to the aqueous side of the adsorbed layer. The addition of the incompatible polysaccharide dextran to a solution of legumin gives rise to an increase in thermodynamic activity of both biopolymers, which therefore behave as if they are in a more concentrated solution. This manifests itself in a shift in the adsorption isotherm position towards lower protein concentration.[5-7] Presumably, however, similar shift may also result from complexing of proteins with lipids (or hydrocarbons).[5,6] It has also been suggested that the difference between foaming and emulsifying properties of a protein may arise from the formation of protein–lipid complexes.[6]

Protein multilayer formation seems to result from phase separation in the aqueous continuous phase. This latter gives rise to micro-encapsulation of dispersed particles.[6] The comparison of the phase diagram with the protein adsorption isotherm for the same legumin–dextran–water system shows that the protein concentration required for multilayer formation is located in the vicinity and below the binodal curve. Presumably, multilayer formation can be regarded as a transition from O/W emulsion to a more stable W/W emulsion, or from the simple O/W emulsion to the multiple decane-in-water-in-water (O/W/W) emulsion.[6] The influence of the addition of polysaccharides on the stability of protein-stabilized emulsions can not be only attributed to an increase in the continuous phase viscosity. It may be assumed that stabilization by polysaccharides is mainly due to the increased viscosity and modulus of the adsorbed protein layer. At the transition from protein monolayer to multilayer coverage, the conformational difference between protein molecules in an external layer and in bulk solution may be reduced. Presumably, this also means that depletion flocculation can occur in emulsions with protein concentration lower than that required for a multilayer formation.[7] Hence, on adding more protein and polysaccharide, the aggregative stability of O/W emulsions can be increased.[6,7]

5 Some Features of Water-in-Water Emulsions

Let us briefly consider some specific features of W/W emulsions. Incompatibility of biopolymers, or more precisely limited compatibility of biopolymers, is a general phenomenon. One should not, therefore, underestimate the role of W/W emulsions in foods. One of the main features of a W/W emulsion is the relatively similar compositions of its phases. The presence of a common solvent, water, and the significant co-solubility of biopolymers, seem to be the main contributory factors to the low interfacial tension of W/W emulsions. Another feature of W/W emulsions is the relatively high viscosity of the phases formed by the aqueous solutions of the macromolecular substances. Both these factors influence the deformability of dispersed droplets in flowing W/W emulsions.[7]

Presumably, W/W emulsions are also notable for the presence of an interfacial or depletion layer with a low biopolymer concentration and viscosity. Its presence around dispersed particles is the direct result of the incompatibility of these biopolymers. The interfacial layer surrounding the dispersed particles can act as a specific lubricator affecting the rheological properties of W/W emulsions. This interfacial layer between the two aqueous polymer solutions may be occupied by water-soluble and insoluble food flavouring ingredients of low molecular weight. Dispersed particles can also obviously distribute themselves between the interfacial layer and the aqueous phases according to their surface properties. Adsorption of dispersed particles of lipid (or a flavouring component) within the interfacial layer may be accompanied by the interaction of the dispersed particles and the formation of a new continuous phase in spite of the low lipid content. Thus, the adsorption and interaction of dispersed particles within the interfacial layer may lead to heterogeneity of the system with respect to chemical composition and physical properties. This chemical and physical heterogeneity seems to be an important factor determining the specific organoleptic features of many low fat foods. Presumably, for this reason, it is possible to replace the lipid continuous phase or dispersed droplets of lipids by aqueous dispersed particles covered (and bound) by lipids. In other words, thin lipid layers between aqueous dispersed particles may form a three-dimensional network, thereby reducing the lipid content in foods whilst producing a lipid continuous phase, *e.g.* in such food products as low-fat spreads.

The degree of chemical heterogeneity reflecting structural and physical heterogeneity of foods seems to be important for the sensory evaluated stimulus. In this connection, the relationship between the sizes and distances between the detecting elements of the tongue, on the one hand, and the dispersed particles of W/W emulsions, on the other, needs to be studied, as well as the role of the relative hydrophilicity of the components of both phases in determining the composition and properties of these phases, including the partition of low-molecular-weight soluble and insoluble components between bulk aqueous phases and the interfacial layer.

Surfactants for W/W emulsions also seem to be an interesting research area of great scientific and practical value. The characteristic properties of the emulsifiers for O/W emulsions are due to the functional groups having an affinity for both the non-polar and water phases. In the same way surface-active agents for W/W emulsions containing proteins and polysaccharides may obviously be compounds containing protein and polysaccharide parts covalently bonded together. This means that the hybrid protein–polysaccharide compounds, which have been considered by Dickinson as emulsifying agents for O/W emulsions,[8,9] may also be regarded as surfactants for W/W emulsions. They are likely to perform a variety of functions in W/W emulsions. Presumably, both natural (*e.g.* proteoglycans) and synthetic protein–polysaccharide hybrids may be used to increase the work of adhesion between two aqueous phases containing the same proteins and polysaccharides, and to control co-solubility of proteins and polysaccharides in aqueous media. One of the functions of proteoglycans and glycoproteins appears to be to increase the stability and homogeneity of mixed biopolymer solutions (*e.g.* in the case of egg-white), as well as to increase the adhesion between structural elements in composite foods.

Presumably, incompatibility of protein and polysaccharide could be shown up in micro-segregation of the protein and polysaccharide parts within the volume of the same hybrid macromolecule. This could result in an increased asymmetry of macromolecules and hence provide some interesting rheological effects. The latter could also result from an increased viscosity of the interfacial layer due to the concentrating protein–polysaccharide surfactants. Because of incompatibility of biopolymers, chemical interactions between dissimilar macromolecules are less probable, both in solution and melt. Below the glass transition temperature, the conformational rearrangements of macromolecules and changes in the distribution of chemical interactions can occur very slowly. Biopolymer complexes or conjugate compounds can obviously be obtained at a temperature exceeding the glass transition temperature, more precisely under conditions allowing mobility of segments, but inhibiting mobility of whole macromolecules, and so preventing segregation of dissimilar macromolecules. Accordingly, cross-linking of incompatible biopolymers proceeds when aqueous solutions of these biopolymers are mixed, frozen (concentrated by solvent freezing), dried, and heated at a low moisture level.[9] The latter governs both the denaturation level and the glass transition temperature.

It should be noted that the effect of different media (*e.g.* solvents containing food ingredients of low molecular weight) on the phase behaviour of biopolymer mixtures is of great practical and theoretical significance. For instance, the study of the phase behaviour of mixtures of proteins and polysaccharides in solutions of sugars (such as sucrose, glucose, fructose, and lactose) is necessary for a better understanding of structure formation processes in many food systems, *e.g.* in ice-cream mixes.

Membraneless Osmosis

A feature of W/W emulsions containing proteins and polysaccharides is that generally the two co-existing phases differ in water content. The greater the difference between biopolymers in molecular weight and hydrophilicity, the higher the asymmetry of the phase diagram.[2,3] The asymmetry of phase diagrams for protein, polysaccharide, and water systems provides the basis for a new method of concentrating protein solutions, called membraneless osmosis.[1-3,7] Unlike conventional osmosis (*i.e.* the transfer of solvent molecules to a solution through a semi-permeable membrane), the membraneless osmosis process involves two immiscible solutions of biopolymers with different chemical potentials of the solvent water. The interface between the two immiscible aqueous solutions is the hypothetical semi-permeable membrane in membraneless osmosis. When the W/W emulsion is made, there is an enormous increase in the total surface area between the two aqueous solutions. Therefore, it only takes a very short time for material equilibrium to be established. Then the two liquid phases can be separated by means of a centrifuge, or simply by standing and decantation. This inexpensive method allows a rapid concentration of a large amount of dispersed protein (*e.g.* skimmed milk) under mild conditions. On mixing a solution of CMC with skimmed milk, the mixture separates into two phases.[2] The phase rich in casein micelles contains 80–90% of the milk proteins with a concentration of 20–40 wt%. Casein concentration from skimmed milk by membraneless osmosis is an example of depletion flocculation. It also should be noted that the membraneless osmosis process presumably governs water partition between the phases in many food systems.

Spinneretless Spinning and Thermoplastic Extrusion

Let us briefly consider some other applications of the incompatibility phenomenon.

Figure 3 shows the general scheme of the spinneretless spinning (or fibre-shaping) process which is presumably responsible for the fibrous and lamellar structure of many foods.[1,3,10] In a flowing W/W emulsion an anisotropic structure may be formed because of deformation and orientation of liquid dispersed particles. The shape of the liquid filaments can be fixed by converting one or both liquid phases into the solid state. Here, the dispersed phase acts as a spinning dope, whilst the dispersion medium functions as a spinneret and coagulating bath simultaneously.

It should be noted that the existence of polymer incompatibility in a solution implies incompatibility in the melt as well. Therefore, it may be supposed that in the thermoplastic extrusion process we are dealing with heterophase mixtures of melted water-plasticized biopolymers. This assumption is confirmed by the results presented in Figure 4. Mixtures of soybean protein isolate with starch have been extruded through a cooled die under conditions preventing explosive water evaporation. Figure 4 shows the change in the expansion of the visco-elastic jet at the exit of the shaping die, and the dispersibility of the extrudate in an excess of water,

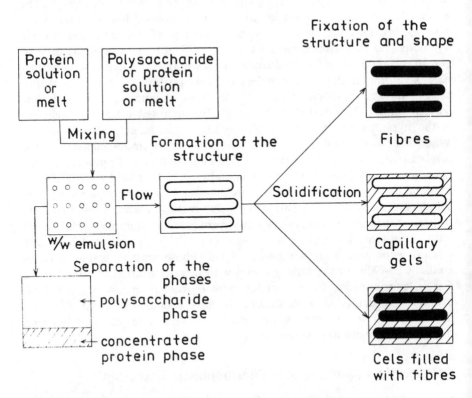

Figure 3 *Schematic representation of the processes of spinneretless spinning and membraneless osmosis*

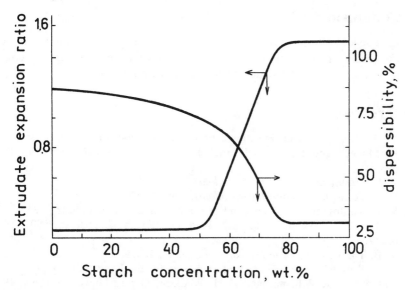

Figure 4 *Thermoplastic extrusion of soybean protein isolate and starch mixtures. The degree of expansion of the extruded jet and the dispersibility of the extrudates in excess water (at 20 °C after 48 hours) are plotted against the starch content. Data were obtained using a Brabender DN Extruder (L/D 20:1) with A4:1 compression screw operating at 20 rpm (zones: conveying, heating, forming die 160 °C, cooled die 2 × 30 × 100 mm 110 °C). The initial water content was 30 wt %*

both as a function of the starch content. These specific changes in properties may reflect phase inversion in the melt mixture of water-plasticized proteins and polysaccharides, arising from incompatibility of the biopolymers in the melt. As the volume fraction of dispersed phase is increased, the emulsion of polysaccharide melt in protein melt changes to an emulsion of protein melt in polysaccharide melt. It is interesting that, as with O/W and W/O emulsions, the extruded systems undergo the process of phase inversion at a volume fraction of dispersed phase exceeding 80–85%. Formation of protein–polysaccharide complexes or protein–carbohydrate compounds by Maillard reaction may also be of importance to the phase behaviour of melt mixtures during thermoplastic extrusion, as well as to the adhesion between the phases in composite extrudates.

It should also be noted that incompatibility in mixtures of different polysaccharides may be of importance for the structure–property relationship in thermoplastic extrusion products. Chinnaswamy and Hanna recently reported[13] that the ratio of amylose and amylopectin in starch determines the jet expansion ratio. The latter is increased as the amylose content increases. This could perhaps be related to the incompatibility of structurally unlike polysaccharides, including amylose and amylopectin in melts.[3,13–15]

6 Conclusion

In conclusion, it should be stressed that the general nature of the phenomenon of thermodynamic incompatibility of biopolymers accounts for its importance in the structure formation of many foods.

References

1. V. B. Tolstoguzov, in 'Functional Properties of Food Macromolecules', ed. J. R. Mitchell and D. A. Ledward, Elsevier Applied Science, London, 1986, p. 385.
2. V. B. Tolstoguzov, *Food Hydrocolloids*, 1988, **2**, 195.
3. V. B. Tolstoguzov, *Food Hydrocolloids*, 1991, **5**, 339.
4. V. B. Tolstoguzov, in 'Gums and Stabilisers for the Food Industry', ed. G. O. Phillips, P. A. Willams, and D. J. Wedlock, IRL Press, Oxford, 1990, Vol. 5, p. 157.
5. V. B. Tolstoguzov, *Food Hydrocolloids*, 1991, **4**, 429.
6. V. B. Tolstoguzov, in 'Gums and Stabilisers for the Food Industry', ed. G. O. Phillips, P. A. Williams, and D. J. Wedlock, Oxford University Press, 1992, Vol. 6, p. 241.
7. V. B. Tolstoguzov, *Int. Food Ingredients*, 1990, **2**, 8.
8. E. Dickinson and S. R. Euston, in 'Food Polymers, Gels and Colloids', ed. E. Dickinson, Special Publication No. 82, Royal Society of Chemistry, Cambridge, 1991, p. 132.
9. E. Dickinson and V. B. Galazka, *Food Hydrocolloids*, 1991, **5**, 281.
10. V. B. Tolstoguzov, in 'Food Structure—Its Creation and Evaluation', ed. J. M. V. Blanshard and J. R. Mitchell, Butterworths, London, 1988, p. 181.
11. V. B. Tolstoguzov, in 'Developments in Meat Science', ed. R. A. Lawrie, Elsevier Applied Science, London, 1991, Vol. 5, p. 159.
12. V. B. Tolstoguzov, *Food Technol. Int.*, 1991, 71.
13. R. Chinnaswamy and M. A. Hanna, *Int. Food Ingredients*, 1990, **3**, 2.
14. M. T. Kalichevsky, P. D. Oxford, and S. G. Ring, *Carbohydr. Polym.*, 1986, **6**, 145.
15. M. T. Kalichevsky and S. G. Ring, *Carbohydr. Res.*, 1987, **162**, 323.

Mixed Biopolymers in Food Systems: Determination of Osmotic Pressure

By M. M. G. Koning, J. van Eendenburg, and D. W. de Bruijne

UNILEVER RESEARCH LABORATORIUM VLAARDINGEN, POSTBOX 114, 3130 AC VLAARDINGEN, THE NETHERLANDS

1 Introduction

Most foods are mixtures of a number of high-molecular-weight proteins and carbohydrates. In food processing we often want to adjust the consistency of the food to preset requirements; one of the routes to achieve this goal is by manipulation of the mixing behaviour of the polymers present.

As far as mixing is concerned, high-molecular-weight substances behave differently from low-molecular-weight substances. As a general rule, dissimilar high-molecular-weight substances do not mix on a molecular scale, while low-molecular-weight substances do mix. This difference in mixing behaviour is due to the gain in entropy on mixing which is significant for small molecules but negligible for large ones. The consequence of non-mixing is that the various biopolymers present in an actual food material will form separate aqueous phases that are mutually immiscible (like oil and water).

Tolstoguzov and his coworkers[1] have demonstrated experimentally that indeed in most ternary systems, consisting of two biopolymers and water as solvent, phase separation occurs at sufficiently high concentrations. Exceptions to this rule are biopolymers which exhibit specific interactions, such as, for example, complex coacervates formed through electrostatic attraction, and the complexes between κ-carrageenan and κ-casein.

In the design of food systems, it is desirable to know the mixing behaviour of possible biopolymer combinations under suitable conditions in order to be able to predict the final consistency of the product. Nowadays the food scientist has a great number of biopolymers available which can be used as ingredients. Measurement of the mixing behaviour of all possible combinations would be a tedious and time-consuming job.

In this work we assume that, when two biopolymers exist in a phase-separated system, the one phase consists only of the one biopolymer in aqueous solution and the other phase contains just the second biopolymer. So complete phase separation of the two biopolymer solutions is assumed. A closer look at the phase diagrams which are available at present[1] indicates that this assumption is close to being valid in many cases. When the two phases are in thermodynamic equilibrium, the chemical potential of the solvent in the two phases is the same. Therefore, measurement of the solvent chemical potential as a function of the concentration of single biopolymers in solution should be sufficient to determine the concentrations at which two biopolymers are in equilibrium in a phase separated system. One of the methods which is suitable for measurement of chemical potentials is measurement of osmotic pressures of biopolymer solutions, since the osmotic pressure is directly related to the solvent chemical potential in a solution.

In this paper osmotic pressure measurements for a number of biopolymer systems are presented. In addition, an indirect method for the measurement of the osmotic pressures based on the use of a semipermeable dialysis tube is described and the results are discussed.

2 Experimental

Materials

Dextran T500 was obtained from Pharmacia. According to the supplier the weight-average molecular weight is 503 000 daltons as measured by light scattering. The number-average molecular weight is 170 300 daltons as determined by end group analysis.

All the materials used in the experiments were commercially available food ingredients. When appropriate, the solutions were heated to the temperature required for swelling or gel formation, and cooled afterwards. In some cases the pH was controlled by buffers.

Osmotic Pressure Measurements

Osmotic pressures were measured for the biopolymer solutions in water at 10 °C with a Jupiter osmometer (model 231) equipped with a RC 52 membrane which has a molecular weight cut-off of 20 000 daltons (ex Schleicher and Schuell).

Dialysis Bag Experiments

Osmotic pressures of very viscous solutions and gels were determined in an indirect way using dialysis bags. Such a bag consists of a piece of tubular dialysis membrane. The dialysis bag is filled with the biopolymer solution of known concentration and its weight is determined. Subsequently the bag is placed in a vessel containing a reference dextran solution of a given concentration and thus of a known osmotic pressure. Both the reference solution and the biopolymer solution inside the bag are buffered. The solution inside the bag will now either take up or release water and other low molecular weight substances until the chemical potential inside the bag equals that in the outside solution. At equilibrium, the concentration of the biopolymer is then determined by carefully weighing the dialysis bag again. Using dextran solutions of different concentrations allows the determination of the equilibrium concentrations of the biopolymer solutions at various osmotic pressures.

3 Results and Discussion

Osmotic Pressure Measurements

In Figure 1a the measured osmotic pressure as a function of concentration is shown for Dextran T500. The measurements have a reproducibility of 2.5%. No deviations from data published in the literature[2,3] for these dextran solutions were observed.

Based on the lattice model developed by Flory, the osmotic pressure Π as function of concentration c can be described by the following third degree polynomial function:[4]

$$\Pi = RT\left[\left(\frac{1}{M_n}\right)c + \left(\frac{1}{2} - \chi\right)\left(\frac{\bar{v}^2}{V_S}\right)c^2 + \left(\frac{\bar{v}^3}{3V_S}\right)c^3\right] \tag{1}$$

In equation (1), R is the gas constant, T is the absolute temperature, M_n is the number-average molecular weight of the polymer, V_S is the molar volume of the solvent, v is the specific volume of the polymer, and χ the Flory–Huggins interaction parameter, which is a measure of the strength of the interaction between the polymer and the solvent. Fitting the data to a third degree polynomial yielded a set of coefficients from which M_n was determined to be 220 000 daltons, the specific volume $v = 0.60 \text{ cm}^3 \text{ g}^{-1}$ and $\chi = 0.48$. The osmotic pressure curve calculated on the basis of these variables is indicated in Figure 1a. The value for the number-average molecular weight agrees quite well with the one given by the manufacturer. The value of $0.6 \text{ cm}^3 \text{ g}^{-1}$ for v is typical for the specific volume of carbohydrates. The value of 0.48 for χ is close to 0.5 where the solvent quality changes from good to poor. The inferred value of χ for this type of dextran in water is possibly so high because this dextran has a high molecular weight and is heavily branched. Both factors reduce the solvent quality of water for this molecule.

A different approach to the interpretation of the data is based on scaling theory.[5] In this semi-empirical approach, certain polymer properties, of which osmotic pressure is one, are related to the concentration and degree of polymerization from a universal point of view. The relations generally are power laws where the polymer characteristic is related to, for example,

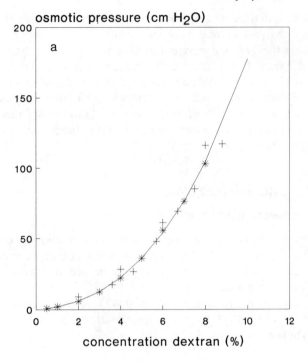

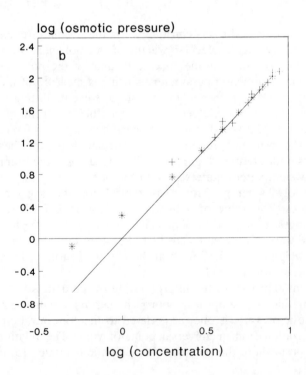

the concentration through a scaling exponent. For the osmotic pressure the following relation is derived for non-ionic polymers in a good solvent:[5]

$$\frac{\Pi}{RT} \approx \left(\frac{c}{c^*}\right)^{3v/3v-1} \approx \left(\frac{c}{c^*}\right)^{9/4} \tag{2}$$

The value for v is $3/5$ in a good solvent, yielding a scaling exponent of $9/4$. This relation holds in the semi-dilute region where the concentrations are above the threshold concentration c^* at which the polymer chains start to overlap. This approach is thus valid when the second quadratic term in equation (1) dominates. The c^* value for this type of dextran is close to a concentration of 2.5 wt%. In Figure 1b a log–log plot of the osmotic pressure *versus* concentration is shown. It is clear from the slope of the curve at the higher concentrations, which is 2.22, that the dextran solution obeys this power-law. Another approach is required for the region below c^*.

Salt Dependency

When dealing with polyelectrolytes the situation is far more complicated than for the case of the neutral dextran molecules. To the parameters characterizing the behaviour of the neutral polymers (concentration, molecular weight, and solvent quality) must be added ionic strength and polyion charge.

The ions associated with a strong electrolyte contribute considerably to the osmotic pressure. In the presence of salt, all the small ions can permeate freely through the membrane and they will tend to distribute themselves in order to produce electrical neutrality. Hence, the distributions of the ions on either side of the membrane, and the measured osmotic pressure, depend on the amount of salt added. This phenomenon is called the Donnan effect. The effect of the addition of salt can be clearly seen in Figure 2 where the osmotic pressure of sodium alginate is shown in the presence of various levels of added NaCl. With no salt added the measured values for the osmotic pressure are quite high, whereas upon addition of salt the values decrease drastically. When a very large amount of salt is present, the Donnan-effect can be completely eliminated and the parameters for the alginate molecules can be obtained from extrapolation of the data to infinite salt concentration.

Figure 1 (*a*) *The osmotic pressure as a function of concentration for dextran T500. Our experimental points are indicated (***); data from the literature are also indicated (+). The measurements were performed at 10 °C with the Jupiter osmometer. The continuous line indicates the fit obtained with equation (1) and parameter values which are mentioned in the text. (*b*) A log–log plot of the same osmotic pressure data against concentration. The straight line represents a least square fit through the last five data points*

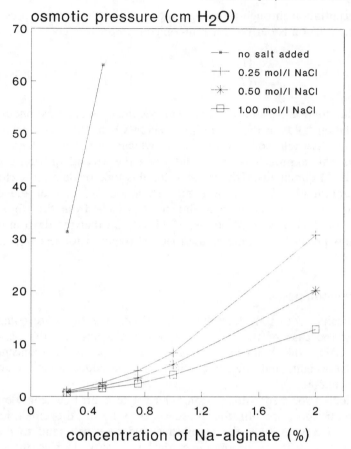

Figure 2 *The effect of salt on the osmotic pressure of sodium alginate. The alginate was dissolved in solutions of different salt content and the osmotic pressures were recorded at 10 °C with the Jupiter osmometer*

Dialysis Experiments

Typical experimental curves obtained with the dialysis bag experiments are shown in Figure 3 for sodium carboxymethylcellulose (CMC) and agar. It is clear that equilibrium between the solution inside the bag and the reference solution outside is reached only after a long time, *i.e.* of the order of weeks. Diffusion of water, and the other small molecules which

Figure 3 *Build-up to equilibrium in the dialysis experiments. (a) The concentration of CMC in the dialysis bag as a function of time in three different reference solutions. Measurements were performed at 10 °C. No buffers or salt were added to the CMC or dextran solutions. (b) The concentration of agar in the dialysis bag as a function of time. The measurements were performed at 10 °C and both the agar and dextran solutions were buffered at pH = 3.5*

concentration Na-CMC (%)

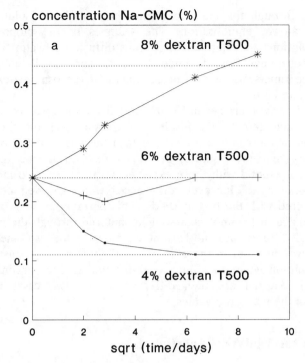

sqrt (time/days)

concentration agar (%)

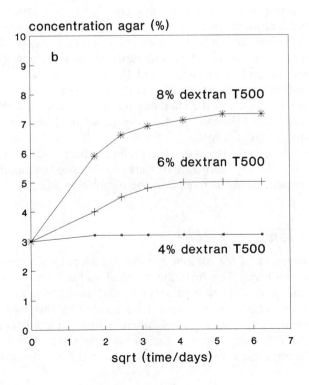

sqrt (time/days)

are present, through the membrane and the biopolymer solutions determines the rate of equilibration. The volumes of the biopolymer and reference solutions are relatively large; thus diffusion of water takes a long time. The osmometer is designed to have very small volumes, and therefore the time-scale of the measurement in the osmometer is of the order of half an hour.

The dotted horizontal lines in Figure 3a indicate the CMC concentrations which are calculated to be inside the dialysis bag on the basis of extrapolation of the osmotic pressure data measured with the osmometer. It is obvious that the equilibrium values reached in the dialysis bag experiment correspond with those expected on the basis of osmometry for CMC. In some cases, however, differences were observed between the values measured with the two methods. This observation was explained in terms of leakage of low-molecular-weight material through the membrane. This material diffuses into the solvent chamber in the osmometer, which has a small volume, but in the case of the dialysis bag it diffuses into the reference solution which has quite a large volume. The concentrations of the diffused material are then different in the two cases leading to differences in the measured values.

Data from the Dialysis Experiments

Having shown that the data from the osmotic pressure measurements reasonably correspond with those obtained from osmometry when the conditions are controlled carefully, we have studied a number of different biopolymer solutions and gels. Some of the results are shown in Figure 4. Clear differences can be observed amongst the various biopolymers. For the solutions, the molecular weight and the quality of the solvent determine these differences. For gelled systems, the elastic contribution of the network also contributes to the chemical potential and should therefore be taken into account. This can be done by using modified Flory expressions.[6,7] Although this approach is widely used, it is not good enough for quantitative predictions. Recently, the scaling concept was also applied to non-ionic gels. Horkay and Zrinyi[8] have developed equations of state where the elastic modulus is related to the osmotic pressure.

4 Conclusions

It is demonstrated that the osmotic pressure of biopolymer systems can be measured in two ways. The first, the classical method with an osmometer has the advantage of a quick measurement and accurate values. However, very viscous or gelled systems cannot be studied in this way. With the second method, *i.e.* dialysis bags filled with a biopolymer solution immersed in a reference solution, the osmotic pressure for systems which could not be measured with the osmometer could be determined. The

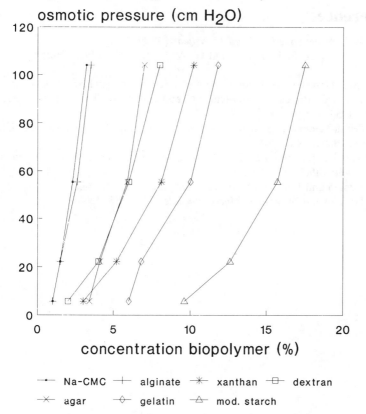

Figure 4 *Osmotic pressures determined at* 10 °C *and* pH 5 *by dialysis. The applied osmotic pressure is shown as a function of equilibrium concentration for several biopolymer systems*

agreement between the two methods is good when the solvent conditions (such as the ionic strength and pH) are controlled carefully, especially in the case of polyelectrolytes.

With these methods it is now also possible to determine which concentrations of biopolymers have the same osmotic pressure. Furthermore, the determined set of equilibrium concentrations allows the possibility of determining the slopes of tie-lines in phase diagrams of mixtures of two biopolymers in aqueous solution where complete phase separation has occurred.

A number of different theoretical models can be used to interpret the data, ranging from the classical Flory–Huggins approach to the recently developed scaling concepts. Most of the biopolymer systems of interest are polyelectrolytes, often in the gelled state; for these systems some theoretical models have already been developed. However, more work on the theoretical interpretation of data is required.

References

1. V. B. Tolstoguzov, *Int. Food Ingredients*, 1990, **2**, 8.
2. H. J. Granger, S. H. Laine, and G. A. Laine, *Microcirc., End., Lymph.*, 1985, **2**, 85.
3. J. Smit, University of Leiden, the Netherlands, personal communication.
4. P. J. Flory, 'Principles of Polymer Chemistry', Cornell University Press, Ithaca, NY, 1953.
5. P.-G. de Gennes, 'Scaling Concepts in Polymer Physics', Cornell University Press, Ithaca, NY, 1979.
6. T. Tanaka, D. Fillmore, S. T. Sun, I. Nishio, G. Swislow, and A. Shah, *Phys. Rev. Lett.*, 1980, **45**, 1636.
7. B. Erman and P. J. Flory, *Macromolecules*, 1986, **19**, 2342.
8. F. Horkay and M. Zrinyi, *Macromolecules*, 1982, **15**, 1306.

Polysaccharide–Surfactant Systems: Interactions, Phase Diagrams, and Novel Gels

By Björn Lindman, Anders Carlsson[1], Stina Gerdes, Gunnar Karlström, Lennart Piculell, Kyrre Thalberg[2], and Kewei Zhang

PHYSICAL CHEMISTRY 1, CHEMICAL CENTER, LUND UNIVERSITY, PO BOX 124, S-221 00 LUND, SWEDEN

1 General Aspects

One of the most important practical aspects of polymer and surfactant systems is the possibility to control rheology over very wide ranges. The basic mechanism involved is illustrated in Figure 1. Surfactant molecules that bind to a polymer chain generally do so in clusters which closely resemble the micelles formed in the absence of polymer. (For reviews of the field, see ref. 1–5). If the polymer is less polar or contains hydrophobic regions or sites, there is an intimate contact between the micelles and the polymer chain. In such a situation the contact between one surfactant aggregate and two polymer segments will be favourable. The two segments can be in the same polymer chain or in two different chains, the former being the typical case for a dilute polymer solution and the latter for more concentrated solutions with extensive chain overlap. The cross-linking of two or more polymer chains can lead to network formation and dramatic rheological effects, as exemplified below.

Surfactant–polymer interactions may be treated in different ways, the appropriateness of the approach depending strongly on the polymer. In one approach, one analyses the surfactant binding to the polymer (for example by surfactant-specific electrodes). The binding isotherms obtained typically indicate a strong co-operativity in binding, the onset of binding is quite well-defined, and a critical association concentration (CAC) is introduced. The CAC varies strongly with the system and decreases, for

[1]Present affiliation: Karlshamns LipidTeknik AB, PO Box 15200, S-104 65 Stockholm, Sweden
[2]Present affiliation: Pharmaceutical R and D, Astra Draco AB, PO Box 34, S-221 00 Lund, Sweden

contraction

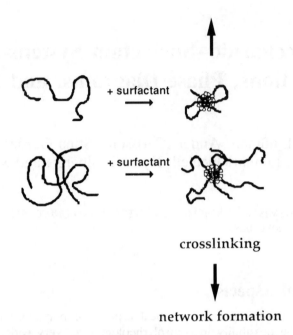

crosslinking

network formation

Figure 1 *Surfactant molecules generally associate with polymer chains in micelle-like clusters. In dilute solutions, contraction of polymer chains can result due to the association of a micelle with two parts of a polymer chain, while for semi-dilute or concentrated solutions, cross-linking of different polymer chains is expected*

example, strongly with increasing chain length of the surfactant. The latter observation suggests the alternative approach, which is to analyse the effect of a polymer on surfactant micellization, the polymer affecting the stability of the micelle by short- or long-range (electrostatic) interactions much in the same way as do low-molecular-weight co-solutes (weak amphiphiles, electrolytes).[6]

The main driving force for surfactant self-assembly in polymer–surfactant systems is generally the hydrophobic interaction between the alkyl chains of the surfactant molecules. Ionic surfactants often interact significantly with both non-ionic and ionic polymers, which can be attributed to the unfavourable contribution to the energetics of micelle formation from the electrostatic effects and their partial elimination due to charge neutralization or lowering of the charge density. For non-ionic surfactants, there is little to gain in forming micelles in the presence of a polymer and they, therefore, in general, do not interact strongly with polymers. If the polymer molecules contain hydrophobic segments or groups, as in block copolymers or so-called hydrophobe-modified polymers, the hydrophobic polymer–surfactant interaction will be significant. Consequences of this are a higher specificity in the interaction leading to a lower CAC, a lower co-operativity, and lower aggregation numbers.

Since the surfactant molecules in a polymer solution in general occur in a strongly self-assembled form, it is a natural starting point to treat the surfactant micelles as a second polymer and use the quite good understanding of polymer plus polymer and solvent systems[7] as a basis, as we recently have reviewed.[8] A typical phenomenon for two polymers in a common solvent is 'polymer incompatibility', resulting in a segregation into two solutions each enriched in one of the polymers. The reason for this is the negligible entropy of mixing for macromolecules, and thus enthalpic effects, which are for most systems repulsive, govern the thermodynamics. It is only for cases of very strong attractive interpolymer interactions that associative interactions will dominate. Schematic phase diagrams for segregating and associating systems are shown in Figure 2. The associative type, characterized by forming one concentrated phase enriched in both polymers and one dilute phase, is primarily found for systems of two oppositely charged polyelectrolytes.

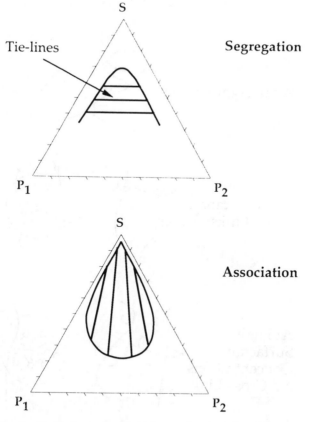

Figure 2 *Schematic illustration of segregative and associative phase separation. We consider the micelles as a second polymer*

2 Cross-linking and Network Formation

For hydrophobe-modified polymers, the cross-linking that may be induced
by surfactant aggregates is visualized in Figure 3. The polymer molecules
show a weak self-association in the absence of surfactant, which may be
strongly enhanced by the formation of surfactant aggregates around hydrophobic centres from two or more polymer molecules. As has been
demonstrated for many systems of a non-ionic polysaccharide plus an ionic
surfactant,[9-11] this may increase the viscosity by several orders of magnitude. However, the viscosity is strongly dependent on the surfactant
concentration with a very marked maximum at intermediate concentrations
and a decrease to quite low values at higher concentrations, where the
viscosity is comparable to that in the absence of surfactant or even lower.

Hydrophobe-Modified Polymer

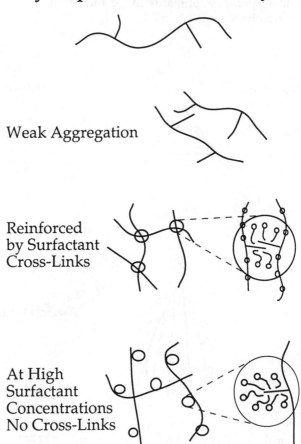

Figure 3 *Schematic illustration of possible association processes in solutions of a
surfactant plus polymer with hydrophobic centres*

This decrease can be understood as depicted in Figure 3. Thus, under conditions of excess of micelles with respect to the hydrophobic centres, the micelles form around individual centres and cross-linking is inhibited.

For a system of a non-ionic cellulose ether, ethyl(hydroxyethyl)cellulose (EHEC), in combination with an ionic surfactant (anionic or cationic), a greatly increased viscosity can be induced by a slight increase in temperature.[12-14] As illustrated in Figure 4, one observes in the absence of surfactant a monotonic decrease of viscosity with increasing temperature, while at relatively low concentrations of surfactant a major increase in apparent viscosity is obtained as the temperature is increased. The maximum can be shifted on the temperature scale by adjusting the degree of substitution on the cellulose backbone.

The rheological characteristics of this 'reversed gel formation' are further illustrated in Figure 5, where it can be seen that, at a low temperature, the loss modulus dominates over the storage modulus. At a high temperature, the elastic properties become strongly dominating.

One possile mechanism of this thermal gelation is sketched in Figure 6. It involves a cross-linking and network formation analogous to that discussed above. (It should be noted, however, that the association between ionic surfactant micelles and the polymer molecules in this case also has the important function of redispersing the polymer owing to the strong intermicellar repulsion. In the absence of surfactant, clouding would occur at the gel temperature.) The temperature variation is induced by a temperature dependent polarity of the polymer chain such that it becomes less polar at a higher temperature and thus a better 'nucleus' for surfactant

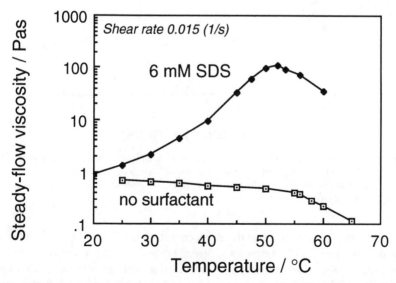

Figure 4 *Temperature dependent apparent viscosity for 1 wt% solutions of ethyl (hydroxyethyl)cellulose with and without added surfactant (sodium dodecylsulfate)*

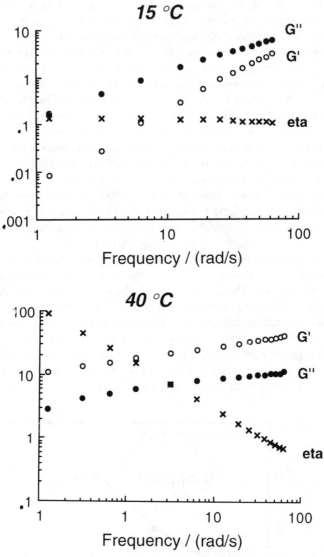

Figure 5 *Rheological behaviour for systems of ethyl(hydroxyethyl)cellulose* (1%) *plus ionic surfactant* (3 mM *sodium dodecylsulfate*) *at two temperatures. The plots show the storage* (G') *and loss moduli* (G") *and the dynamic viscosity* (eta)

micellization. This was directly demonstrated by Zana *et al.*[15] who observed that the CAC decreases strongly with small increases in temperature; it was, furthermore, observed that the micelle size decreases markedly as the temperature increases. Therefore, under suitable conditions, a temperature increase for a solution of EHEC plus an ionic surfactant can lead to the formation of a much larger number of micelles,

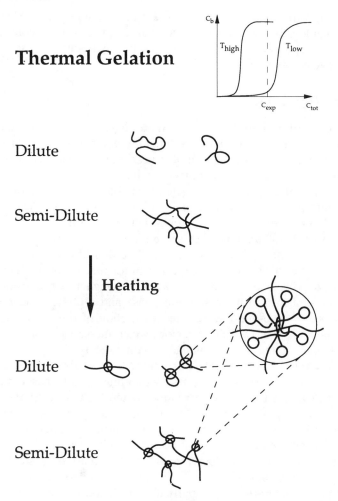

Figure 6 *Hypothetical mechanism of thermal gelation in systems of a non-ionic polysaccharide* (EHEC) *plus an ionic surfactant*

which, because of the low polarity of the polymer, are intimately associated with the polymer chains and can act as cross-links between polymer chains. At higher temperatures and at higher concentrations of surfactant, the cross-linking ability is lost in line with the reasoning given above. It should be stressed that this explanation in terms of cross-linking has received direct support from determinations of the intrinsic viscosity by Sundelöf *et al.*,[16] but that direct structural studies still remain to be performed.

This (reversible) thermal gelation should have a number of applications (for example, pharmaceutical) as described elsewhere.[17,18] One application is as an *in situ* forming drug-delivery matrix compatible with a wide range of modes of administration; another is as a 'liquid fibre' with potential to affect eating patterns. Both animal and clinical studies have shown that easily drinkable solutions of EHEC and surfactant of a total concentration of *ca*. 1% can spontaneously form a gel in the gastro-enteric system and so retard gastric emptying dramatically.

3 Phase Behaviour

It has been appreciated for a long time that, for example, the phase separation pattern of a non-ionic polymer in water is strongly affected by an ionic surfactant and that a complex precipitation–dissolution sequence may occur for a mixture of a polyelectrolyte and an oppositely charged surfactant. But attempts to obtain more complete phase diagrams, or to put them into a broader context, are few. A detailed review can be found elsewhere.[8] Just a few selected examples will be given here.

As an example of an ionic polysaccharide, let us consider hyaluronan (Hy, the sodium salt of hyaluronic acid). Addition of a cationic surfactant leads to micelle formation at concentrations much lower than in the absence of a polyelectrolyte and, at a slightly higher concentration, to a phase separation into a dilute solution phase and a phase concentrated in both solutes.[19,20] A pseudo-ternary representation of the phase diagram is given in Figure 7. The two-phase region, which increases strongly in extent with the length of the surfactant alkyl chain (and is absent for a shorter chain than decyl)[21] is completely surrounded by a single one-phase region. The concentrated phase of the phase-separating sample has a very high viscosity, which can mainly be attributed to the high concentration of the high-molecular-weight polymer. In the gel-like phase, the surfactant molecules occur in micellar clusters with aggregation numbers and dynamic properties closely similar to those of normal micelles.[22,23]

This phase behaviour is closely analogous to that for systems of two oppositely charged polyelectrolytes; it can be understood from simple electrostatic considerations, in particular the contribution to the free energy from the entropy in the counterion distribution. As can be inferred from Figure 7, addition of an electrolyte dramatically shrinks the phase separation region, and in a wide range of electrolyte concentrations phase separation does not occur.[24] At high enough salt concentrations we observe another type of phase separation, where the surfactant and the polysaccharide are separating into different solutions. This behaviour is not of the common segregative type mentioned in the introduction. The occurrence of this at higher salt concentrations demonstrates that the polymer plus surfactant system is of an intrinsically segregating type, as segregation results when the electrostatic effects have been reduced. A strong segregation is obtained between Hy and a surfactant of similar charge as illustrated in Figure 8. On addition of inorganic electrolyte, segregation is accentuated.

For systems of a non-ionic polysaccharide plus a surfactant, phase separation may be strongly influenced by temperature. Systems of EHEC

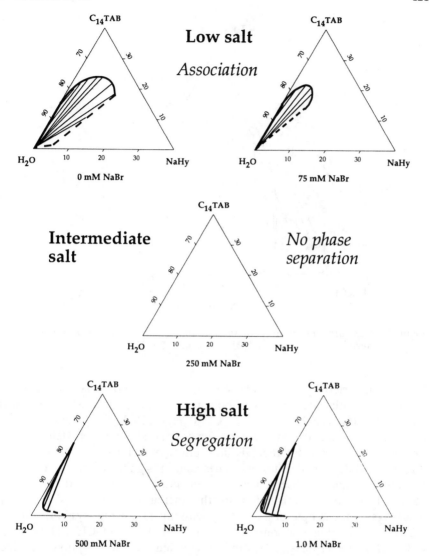

Figure 7 *Salt-dependent phase diagrams for tetradecyltrimethylammonium bromide plus hyaluronan*

plus sodium dodecylsulfate are of an associative type, with the tendency towards phase separation generally increasing (the cloud point decreases) in the presence of salt.[25,26] Ternary phase diagrams have been established which demonstrate an associative phase separation, although the two-phase region is quite asymmetrical as it extends to the EHEC/water axis.

In general, mixtures of non-ionic polymers and non-ionic surfactants show no strong mutual attraction, but it could be expected that sufficiently concentrated systems should show a segregative type of phase separation.

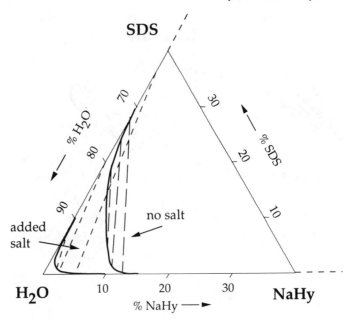

Figure 8 *Phase diagram at* 40 °C *of sodium dodecylsulfate plus hyaluronan* (9 × 10⁴ *daltons*): — —, *no added salt*; - - - -, *with* 1 M NaBr

This is supported by recent observations in our laboratory on the phase behaviour with dextran as polymer (Figure 9). Mixtures with $CH_3(CH_2)_{11}(OCH_2CH_2)_5OH$ or $CH_3(CH_2)_{11}(OCH_2CH_2)_8OH$ show a segregation similar to that observed for dextran plus poly(ethylene glycol). For the latter surfactant there is little effect of temperature on the phase diagram, while for the former surfactant a significant increase in the two-phase region is observed as the temperature is increased from 10 to 25 °C. This can be attributed to the increasing micellar size in this case — again an example of an analogy with mixed polymer systems.

For systems of a non-ionic surfactant, $CH_3(CH_2)_{11}(OCH_2CH_2)_4OH$ or $CH_3(CH_2)_{11}(OCH_2CH_2)_8OH$, and a clouding non-ionic polysaccharide, EHEC, the phase separation was found, as illustrated in Figure 10, to be associative rather than segregative.[27] The polymers of the latter study contain hydrophobic regions and the association of surfactants with these regions seems to have a dominating influence on the phase behaviour. As Figure 10 shows, the segregation may increase very strongly with minor temperature increases in such a system.

4 Conclusions

In this brief review of polysaccharide–surfactant systems, we have emphasized the cross-linking properties and the phase behaviour. Surfactants can

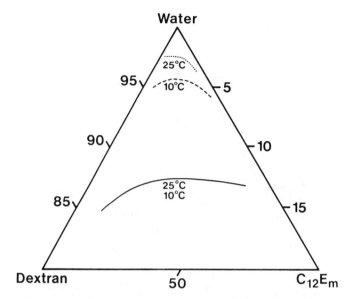

Figure 9 *Binodals for aqueous mixtures of dextran with* $CH_3(CH_2)_{11}(OCH_2CH_2)_8OH$ *at* 10 °C *or* 25 °C (——; *no significant temperature dependence was detected), or with* $CH_3(CH_2)_{11}(OCH_2CH_2)_5OH$ *at* 10 °C (----) *or* 25 °C (· · ·)

be employed effectively to control the rheology of polysaccharide systems, due to a cross-linking by surfactant micelles. Depending on whether the polysaccharide is charged or not, the rheological properties will be sensitive to salt concentration or temperature, respectively. For polysaccharides with decreasing polarity at a higher temperature, novel gels which form on heating and melt on cooling are found.

Phase diagrams show a close analogy with mixed polymer systems. For certain systems of an ionic polysaccharide and an oppositely charged surfactant, an associative phase separation may be converted into homogeneous mixing or segregative phase separation on addition of salt. Mixtures of an ionic polysaccharide and a similarly charged surfactant have been found to show segregation irrespective of added electrolyte. For mixtures of two non-ionic solutes, either a segregative or an associative phase separation can result, the latter only being expected for a very strong (normally hydrophobic) interaction between the polysaccharide and the non-ionic surfactant.

Acknowledgements

This work was supported by the Swedish National Board for Industrial and Technical Development, Berol Nobel AB, and Kabi Pharmacia AB.

124   *Polysaccharide–Surfactant Systems*

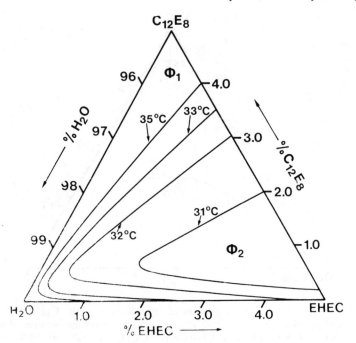

Figure 10 *Phase diagram of non-ionic polymer plus non-ionic surfactant. The two-phase region (Φ_2) of mixtures of EHEC plus $CH_3(CH_2)_{11}(OCH_2\text{-}CH_2)_8OH$ strongly increases with increasing temperature and its shape gives evidence of an associative interaction*

References

1. E. D. Goddard, *Colloids Surf.*, 1986, **19**, 255, 301.
2. E. D. Goddard and K. P. Ananthapadmanabhan (ed.), 'Polymer-Surfactant Interactions', CRC Press, Boca Raton, 1992, in press.
3. B. Cabane and R. Duplessix, *Colloids Surf.*, 1985, **13**, 19; *J. Physique (Paris)*, 1982, **43**, 1529.
4. K. Kayakawa and J. C. T. Kwak, in 'Cationic Surfactants, Physical Chemistry', ed. D. N. Rubingh and P. M. Holland, Marcel Dekker, New York, 1991, p. 189.
5. B. Lindman and K. Thalberg, in Ref. 2, Chapter 5.
6. B. Lindman and H. Wennerström, *Top. Curr. Chem.*, 1980, **87**, 1.
7. L. Zeman and D. Patterson, *Polym. Incompatibility*, 1972, **5**, 513. C. C. Hsu and J. M. Prausnitz, *Macromolecules*, 1974, **7**, 320. P. Å. Albertsson, 'Partition of Cell Particles and Macromolecules', 3rd Edn, Wiley, New York, 1986.
8. L. Piculell and B. Lindman, *Adv. Colloid Interface Sci.*, 1992, **41**, 149.
9. R. Tanaka, J. Meadows, G. O. Phillips, and P. A. Williams, *Carbohydr. Polym.*, 1990, **12**, 443.
10. R. A. Gelman, 1987 International Dissolving Pulps Conference, *TAPPI Proc.*, 1987, p. 159. F. M. Winnik, *Langmuir*, 1990, **6**, 522. C. A. Steiner and R. A. Gelman, in 'Cellulosics Utilization, Research and Rewards in Cellulosics', ed.

H. Inagaki and G. O. Phillips, Elsevier Science Publishers, London, 1989, p. 131.

11. K. Sivadasan and P. Somasundaran, *Colloids Surf.*, 1990, **49**, 229. P. Zugenmaier and N. Aust, *Macromol. Chem., Rapid Commun.*, 1990, **11**, 95.
12. A. Carlsson, G. Karlström, and B. Lindman, *Colloids Surf.*, 1990, **47**, 147.
13. A. Carlsson, B. Lindman, G. Karlström, and M. Malmsten, in 'Cellulose Sources and Exploitation', ed. J. F. Kennedy, G. O. Phillips, and P. A. Williams, Ellis Horwood, Chichester, 1990, p. 317.
14. B. Nyström, J. Roots, A. Carlsson, and B. Lindman, *Polymer*, 1992, **33**, 2875.
15. R. Zana, W. Binana-Limbele, N. Kamenka, and B. Lindman, *J. Phys. Chem.*, 1992, **96**, 5461.
16. C. Holmberg, S. Nilsson, S. K. Singh, and L. O. Sundelöf, *J. Phys. Chem.*, 1992, **96**, 871.
17. B. Lindman, A. Carlsson, K. Thalberg, and C. Bogentoft, *L'actualité Chimique*, 1991, 181.
18. B. Lindman, J. Tomlin, and A. Carlsson, Proceeding of Cellulose '91, in press.
19. K. Thalberg and B. Lindman, *J. Phys. Chem.*, 1989, **93**, 1478.
20. K. Thalberg, B. Lindman, and G. Karlström, *J. Phys. Chem.*, 1990, **94**, 4289.
21. K. Thalberg, B. Lindman, and G. Karlström, *J. Phys. Chem.*, 1991, **95**, 3370.
22. K. Thalberg, J. van Stam, C. Lindblad, M. Almgren, and B. Lindman, *J. Phys. Chem.*, 1991, **95**, 8975.
23. T. C. Wong, K. Thalberg, and B. Lindman, *J. Phys. Chem.*, 1991, **95**, 8850.
24. K. Thalberg, B. Lindman, and G. Karlström, *J. Phys. Chem.*, 1991, **95**, 6004.
25. G. Karlström, A. Carlsson, and B. Lindman, *J. Phys. Chem.*, 1990, **94**, 5005.
26. K. Zhang, G. Karlström, and B. Lindman, *Progr. Colloid Polym. Sci.*, 1992, **88**, 1.
27. K. Zhang, G. Karlström, and B. Lindman, *Colloids Surf.*, 1992, in press.

Physical Properties of Starch Products: Structure and Function

By T. R. Noel, S. G. Ring, and M. A. Whittam

AFRC INSTITUTE OF FOOD RESEARCH, NORWICH RESEARCH PARK, COLNEY
LANE, NORWICH NR4 7UA, UK

1 Introduction

Starch is a valued component of many foods and has an important role
both as a macronutrient and as a structural component. The detailed
chemical structure and composition of starch are primarily dependent on its
botanical origin. Usually it is processed by a moist heat treatment which
results in the disruption of the native granular structure. This processing
eventually leads to the development of a new microstructure which
determines both the mechanical properties of the material, and the rate
and extent of bio-erosion. In this article we consider the relationships
between the structure and composition of a starch and the physical
properties of processed starch products. Rather than consider specific
products, such as an extruded snack product, we will examine the
relationship between molecular and mechanical behaviour for three distinct
types of material—a paste, a gel, and a film. All three types are produced
after melting of the granular structure. On cooling to ambient temperature,
phase separation/crystallization of the starch polysaccharides leads to
time-dependent changes in the material behaviour.

2 Starch Structure and Granule Organization

Starch usually consists of a mixture of two polysaccharides, amylose and
amylopectin, which are based on 1–4 linked α-D-glucan chains. Some
starches consist solely of amylopectin, e.g. waxy maize, while others such
as amylomaize have a high amylose content of 60–70%. Most starches
have an amylose content of between 20 and 30%. Although both polysaccharides show variations in structure, it is possible to identify characteristics typical of amylose and amylopectin. Amylose is an essentially linear
polymer which gives the well-known intense blue colour with iodine in
aqueous solution. It can be quantified in relation to an iodine binding
capacity of 20 wt% under standard conditions. From measurements of

iodine binding behaviour, it is then possible to determine the amylose content of a starch, providing that the amylopectin binds essentially no iodine under the same conditions. The determination of amylose content in this way, however, is not necessarily straightforward. For example, it has been found[1] that some rice amylopectins can bind up to 2.67 wt % iodine; as a result, such starches have a high apparent level of amylose.

It has been known for some time that, as a result of some limited branching, amylose may not be quantitatively converted to maltose by the exo-acting enzyme β-amylase. Recently, the heterogeneity of amyloses has been investigated in more detail.[2,3] For maize amylose, the linear polymer is the most abundant ($\sim$70%), while the main branched fraction has a degree of polymerization (DP) of $\sim$2200, and 6 side-chains on average per molecule. These consist of short chains (DP = 18) and very long chains (DP > 370). Further fractionation of the amylose has yielded a more branched polysaccharide with, on average, 20 chains per molecule, consisting of short (DP = 18), long (DP = 230), and very long chains (DP > 2730). These branched polysaccharides have structural features in common with amylopectin, but their role in starch biosynthesis still remains to be assessed.

Amylopectin is a highly branched molecule with, on average, one 1–4–6 linked branch point for every 20–25 straight chain residues. The constituent chain profile of amylopectin may be obtained, following enzymic debranching, using gel permeation chromatography. Two main populations are observed, one with DP $\sim$ 60 and the other, more abundant, with DP $\sim$ 15. Improved separation techniques are refining this picture, with a polymodal distribution of constituent chains usually being observed,[4] with the additional presence in some amylopectins of longer chains of DP in the range 85–180 which contribute to the high iodine binding capacity of these amylopectins. In schematic representations of amylopectin structure the short chains are arranged in clusters on the longer chains. Current research is determining the variation in these chain populations with botanical origin, and defining how they are linked to the rest of the molecule.[5] Both polysaccharides are of high molecular weight, with values of 10^{5-6} daltons and 10^8 daltons, respectively, being reported for amylose and amylopectin.

The starch granule is an ordered structure which is both partially crystalline and birefringent. The birefringence of the granule indicates that the polysaccharides have a preferred radial orientation. Estimates of the crystallinity of common starch granules are generally in the range 30–50%. Three different crystalline polymorphs, designated A, B, and C, can occur in granular starches. All are based on the packing of double helices. The A form is found in most cereal starches, the B form in tuber starches, and the C form in legume starches. Recent research is distinguishing between the extent of double helical and crystalline order. The relationship between chain length distributions and the organization of amylopectin chains in the crystallite remains to be determined. The observation that starches which give an A-type diffraction pattern have amylopectins with shorter short chains than those that give the B-type pattern[6] provides a link between structure and granule organization. Comparable behaviour is observed on isolated linear chains;[7,8] short chains with $9 \leqslant DP \leqslant 12$ crystallize in the A-form from concentrated aqueous solution, while longer chains crystallize in the B-form.

3 Melting and Glass Transition Behaviour

While the melting and glass transition temperatures, T_m and T_g, of synthetic polymers can be found in standard texts, it is only relatively recently that estimates of these parameters for starch polysaccharides have become available. These parameters affect both the processing and the end use to which a particular polymer may be put. Both parameters can be conveniently estimated by differential scanning calorimetry, T_m from the temperature of the melting endotherm, and T_g from the sharp increase in heat capacity at the glass transition of the amorphous material.

The starch granule is a partially crystalline structure; in the absence of a diluent, however, it is not possible to melt the granule without causing thermal degradation. A diluent such as water may be added to depress the melting temperature into a temperature region where thermal degradation does not occur. For the determination of the equilibrium melting temperature of starch crystals, as a function of diluent content, it is necessary to prepare extended chain single crystals of macroscopic size.[9] This has not so far proved possible, although it is possible to prepare highly crystalline spherulitic material, consisting of assemblies of single crystals, from the crystallization of short amylose chains ($15 \leqslant DP \leqslant 20$) from aqueous solution.[10] By varying the solvent conditions, it is possible to prepare both the A- and B-type crystalline polymorphs. In a recent study the melting temperatures of A- and B-type crystalline materials have been studied as a function of water content.[11] At high water contents, the B spherulites melted at 348 K. With decreasing water content, the melting temperature increased, and it was 372 K at a water content of 40 wt %. At a fixed water content, the A polymorph was found to melt at temperatures approximately 20 K higher. Using the Flory–Huggins relationship for polymer crystal melting in the presence of a diluent, it was possible to predict a value of 530 K for the ideal melting temperature of the dry crystals. The melting of the starch granule shows a similar general dependence on diluent content.[12] But, as it is formed biosynthetically, and is only partially crystalline, its melting behaviour is more influenced by non-equilibrium factors.

For a polymeric material, the glass transition is typically described as a transition from brittle to rubbery behaviour and occurs over a small temperature range. Glassy and rubbery polymers clearly have different uses. For a low-molecular-weight carbohydrate liquid, the viscosity of the liquid increases with decreasing temperature below T_m, providing crystallization does not intervene. The form of the increase in shear viscosity η follows an empirical relationship of the form

$$\eta = A \exp[B/(T - T_0)] \tag{1}$$

where A, B, and T_0 are constants. The process of relaxation is the response in any property after a perturbation, the perturbation being relaxed as a result of diffusive motions within the liquid. It is expected that the rate of diffusion of a component in a liquid is linked to shear viscosity, with an increase in viscosity slowing down the diffusion. It is possible to calculate a shear relaxation time, τ_s, from

$$\tau_s = \eta/G_\infty \tag{2}$$

At T_g, the shear viscosity is in the region of 10^{12} Pa s, and the high frequency limit of the shear modulus, G_∞, is of the order 10^{10} N m^{-2}, leading to a value of τ_s of the order of 100 seconds. For observations lasting seconds or fractions of seconds, the stress does not relax significantly and the material behaves as a solid. Over longer periods of time liquid-like behaviour is observed. In a calorimetric experiment the sharp change in heat capacity at the glass transition occurs because the liquid cannot explore available configurations within the timescale imposed by the experiment. As a result a sharp fall in heat capacity occurs as the material is cooled through the glass transition.

From calorimetric experiments on malto-oligomers, it is possible to obtain an estimate of the high-molecular-weight limit of T_g by extrapolation.[13] The T_g of the malto-oligomer series increases from 364 K for maltose to 448 K for maltohexaose, giving an estimate of 500 K for the high-molecular-weight limit by extrapolation. Water is a powerful plasticizing agent of amorphous starch films, the T_g value falling to 400 K on the addition of just 6 wt % water.[14] The effect of water content on the glass transition and melting behaviour of starch is shown in Figure 1.

The glass transition temperature of an amylose of DP $\sim$ 15 is estimated as 470 K. The calculated ratio of T_g/T_m of 0.89 is outside the expected range of 0.5–0.8, and this probably indicates that there is a relatively large error in the estimate of T_m arising as a result of the extended extrapolation. At a fixed water content of 40 wt %, the T_m value of the B polymorph is 372 K, that of the A polymorph is 402 K, and T_g is estimated as 230–240 K, giving a ratio of $T_g/T_m \sim 0.6$ for the A polymorph, which is within the expected range. Further research will undoubtedly lead to a refinement of these values and establish the composition dependence of T_g and T_m as a function of water content and degree of polymerization.

For amorphous polymeric materials, the form of the increase in local viscosity with decreasing temperature is often comparable to that of a low-molecular-weight analogue. At the glass transition there is a sharp change in material properties from a material which is glassy and brittle to

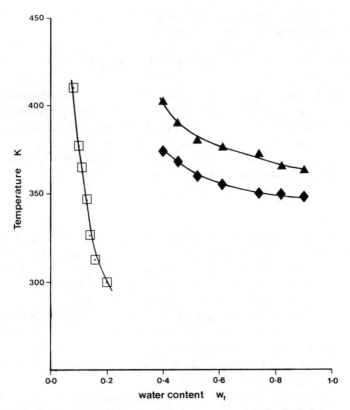

Figure 1 *Plot of glass transition temperature* (T$_g$) *and melting temperature* (T$_m$) *for amorphous and crystalline forms of starch as a function of water content* w$_f$: □, T$_g$, *for starch (K. J. Zeleznak and R. C. Hoseney*, Cereal Chem., *1987*, **64**, *121*); ▲, T$_m$ *for A-type spherulites*, ◆, T$_m$ *for B-type spherulites (M. A. Whittam, T. R. Noel, and S. G. Ring*, Int. J. Biol. Macromol., *1990*, **12**, *359)*

a rubbery material with a typical fall in shear modulus of *ca.* 10^5 N m^{-2}. The temperature of the observed transition is again dependent on the timescale of the measurement: immediately above the glass transition, the local viscosity is such to permit partial relaxation of the applied strain, within the experimental timescale, to reveal the rubbery behaviour of the polymer. In a study[15] on the glass transition behaviour of amylopectin films, plasticized with water, an approximately hundred-fold decrease in modulus was observed on heating through the glass transition. Even a limited partial crystallization (2–5%) of the amylopectin in the film broadened the transition and decreased the observed heat capacity increment at T_g. Figure 2 depicts the mechanical behaviour as a function of temperature of an amylopectin film plasticized with sorbitol. In addition to the effect on the small-strain behaviour, the fracture properties of starchy materials are also dependent on the proximity to T_g. Below T_g, typical

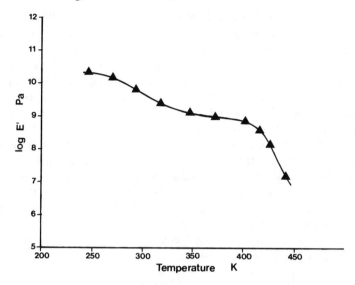

Figure 2 *Plot of logarithm of tensile modulus* (E') *versus temperature for an amylopectin film plasticized with sorbitol scanned from* 240 K *to* 440 K *at a heating rate of* 2 K min^{-1}

brittle fracture, comparable to that of synthetic glassy polymers, is observed.[16]

As a material is cooled below T_m, the driving force favouring crystallization increases, but, at the same time, the rate at which molecules or polymer segments can diffuse to the growing crystal surface is retarded by the increasing viscosity. Crystallization below the T_g can take several years. It is expected that the maximum rate of crystallization should occur between T_g and T_m. There have been few observations on the rate of crystallization of the starch polysaccharides as a function of temperature, in the temperature interval between T_m and T_g, and of the changes in mechanical properties which occur as a result. It is also to be expected that the change in local viscosity with temperature would have an effect on the rate of diffusion of small molecules through the polymer matrix. With the determination of T_g becoming more routine, it is expected that attention will turn to how the proximity to T_g affects local diffusion and the functional behaviour of the matrix, *e.g.* during drying.

4 Starch Pastes and Gelatinization

The gelatinization of starch is comparable to the non-equilibrium melting of partially crystalline synthetic polymers.[17] In excess water it usually occurs over a few degrees—in the temperature range from 55 to 70 °C for starches of normal amylose content (20–30%).[18,19] The gelatinization temperatures for a range of starches as determined by differential scanning calorimetry are given in Table 1. Recently, the extent of different types of order—double-helical, crystalline, and orientational—have been measured as a function of temperature during gelatinization.[20] It was found that the loss of order occurs in a sequence starting with the melting of double helices. True solution of the starch polysaccharides can only be approached by heating to temperatures in excess of 120 °C in a shear field. The rheological properties of processed starches are often dependent on the particulate nature of the dispersion.

The starch paste usually consists of swollen gelatinized granules or, occasionally, granular fragments dispersed in a predominantly amylose solution. To describe the viscous behaviour of this system, it is necessary to evaluate the separate contributions to viscosity from the particulate and soluble material.[21-23] There is a substantial literature on the rheological properties of particulate suspensions.[24-26] It has been found that starch pastes behave as power law fluids[21] where the apparent viscosity is related to the shear rate γ by the expression

$$\eta_{app} = K'\gamma^m \tag{3}$$

The power law constant m ranges between -0.6 and -0.4 for gelatinized maize and wheat starch suspensions.[27] The viscosity of these suspensions is dependent primarily on the fraction of the volume occupied by the particles. As volume fraction increases suspension viscosity progressively increases. At low volume fractions particles are relatively remote from each other, but as the volume fraction increases particle–particle interactions become important and the observed viscosity shows an increasing dependence on particle volume fraction. The increase in relative viscosity of an aqueous suspension of gelatinized starch granules with increasing volume fraction is shown in Figure 3. There are a number of theoretical and empirical relationships which can describe this behaviour; one due to Krieger[28] is of the form

Table 1 *Gelatinization temperatures for starches from different sources*

	Gelatinization Temp. °C	
	(*a*)	(*b*)
Potato	63	65
Pea	65	64
Wheat	58	60
Maize	66	71
Rice	55	

[a] M. G. E. Wolters and J. W. Cone, *Starch/Stärke*, 1992, **44**, 14.
[b] P. D. Orford, S. G. Ring, V. Carroll, M. J. Miles, and V. J. Morris, *J. Sci. Food Agric.*, 1987, **39**, 169.

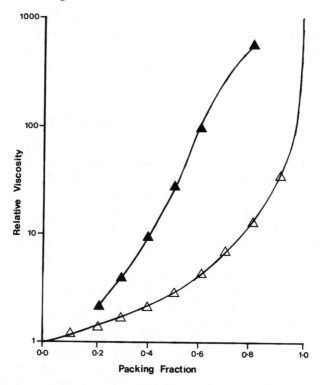

Figure 3 *Plot of relative viscosity* versus *packing fraction for an aqueous suspension of swollen gelatinized granules from wheat starch* (▲). *The predicted behaviour for rigid spherical particles is included for comparison* (△), *calculated from equation (4) with* K = 1/0.64

$$\eta_r = (1 - K\varphi)^{-5/(2K)}, \tag{4}$$

where the relative viscosity η_r is dependent on particle volume fraction, and where K is a crowding factor which is usually set at $1/\varphi_m$, where φ_m is the maximum packing fraction. This relationship usefully describes the behaviour of suspensions of rigid spheres and it also gives an insight into the rheology of starch pastes. However, gelatinized starch granules are not rigid, nor are they spherical. At a fixed volume fraction, an anisometric particle will produce a greater suspension viscosity than a spherical particle. Particle deformability has an effect at higher packing fractions when the particles 'bump into each other' and thereby deform, resulting in a flattening of the dependence of relative viscosity on particle volume fraction.[27] It has been found that the viscosity of starch pastes is dominated by the particulate fraction, the dissolved starch polysaccharides making a relatively minor contribution. Any process or interaction which alters particle volume fraction can have a major impact on viscosity, particularly at high packing fractions. One such interaction is retrogradation, as the retrogradation of the solubilized amylose produces more particulate material. The retrogradation of amylopectin can lead to a decrease in the size of gelatinized granules on storage,[29] which in turn can lead to a reduction in viscosity and the separation of a watery layer.

5 Gels of Amylose, Amylopectin, and Starch

Aqueous solutions of amylose are unstable. On cooling a dilute solution to room temperature, a precipitate eventually separates out with time, while from more concentrated solutions—above ~1.5 wt%—an opaque white gel forms. Gelation occurs as a result of the formation of an interconnected network. By following the development of gel stiffness with time, and by comparing it with the development of crystallinity and the extent of aggregation of polymer chains, it was proposed[30] that gel formation occurred as a result of a phase separation which produced an interconnected polymer-rich phase. The development of crystallinity as assessed by the appearance of an X-ray diffraction pattern occurred at a slower rate, and occurred subsequent to gelation. Evidence has since been obtained from solid state NMR measurements indicating that the conformation of amylose chains in the gel can be double-helical.[31] The inter-connectedness in amylose gels is, however, not a result of double-helix formation connecting individual molecular strands. The gel is opaque indicating the formation of polymer aggregates. Additionally the network strands of the amylose gel, as visualized by electron microscopy, consist of assemblies of many chains.[32] Amylose gelation can be reversed by heating to 160 °C. To initiate gelation on practical timescales, for concentrations in the range 2–10 wt%, it is necessary to quench to temperatures below 40–50 °C. The phase separation producing the network is therefore occurring at temperatures far removed from the liquid–liquid phase separation boundary. The deepness of the quench needed to obtain a practical indication of a phase separation, such as an increase in turbidity, has also excluded the possibility of obtaining a phase diagram for amylose plus water mixtures.

Amylose gels behave as elastic solids at small deformations, the shear modulus of the gel showing a very strong power-law dependence on concentration ($G' \propto C^7$) in the concentration range 1.5–7 wt%. The properties of a starch gel will therefore be very dependent on the amount of amylose solubilized during gelatinization.

Concentrated amylopectin solutions (above 20 wt%) also form gels on cooling to below 10 °C. While the stiffness of an amylose gel may only take minutes or at most a few hours to reach a plateau value, the gelation of amylopectin at these concentrations takes several weeks to reach a plateau modulus. Amylopectin gelation can be reversed by heating to temperatures below 100 °C. With differential scanning calorimetry, a broad endotherm is observed on heating, which is associated with the melting of the gel structure. Typically the endotherm spans 20 °C with a mid-point at 50 °C.

The size of the endotherm is similar to that of the gelatinization endotherm indicating that the extent of reordering in the gel is roughly equivalent to that in the starch granule.

A similar experimental approach to that used to study amylose gelation was used to study amylopectin gelation.[33] For amylopectin gelation, the separation of a partially crystalline amylopectin-rich phase was closely associated with the development of gel stiffness. Through heterogeneous acid hydrolysis of the gel it was possible to isolate the fragments of amylopectin involved in the aggregated regions. The insoluble product of this hydrolysis was found to be a branched fragment composed of short chains of average DP ~ 15. From these experiments it was concluded that chain association in amylopectin gels involves these short regions.

Amylopectin gel stiffness does not show a strong dependence on concentration, but it does show a strong dependence on amylopectin structure. An association has been noted between the constituent chain length of amylopectin and the rate of gelation, amylopectins with shorter short chains (*e.g.* wheat and barley) gelling at a slower rate than amylopectins with longer short chains (*e.g.* potato and pea).[34] It is to be expected that the driving force towards crystallization would increase with increasing chain length. The shear modulus of the gel, and the extent of chain association within the gel, does not, however, show any obvious correlation with the fine structure of the amylopectin. With increasing exploitation of genetic variations in starch structure and composition, a wider range of amylopectin structures will become more readily available; this should permit refinement of these initial observations.

Starch gels are readily formed when a concentrated gelatinized paste (6–50 wt%) is cooled to room temperature. At lower concentrations visco-elastic pastes are formed which flow on the application of a small deformation. At higher concentrations 'homogeneous' gelatinization of all the granules in the population is difficult to achieve. On cooling a paste to room temperature, the amylose solubilized during gelatinization forms aggregates, and, if it is sufficiently concentrated, it forms an interpenetrating network, within which the gelatinized granules are embedded.

Amylose gels are free-standing. At small strains the material behaves as a Hookean solid. The starch gel is a composite structure in which gelatinized starch granules are embedded in, and reinforce, an amylose gel matrix. Composite systems are relatively common in material science, and there is a substantial literature on the reinforcement of elastomeric materials by filler particles.[35] In these systems rigid particles modify the properties of the elastomeric matrix. In composite gel systems,[36] rigid filler particles stiffen the gel matrix, the extent of this depending on the volume fraction occupied by the filler particles and their shape. The extent of stiffening of gels filled with deformable particles depends on particle deformability. At a fixed volume fraction, the more rigid the particle, the greater the stiffening. At large deformations the presence of rigid particles can more easily lead to failure, as a stress localization occurs at the filler–matrix interface. At these large deformations, interactions between the filler matrix and the gelling polymer become important; the greater the adhesion between matrix and filler, the greater the strength of the material. At gelling concentrations, the gelatinized starch granules are 'close packed'. The stiffness of the starch gel is dependent on the stiffness of an amylose matrix gel and the deformability of the gelatinized granules.[36] Any transitions or interactions which affect these parameters will therefore have an effect on gel mechanical properties.

References

1. Y. Takeda, S. Hizukuri, and B. O. Juliano, *Carbohydr. Res.* 1987, **168**, 79.
2. Y. Takeda, K. Shirasaka, and S. Hizukuri, *Carbohydr. Res.*, 1984, **132**, 83.
3. Y. Takeda, T. Shitaozono, and S. Hizukuri, *Carbohydr. Res.*, 1990, **199**, 207.
4. S. Hizukuri, *Carbohydr. Res.*, 1986, **147**, 342.
5. S. Hizukuri and Y. Maehara, *Carbohydr. Res.*, 1990, **206**, 145.
6. S. Hizukuri, *Carbohydr. Res.*, 1985, **141**, 295.
7. B. Pfannemuller, *Int. J. Biol. Macromol.*, 1987, **9**, 105.
8. M. J. Gidley and P. V. Bulpin, *Carbohydr. Res.*, 1987, **161**, 291.
9. B. Wunderlich, 'Macromolecular Physics', Academic Press, London, 1980, Vol. 3, Chapter 8.
10. S. G. Ring, M. J. Miles, V. J. Morris, R. Turner, and P. Colonna, *Int. J. Biol. Macromol.*, 1987, **9**, 158.
11. M. A. Whittam, T. R. Noel, and S. G. Ring, *Int. J. Biol. Macromol.*, 1990, **12**, 359.
12. J. W. Donovan, *Biopolymer*, 1979, **18**, 263.
13. P. D. Orford, R. Parker, S. G. Ring, and A. C. Smith, *Int. J. Biol. Macromol.*, 1989, **11**, 91.
14. K. J. Zeleznak and R. C. Hoseney, *Cereal Chem.*, 1987, **64**, 121.
15. M. T. Kalichevsky and J. M. V. Blanshard, *Carbohydr. Polym.*, 1992, **18**, 77.
16. A.-L. Ollett, R. Parker, and A. C. Smith, *J. Mater. Sci.*, 1991, **26**, 1351.
17. H. Liu, J. Lelievre, and W. Ayoung-Chee, *Carbohydr. Res.*, 1991, **210**, 79.
18. P. D. Orford, S. G. Ring, V. Carroll, M. J. Miles, and V. J. Morris, *J. Sci. Food Agric.*, 1987, **39**, 169.
19. M. G. E. Wolters and J. W. Cone, *Starch/Stärke*, 1992, **44**, 14.
20. M. J. Gidley and D. Cooke, *Biochem. Soc. Trans.*, 1991, **19**, 551.
21. I. D. Evans and D. R. Haisman, *J. Text. Stud.*, 1979, **10**, 347.
22. D. D. Christianson and E. B. Bagley, *Cereal Chem.*, 1983, **60**, 116.
23. J. L. Doublier, *J. Cereal Sci.*, 1987, **5**, 247.
24. J. W. Goodwin, *Colloid Sci.*, 1975, **2**, 246.
25. R. C. Ball and P. Richmond, *Phys. Chem. Liq.*, 1980, **9**, 99.
26. C. R. Wildemuth and M. C. Williams, *Rheol. Acta.*, 1984, **23**, 627.
27. H. S. Ellis, S. G. Ring, and M. A. Whittam, *J. Cereal Sci.*, 1989, **10**, 33.
28. I. M. Krieger, *Adv. Coll. Sci.*, 1972, **3**, 111.
29. H. S. Ellis, S. G. Ring, and M. A. Whittam, *Food Hydrocolloids*, 1988, **2**, 321.
30. M. J. Miles, V. J. Morris, and S. G. Ring, *Carbohydr. Res.* 1985, **135**, 271.
31. M. J. Gidley, *Macromolecules*, 1989, **22**, 351.

32. V. M. Leloup, S. G. Ring, K. Roberts, B. Wells, and P. Colonna, *Carbohydr. Polym.*, 1992, **18**, 189.
33. S. G. Ring, P. Colonna, K. J. I'Anson, M. T. Kalichevsky, M. J. Miles, V. J. Morris, and P. D. Orford, *Carbohyr. Res.*, 1987, **162**, 277.
34. M. T. Kalichevsky, P. D. Orford, and S. G. Ring, *Carbohydr. Res.*, 1990, **198**, 49.
35. Y. Lipatov, *Adv. Polym. Sci.*, 1977, **22**, 1.
36. G. J. Brownsey, H. S. Ellis, M. J. Ridout, and S. G. Ring, *J. Rheol.*, 1987, **31**, 635.

A Rheological Description of Amylose–Amylopectin Mixtures

By J.-L. Doublier and G. Llamas

LABORATOIRE DE PHYSICOCHIMIE DES MACROMOLÉCULES, INRA, BP 527,
44026 NANTES CÉDEX 03, FRANCE

1 Introduction

Starch is composed of two macromolecular components which differ in their chemical structure—amylose (A) is almost totally linear, whereas amylopectin (AP) is highly branched. Amylose accounts for 18–33% of normal starch, depending upon the botanical origin, and it is known to play a major role in starch functionality. It is remarkable, for instance, that cereal starches behave in a totally different way from potato starch despite the amylose content differing only slightly, namely, 27–28% for the former against 22% for the latter.

It is also noteworthy that each starch component behaves in a different way in aqueous solution. Amylopectin alone is easily solubilized in neutral aqueous solution, and it yields essentially stable solutions. Amylose, on the other hand, can be solubilized only under drastic conditions, in alkali (1 M) or at an elevated temperature (>130 °C) at neutral pH. Moreover, opaque elastic amylose gels are produced upon neutralization or cooling. Another peculiarity of these two polysaccharides is their thermodynamic incompatibility in aqueous medium; this leads to phase separation, one phase being enriched with amylose while the other contains mostly amylopectin.[1] The practical consequence of this phenomenon in starch gelation is composite material formation.[1-3] Of particular interest is the range of amylose contents lower than 30%. Starting from 0% (amylopectin alone), it is possible to increase gradually the amylose content so that this latter component plays an increasing role until reaching a value where amylose replaces amylopectin in the continuous phase, the latter component then being located in the dispersed phase. This critical point may be defined as a phase inversion point as in emulsions. The position of this point has been estimated to correspond to an amylose:amylopectin (A:AP) ratio of the order of 30:70, on the basis of rigidity measurements in compression experiments.[2] It is thus clear that a low amylose content determines strongly the properties of amylose plus amylopectin mixtures. The object of the present work is to describe more accurately the visco-elastic properties of such mixed systems. The overall aim is to understand the physical basis for the properties of starch.

2 Experimental

Materials

Potato amylose was obtained from Avebe (The Netherlands) and has been characterized in a previous investigation ($[\eta] = 70 \, \text{ml g}^{-1}$ in KCl (0.33 M); $M_v = 390\,000$).[4] Amylopectin was prepared from a waxy maize starch. Both crude materials were purified by dispersion in 95% dimethylsulfoxide (DMSO), followed by precipitation using pure ethanol. Amylose was then dried under vacuum, while amylopectin was redispersed in boiled water before being freeze-dried.

Preparation of Mixed Systems

Amylose was dispersed in hot water (150 °C) as described previously.[5] Amylopectin was solubilized in water at 95 °C while stirring. Weighed amounts of amylose and amylopectin solutions were mixed at 50–60 °C in order to yield a final total concentration of 4 wt% and immediately transferred to the measuring system of the rheometer.

Measurements of Turbidity

Development of turbidity was monitored by measuring absorbance variations at 25 °C as a function of time at 600 nm, using a UV/visible spectrophotometer.

Rheology

Small-amplitude oscillatory shear experiments were performed at 25 °C with a Rheometrics Fluids Spectrometer (RFS II) using the cone-and-plate geometry (diameter, 5 cm; angle, 0.04 radians). For gel-cure experiments, measurements were performed at $1 \, \text{rad s}^{-1}$ for 8 hours, the strain amplitude being fixed at 0.05 or lower. The characteristics of the final system were described by the mechanical spectrum plotted between 0.01 and $100 \, \text{rad s}^{-1}$ at the same strain amplitude.

3 Results and Discussion

Figure 1 illustrates the visco-elastic behaviour of amylopectin and amylose at concentrations of 4 wt% and 1 wt%, respectively. These concentrations have been chosen since they correspond to the concentrations employed in the present work. The amylose content in the mixture varied between

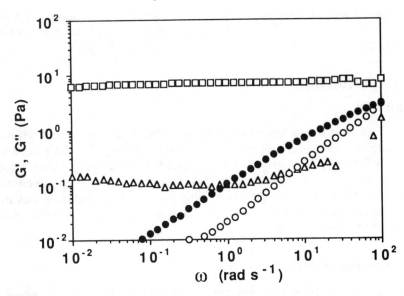

Figure 1 *Mechanical spectra of a* 4 wt % *amylopectin solution* (●, G'; ○, G") *and a* 1 wt % *amylose gel* (□, G'; △, G"). *Moduli are plotted against frequency,* ω

0.02 wt % (A:AP = 5:95) and 1.08 wt % (A:AP = 27:73) and hence never exceeded 1.1 wt % (see below). The total polysaccharide concentration was fixed at 4 wt %. The mechanical spectrum exhibited by amylopectin is characterized by a strong variation in the storage modulus (G') and in the loss modulus (G'') as a function of frequency, with $G'' \gg G'$ at low frequency, and a trend to a crossover of the curves at high frequency (*ca.* 100 rad s^{-1}). Such behaviour is typical of a macromolecular solution, and it indicates that there are no strong interactions between the macromolecules apart from topological entanglements, as has been reported for other non-gelling polysaccharides.[6] Amylose, on the other hand, exhibited, after ageing for 8 hours, the behaviour typical of a gel, with G' independent of frequency, and $G' > 10G''$.[7] It is to be emphasized that the final value of G', at 1 rad s^{-1}, is 6.3 Pa suggesting that we are quite close to the lowest concentration (C_0) where gelation is possible. This result is consistent with a previous investigation on the same sample, where a C_0 value of 0.9 wt % was found, although experimental conditions for the preparation of gels were different.[4,8]

Figure 2 shows the evolution of G' and G'' as a function of time for an amylose gel at 1 wt %. These rheology traces are plotted together with the turbidity evolution. Details of the rheology of amylose gelation have been previously reported.[4,8] The present figure shows clearly that gelation takes place after a lag period. The gelation time, ~77 min in the present example, is defined as usual by the point $G'(\omega)$ and $G''(\omega)$ crossover.

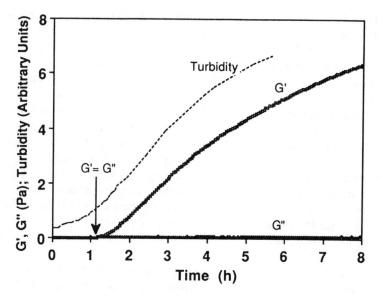

Figure 2 *The storage modulus G' and the loss modulus G" as a function of time for a 1.0 wt % amylose gel. Also shown is the evolution of the turbidity*

Beyond this time, turbidity and *G'* increase simultaneously as a result of the phase separation process.[4,9]

Figure 3 shows what is obtained for a mixture with a 22:78 A:AP ratio. The shape of the *G'* trace is comparable to that for amylose in Figure 1

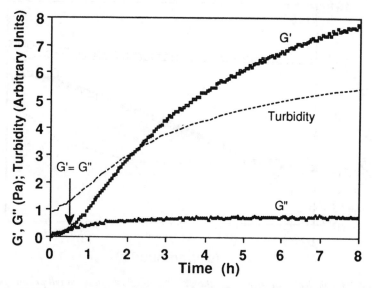

Figure 3 *Plots of G' and G" as a function of time for a 22.:78 amylose–amylopectin gel (total concentration = 4 wt %). Also shown is the evolution of turbidity*

despite the actual concentration being lower, *i.e.* 0.88 wt% as against 1.0 wt%. The only difference is that gelation took place much earlier, 32 min as against 77 min. In contrast, the turbidity trace is different. The system became turbid as soon as the mixture was prepared (initial absorbance = 0.15, as against 0.06 for 1 wt% amylose) and turned progressively opaque. However, the final turbidity of the 8-hour old gel was slightly lower than with amylose alone (absorbance = 0.8 for the mixture, compared to 1.2 for amylose). Figure 4 shows the mechanical spectrum of the 22:78 mixture as compared to that of amylose. It is clear that both systems display the characteristics of a gel. However, although G' is of the same order of magnitude in both cases, the $G''(\omega)$ variations are completely different. This parameter was found to remain almost constant and very low for amylose alone, whereas a steady increase was seen in the case of the mixture, the slope of the curve being of the order of 0.5.

From the comparison of Figures 2, 3, and 4, we infer that phase separation occurs as soon as both macromolecular components are mixed. This is ascribed to the thermodynamic incompatibility of amylose and amylopectin, which leads to the formation of two phases, one enriched with amylose, the other with amylopectin. The fact that the system exhibits unambiguously the visco-elastic behaviour of a gel means that it is amylose that is in the continuous phase. We are thus beyond the phase inversion point. However, the frequency dependence of G'' suggests that amylopectin also contributes to the visco-elasticity of the system. Final values of G' are of the same order of magnitude as in 1 wt% amylose gel; this means that amylose has been concentrated beyond C_0 in the continuous phase.

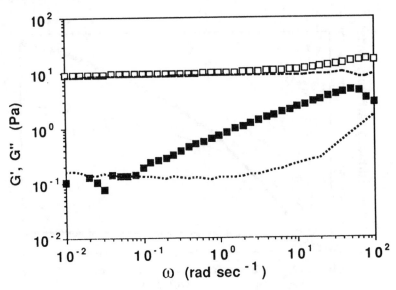

Figure 4 *Mechanical spectrum of the* 22:78 *amylose and amylopectin mixutre* ($\square$, G'; $\blacksquare$, G'') *in comparison with that of a* 1 wt% *amylose gel (dashed lines; data from Figure 1)*

A comparison was made of behaviour at A:AP ratios in the range from 5:95 to 27:73. Two examples of the mechanical spectra are shown in Figure 5. At a low ratio (5:95) (not shown), the $G'(\omega)$ and $G''(\omega)$ traces did not differ from those of amylopectin alone. It is only beyond a 10:90 ratio that the presence of amylose is noticeable. The first effect is a dramatic increase in G' particularly in the low frequency range. The shape of the $G'(\omega)$ curve is modified with a flattening towards the low frequency range. This is classically interpreted as a manifestation of an elastic plateau as a result of network formation. In contrast, the $G''(\omega)$ curve is modified much more progressively, its shape being similar to that illustrated in Figure 4. Similar trends were exhibited at 15:85 and 17:83 A:AP ratios. It is also to be mentioned that these systems remained fluid when submitted to shearing, their flow behaviour being typical of suspensions with the appearance of a yield stress. Hence, despite these mixed systems remaining fluid-like, they exhibited the visco-elastic behaviour typical of a gel as the A:AP ratio was increased.

Regarding the kinetics of gelation, the same effects as illustrated in Figure 2 were seen. The mixed systems became cloudy more rapidly than with amylose alone at an equivalent concentration, and gelation, as estimated by the $G'-G''$ crossover, took place earlier as the A:AP ratio increased. Table 1 illustrates the dramatic effects of the A:AP ratio on gelation time and G'. The former parameter decreases drastically as the ratio increases from 15:85 to 27:83, namely from ~180 min to 5 min; meanwhile G' increases from 0.5 to 25 Pa. These dependencies could be expressed as a function of the amylose content using power-law equations with exponents of -3.0 and 6.5 for gelation time and G', respectively.

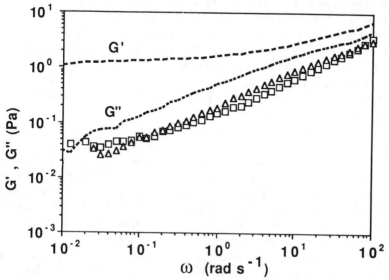

Figure 5 *Mechanical spectra of two amylose and amylopectin mixtures (A:AP = 17:83, dashed lines; A:AP = 10:90, □, G'; △, G'')*

Table 1 *Storage modulus* G' *and gelation time as a function of the amylose:amylopectin (A:AP) ratio*

A:AP	Amylose (wt %)	G'[1] (Pa)	Gelation time (min)
100:0	100	6.4	77
0:100	0	0.104	–
5:95	0.20	0.104	–
10:90	0.40	0.196	–
15:85	0.60	0.50	179
17:83	0.68	1.21	116
20:80	0.80	2.46	32
22:78	0.88	7.74	32
25:75	1.00	14.0	17
27:73	1.08	25.0	5

[1] after 8 h ageing (measured at 1 rad s^{-1})

Such dependencies are comparable with data reported for amylose alone.[4,10] This suggests that the effects are merely to be ascribed to the presence of the amylose component. Moreover, the point of phase inversion can be estimated as being close to the A:AP ratio of 15:85, corresponding to an amylose content of 0.60 wt %. It is thus clear that the presence of amylose induces gelation in the mixture at a content that is much lower than C_0 of amylose alone.

These results confirm that the behaviour of amylose and amylopectin mixtures is governed by a phase separation process. Amylopectin, the major component which accounts for more than 73 wt % of the total polysaccharide content, plays a minor role in the rheology of the system. Its presence, however, effects the $G''(\omega)$ curve, and it induces cloudiness much earlier due to the onset of phase separation. Amylose, on the other hand, plays a major role as soon as the point of phase inversion is reached (A:AP ~ 15:85). It forms a continuous phase which entraps amylopectin molecules, yielding a gel, the properties of which strongly depend upon the A:AP ratio. A weak gel is experienced below A:AP = 22:78, whereas quite strong gels are seen beyond this ratio.

These results are to be compared with those obtained with starch, as in Figure 6 for tapioca starch (A:AP ~ 17:83) and maize starch (A:AP ~ 27:73) gels at a comparable concentration (4 wt %). Both samples were prepared by heating a starch suspension at 95 °C under conditions of slow stirring and immediately transferring the sample to the measuring system of the rheometer. The gel was aged at 25 °C for 8 hours, and then a mechanical spectrum was recorded. It is clear that these starch gels do not compare directly to the mixtures. The spectrum of tapioca starch is characterized by a strong dependence of G' and G'' upon frequency and by $G' \sim G''$ over the whole frequency range. This is similar to the data for the amylose and amylopectin mixture at a 10:90 ratio. Such a result suggests that only part of amylose in the starch is effective in forming a gel.

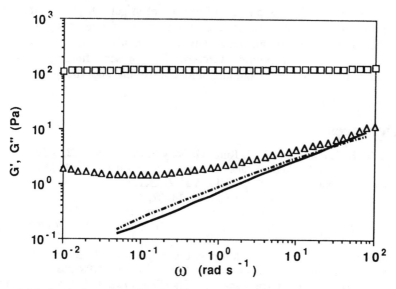

Figure 6 *Mechanical spectra of a* 4 wt% *maize starch gel* (□, G'; △, G") *and a* 4 wt% *tapioca starch gel* (—, G'; – –, G")

Probably, the phase separation is not completed as effectively in the starch as in the mixture owing to an incomplete separation of components during the pasting process. The data obtained with maize starch, on the other hand, are typical of a gel. The behaviour differs, however, from that of the 27:73 mixture in having a higher *G'* value (of the order of 100 Pa against 25 Pa in the mixture) and a low value of *G"* which did not depend upon frequency as in the mixture. A maize starch gel can therefore be described as a composite, its continuous phase being enriched with amylose.[11-13] The visco-elastic behaviour of the 27:73 mixed system also reflects the properties of a composite. However, though it appears evident that phase separation does occur in both cases, the redistribution of the components seems to take place in different ways, and to depend upon the preparation procedure.

4 Conclusion

We confirm that the properties of starch systems strongly depend upon the amylose:amylopectin ratio. These properties are mainly governed by a phase separation process due to the thermodynamic incompatibility of amylose and amylopectin. On mixing, a phase separation takes place which results in a composite system. The critical range in terms of A:AP ratio is between 15:85 and 22:78, and the point of phase inversion is around 15:85. Below 15:85, the continuous phase contains predominantly amylopectin playing the major role. Beyond a 22:78 ratio, the amylose is concentrated enough in the continuous phase to form a strong gel. Of particular interest is the intermediate range, where the properties are far removed from those of either amylopectin or amylose. This corresponds, in fact, to the A:AP ratios of tapioca and potato starches. The present findings may thus provide a basis for understanding and controlling the rheology of these starches.

References

1. M. T. Kalichevsky and S. G. Ring, *Carbohydr. Res.*, 1987, **162**, 323.
2. V. M. Leloup, P. Colonna, and A. Buléon, *J. Cereal Sci.*, 1991, **13**, 1.
3. M. L. German, A. L. Blumenfeld, Ya. V. Guenin, V. P. Yuryev, and V. B. Tolstoguzov, *Carbohydr. Polym.*, 1992, **18**, 27.
4. J.-L. Doublier and L. Choplin, *Carbohydr. Res.*, 1989, **193**, 215.
5. J.-L. Doublier, I. Coté, G. Llamas, and G. Charlet, *Progr. Colloid Polym. Sci.*, 1992, in press.
6. G. Robinson and S. B. Ross-Murphy, *Carbohydr. Res.*, 1982, **107**, 17.
7. A. H. Clark and S. B. Ross-Murphy, *Adv. Polym. Sci.*, 1987, **83**, 57.
8. J.-L. Doublier, G. Llamas, and L. Choplin, *Makromol. Chem. Macromol. Symp.*, 1990, **39**, 171.
9. M. J. Miles, V. J. Morris, and S. G. Ring, *Carbohydr. Res.*, 1985, **135**, 257.
10. H. S. Ellis and S. G. Ring, *Carbohydr. Polym.*, 1985, **5**, 201.
11. M. J. Miles, V. J. Morris, P. D. Orford, and S. G. Ring, *Carbohydr. Res.*, 1985, **135**, 271.
12. S. G. Ring, *Starch*, 1985, **37**, 80.
13. J.-L. Doublier, G. Llamas, and G. Le Meur, *Carbohydr. Polym.*, 1987, **7**, 251.

Stringiness of Aqueous Starch Pastes

By P. A. M. Steeneken and A. J. J. Woortman

NETHERLANDS INSTITUTE FOR CARBOHYDRATE RESEARCH—TNO,
ROUAANSTRAAT 27, 9723 CC GRONINGEN, THE NETHERLANDS

1 Introduction

Apart from its nutritive value, starch has numerous industrial applications in products like foodstuffs, paper, textiles, drilling muds, bioplastics, and pharmaceuticals.[1-2] In the food sector, one of the main uses of starch is as a thickener and binding agent in sauces, soups, dairy desserts, and pie fillings. The performance of such starch products depends on their functional properties, which are determined largely by rheological and other textural characteristics. An important textural parameter of aqueous starch pastes is 'stringiness', the ability to form long threads. Whereas a 'long' texture is a desired attribute in adhesives, it is seldom wanted in foods, where it is associated with a gluey or slimy mouthfeel. For example, dairy desserts must be 'short'.

In the laboratory, stringiness is measured by allowing a material to flow from a spoon or a funnel, and estimating visually the length of the thread.[3] A ranking can be made with the help of some starch pastes used as a reference. Results are semi-quantitative at best and trained staff are required to assure reliability.

In this contribution we address two aspects of the assessment of stringiness. Firstly, a new method is presented which allows a quantitative estimation of thread lengths. Secondly, the relationship of thread length to some rheological properties is examined with the aim of identifying structural features of starch pastes that are responsible for stringiness and so providing a convenient experimental method for the prediction of stringiness.

2 Some Physical Properties of Starch Pastes

Starch is deposited in plants as small semi-crystalline and birefringent granules which are insoluble in cold water.[4] On heating an aqueous suspension, the granules lose their birefringence and swell enormously. This process is known as gelatinization and the resulting suspension of swollen granules is referred to as a paste. In the course of heating, macromolecular material, mainly amylose, is released from the swollen granules; under conditions of low shear and at temperatures below 95 °C, the amount released is limited to 2–40%, depending on the type of starch.[5-7]

Aqueous starch pastes behave rheologically as soft gels, exhibiting a yield stress,[8,9] and a dynamic storage modulus G' that is not very sensitive to frequency and always higher than the loss modulus G''.[8,10,11] The limiting zero-shear viscosity characteristic of polymer solutions is not observed, not even at very low shear rates.[8] The concentration dependence of the viscosity can be explained from the degree of swelling and the rigidity of the gelatinized starch granules.[6] These properties are by no means unique to aqueous starch systems: they are observed for many types of microgel suspensions.[12-15]

3 Experimental

Method for the Estimation of Stringiness

Problems with the afore-mentioned spoon or funnel tests arise because the velocity of the falling liquid-like thread increases with its length and the process of rupture occurs in a very short time (< 0.1 s). An alternative method[16] involves the measurement of the length of threads that are formed when a steel ball is slowly pulled out from a starch solution.

Thread lengths of flowing starch pastes can be measured conveniently and accurately by means of a video technique (Figure 1).[17] The paste is allowed to flow from a liquid delivery device along a graduated scale. The flow is monitored continuously with a video camera and replayed with a variable-speed video recorder in slow motion. The liquid delivery device can be chosen at will and adapted to flow conditions encountered in practice. For fundamental studies, capillary flow from a pump-driven syringe is preferred because it allows a wide range of concentrations and flow-rates to be studied. Measurable quantities are the drop length l, the drop time t, and the drop mass m, from which the average drop diameter d is calculated, i.e.

$$d^2 = 4Qt/\pi l = 4m/\pi l \rho, \tag{1}$$

where Q and ρ are the volume flow-rate and the density of the paste, respectively.

The use of high-speed film to study the flow of liquid jets has been described previously.[18-20]

Other Experimental Methods

Starch pastes were prepared by rotating appropriate amounts of starch and water in a round-bottom flask at 180 r.p.m. on a water bath at 92 °C

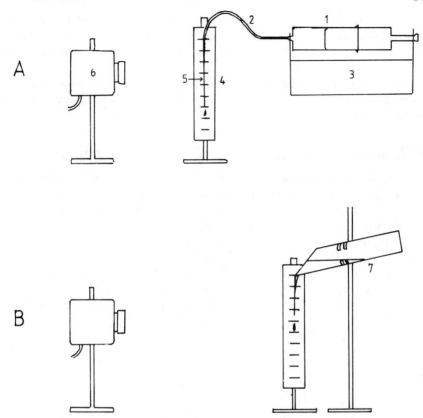

Figure 1 *Experimental set-up for the measurement of the thread length of starch pastes. A = capillary flow from a pump driven syringe; B = flow from an inclined tube. 1, syringe (50 ml); 2, capillary tubing; 3, pump; 4, graduated scale; 5, liquid thread; 6, camera; 7, inclined perspex tube* [Reprinted from ref. 17 with permission from IRL Press Ltd.]

(potato starches) or 97 °C (cereal starches) for 30 min. Steady shear viscosity η (at $1\,s^{-1}$) and dynamic storage modulus G (at $1\,rad\,s^{-1}$)* were measured in a Deer rheometer (a constant-stress apparatus) equipped with a concentric cylinder geometry at 60 °C as described elsewhere.[6]

4 Some Results

The thread length has been found to increase with starch concentration and with volume flow-rate, but it is independent of the diameter of the tubing. The average thread diameter decreases with starch concentration, decreases slightly with flow-rate, and increases with tubing diameter. The

---

*In the rest of this paper, 'modulus' means 'dynamic storage modulus'. (Note the dropping of the prime symbol.)

most noteworthy result is the large variation in stringiness between different starches. The results depicted in Figure 2 agree with our expectations: potato starch is known to be long, whilst cereal starches and cross-linked starches are short. The peculiar behaviour of highly cross-linked potato starch which has a short texture, but which was found here to produce fairly long threads with a spaghetti-like appearance, will be discussed later.

5 Relation of Stringiness to Rheological Properties

A study into the relationship between stringiness and rheological properties serves two intended purposes: to provide a simple measurement for the prediction of stringiness, and to identify the structural features of the starch paste that are responsible for stringiness.

Formation of liquid threads involves large-scale deformations, and so it seems appropriate to attempt to relate stringiness to large-scale deformation behaviour, for example in shear. On the other hand, small-scale deformation properties like the dynamic moduli can be well correlated with the physical structure of visco-elastic materials[21] and also can be measured

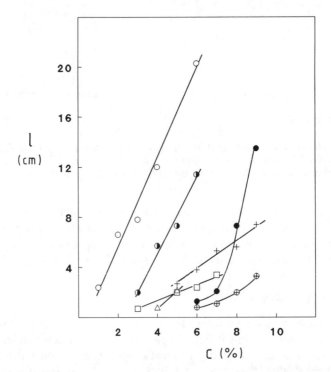

Figure 2 *Thread length* l *of starch pastes as a function of concentration* c. *Capillary tube flow-rate* = 30 ml h^{-1}; *tubing diameter* = 2 mm. *Starches:* ○, *potato*; ◐, *potato, lightly cross-linked;* ●, *potato, highly cross-linked*; +, *waxy maize*; ⊕, *waxy maize, highly cross-linked;* △, *maize*; □, *wheat*

conveniently. The stringiness, the modulus G (at $1\,\mathrm{rad\,s^{-1}}$), and the steady shear viscosity η (at $1\,\mathrm{s^{-1}}$) of 56 starch pastes prepared from 27 different starch products were measured. For 24 pastes, measurement of stringiness was performed at two different flow-rates. Pastes of highly cross-linked potato starch, which behave anomalously, are not included in the following analysis.

In a preliminary study, it was established that, at a specified viscosity, the texture of the paste is shorter if the modulus is higher. This prompted us to explore a relationship of the type $l \sim \eta^a/G^b$. For a double-logarithmic plot of the data, a linear regression coefficient $r = 0.84$ was found for $a = 1.2$ and $b = 0.83$, which is insufficient for an accurate determination of the thread length (Figure 3). At a specified value of the abscissa variable the thread length may vary by a factor of 4! It was interesting to note that the correlation was good for pastes prepared from the same starch, but of different concentration (Figure 4). This suggests that, apart from viscosity and modulus, there is another starch characteristic that determines thread length. An attempt to improve the correlation by including the swelling power q was not successful.

A relationship between thread length and thread volume, $V = Qt$, would seem to provide a useful way to predict stringiness. For a plot of $\log l$ versus $\log V$, r was found to be only 0.77. This suggests that, at a given value of l, the average thread diameter may vary for different starches. This assumption was tested for a number of starch pastes by plotting the real thread length versus a (hypothetical) thread length for plug flow $l_0 = 4\,V/\pi d_0$, where d_0 is the diameter of the tubing used in pump flow (Figure 5). The slope of the plot of l versus l_0 was found to be a

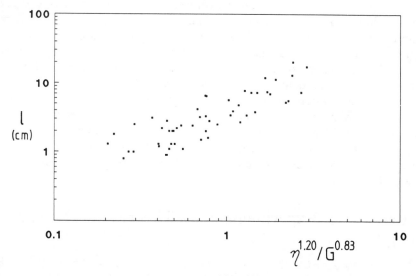

Figure 3 Thread length l as a function of $\eta^{1.2}G^{-0.83}$ plotted on a double logarithmic scale (η = viscosity; G = modulus)

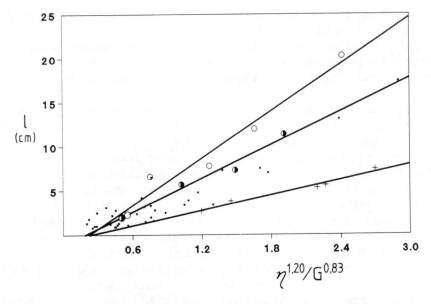

Figure 4 *Thread length* l *as a function of* $\eta^{1.2}G^{-0.83}$ *on a linear plot* ($\eta =$ viscosity; G = modulus). *Starches:* $\bigcirc$, *potato*; $\pmb{\mathbb{O}}$, *potato, lightly cross-linked;* +, *waxy maize*; $\blacksquare$, *other starches*

characteristic property of each starch; hence it can be considered as a measure of extensibility. For highly cross-linked potato starch this slope is only slightly larger than unity. Pastes of this starch are hardly extended, but are extruded from the capillary in plug flow, affording long threads which apparently break under their own weight. Other starches show much steeper slopes and are extended appreciably; the highest value of the slope (approximately 8) was observed with unmodified potato starch.

It can be concluded that the average extension ratio d_0^2/d^2 at a specified thread length varies for different starches. In order to correct for the variation in tensile deformation exerted on threads of different starch pastes, l was plotted as a function of $\eta^e G^f d^{-2}$, where e and f are adjustable parameters. In so doing the regression coefficient of the log–log plot increased from 0.84 (Figure 3) to 0.97 with $e = 0.3$ and $f = -0.23$ (Figure 6). Unfortunately, it seems impossible to measure d in an easier way than by the video method; hence this relationship is of limited practical value.

6 Relation of Experiments with Theories for Thread Length

To our knowledge, the theory of Ziabicki and Takserman-Krozer[18,22-25] is the only one in which the thread length of a fluid is related to the physical properties of that fluid, and to the experimental conditions, in particular

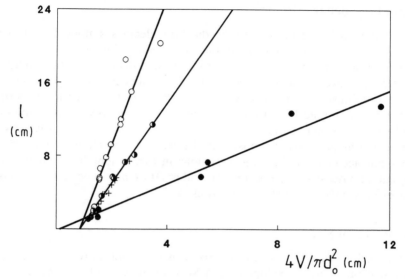

Figure 5 *Thread length l as a function of the hypothetical thread length for plug flow $4V/\pi d_0$ (V = thread volume; d_0 = tubing diameter). Starches: ○, potato; ◑, potato, lightly cross-linked; ●, potato, highly cross-linked; +, waxy maize*

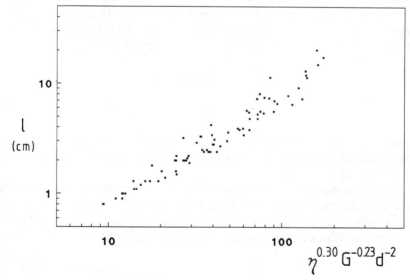

Figure 6 *Thread length l as a function of $\eta^{0.3}G^{-0.23}d^{-2}$ (η = viscosity; G = modulus; d = average thread diameter)*

the flow-rate. Newtonian as well as non-Newtonian fluids can form liquid jets. Two mechanisms taken into account in the theory are discussed below.

Capillary break

Due to extension of the thread the diameter decreases along the jet. Thus the surface energy per unit of volume is increased. This is counteracted by the creation of a sinusoidal periodicity in thread diameter. If the amplitude of this wave exceeds a critical value, the jet breaks up. The development of these instabilities is slowed down by a high viscosity and/or a low surface tension. The thread length increases with increasing viscosity and flow-rate and with decreasing surface tension. This prediction is in accordance with our experimental observations, but the capillary break mechanism[18,22,24] does not account for the large variations in thread length of starch pastes of similar viscosity but different modulus. It is not expected that starch pastes vary widely in surface properties.

Cohesive break

This mechanism[18,22,23] is relevant for elastic liquids only. The theory of cohesive break was developed for a Maxwell fluid.[21] On extension of the thread, a tensile stress is built up. This stress may relax by viscous dissipation at a rate $1/\tau$, where τ is the relaxation time of the fluid. If the stress relaxation is faster than the elongation rate $\dot{\varepsilon}$, the jet is stable. If the stress builds up faster than it can relax, a limiting value is reached (the tensile 'strength' of the jet) and the thread breaks. Ziabicki and Takserman-Krozer derived an explicit expression for the thread length as a function of viscosity, modulus, and flow-rate. For the present purpose it is sufficient to formulate a critical condition for rupture of the jet:

$$1/\tau < \dot{\varepsilon} \text{ (jet breaks)}$$

$$1/\tau > \dot{\varepsilon} \text{ (jet is stable)} \tag{2}$$

Because we have $1/\tau = G/\eta$, a high modulus and/or a low viscosity tend to stabilize the jet. In addition, it was derived that the thread length increases with decreasing flow rate.

All these predictions are *opposite* to our observations. The failure of the Ziabicki–Takserman-Krozer theory may be due to the fact that we are not dealing with Maxwell liquids, but with materials that are particulate in nature and which exhibit solid-like rather than liquid-like character.

Fracture by gravity

In addition to capillary break and cohesive break, we should also consider a third mechanism: fracture by gravity. In this case threads are extruded in plug flow and break under their own weight. An example is highly cross-linked potato starch at high concentrations in water, which shows spaghetti-like behaviour. As the tensile deformation tends to zero ($d \approx d_0$), gravity is the dominant force acting on the thread. An effective strength of the thread, T, can be defined as

$$T = \rho l g \tag{3}$$

where g is the gravitational constant. Prerequisites for fracture by gravity are a high rigidity of the swollen granules and a large coherence (or friction) between them. Such pastes have an appreciable yield stress.

It must be concluded that there is no theory available at present that adequately describes the stringiness of starch pastes. Depending on the starch type and concentration, more than one mechanism may be involved. A satisfactory theory of stringiness of starch pastes should comply with the following observations:

(i) For a particular starch the thread length increases with flow-rate and with concentration; therefore it seems likely that a certain coherence between the swollen granules is required for the formation of long threads.

(ii) At a specified viscosity, the thread length decreases with the rigidity of the granules.

(iii) At a specified thread length, the average extension ratio d_0^2/d^2 varies for different starches.

(iv) In the limiting case of rigid granules and a large friction between them, the extension ratio falls to zero and fracture by gravity occurs.

References

1. 'Starch: Chemistry and Technology', ed. R. L. Whistler, J. N. BeMiller, and E. F. Paschall, 2nd Edn., Academic Press, Orlando, 1984.
2. O. B. Wurzburg, 'Modified Starches: Properties and Uses', CRC Press, Boca Raton, 1986.
3. P. Woldendorp and K. G. de Noord, *Stärke*, 1966, **18**, 293.
4. D. French, in Ref. 1, p. 183.
5. J.-L. Doublier, G. Llamas, and M. Le Meur, *Carbohydr. Polym.*, 1987, **7**, 251.
6. P. A. M. Steeneken, *Carbohydr. Polym.*, 1989, **11**, 23.
7. H. S. Ellis, S. G. Ring, and M. A. Whittam, *J. Cereal Sci.*, 1989, **10**, 33.
8. I. D. Evans and D. R. Haisman, *J. Texture Studies*, 1979, **10**, 347.
9. J.-L. Doublier, *Stärke*, 1981, **33**, 415.
10. R. B. K. Wong and J. Lelievre, *Rheol. Acta*, 1981, **20**, 299.
11. K. Svegmark and A.-M. Hermansson, *Carbohydr. Polym.*, 1990, **13**, 29.
12. N. W. Taylor and E. B. Bagley, *J. Appl. Polym. Sci.*, 1974, **18**, 2747.
13. N. W. Taylor and E. B. Bagley, *J. Appl. Polym. Sci.*, 1977, **21**, 113.
14. R. J. Ketz, Jr., R. K. Prud'homme, and W. W. Graessley, *Rheol. Acta*, 1988, **27**, 531.
15. M. S. Wolfe and C. Scopazzi, *J. Colloid Interface Sci.*, 1989, **133**, 265.
16. R. Takahashi, T. Ojima, and M. Yamamoto, *Stärke*, 1969, **21**, 315.
17. P. A. M. Steeneken and A. J. J. Woortman, *Food Hydrocolloids*, 1991, **5**, 147.

18. A. Ziabicki and R. Takserman-Krozer, *Kolloid Z. Z. Polym.*, 1964, **198**, 60.
19. W. M. Jones and I. J. Rees, *J. Non-Newtonian Fluid Mech.*, 1982, **11**, 257.
20. J. E. Matta and R. P. Tytus, *J. Non-Newtonian Fluid Mech.*, 1990, **35**, 215.
21. J. D. Ferry, 'Visco-elastic Properties of Polymers', 3rd Edn., Wiley, New York, 1980.
22. A. Ziabicki and R. Takserman-Krozer, *Roczniki Chemii*, 1963, **37**, 1503.
23. A. Ziabicki and R. Takserman-Krozer, *Roczniki Chemii*, 1963, **37**, 1511.
24. A. Ziabicki and R. Takserman-Krozer, *Roczniki Chemii*, 1963, **37**, 1607.
25. A. Ziabicki and R. Takserman-Krozer, *Kolloid Z. Z. Polym.*, 1964, **199**, 9.

Evidence for Protein–Polysaccharide Complex Formation as a Result of Dry-heating of Mixtures

By K. Jumel, S. E. Harding, J. R. Mitchell, and Eric Dickinson[1]

DEPARTMENT OF APPLIED BIOCHEMISTRY AND FOOD SCIENCE, UNIVERSITY OF NOTTINGHAM, SUTTON BONINGTON, LOUGHBOROUGH, LEICESTERSHIRE LE12 5RD, UK
[1]PROCTER DEPARTMENT OF FOOD SCIENCE, UNIVERSITY OF LEEDS, LEEDS LS2 9JT, UK

1 Introduction

The complexes produced by the dry-heating of the globular protein bovine serum albumin (BSA) and the non-ionic polysaccharide dextran have been shown recently to have excellent emulsion stabilizing properties.[1,2] This short paper presents experimental data on the molecular weight of complexes formed by dry-heating mixtures of BSA and dextran T40 (4×10^4 daltons) as determined by gel permeation chromatography with light-scattering detection (GPC/MALLS) and ultracentrifugation.

2 Experimental

Protein–polysaccharide complexes were prepared by dry heating various molar ratios of BSA and dextran T40 at 60 °C for 3 weeks as described previously.[1] The samples were dissolved in phosphate/chloride buffer (pH 7.0, ionic strength 0.1 M). Weight-average molecular weights of samples of various molar ratios were determined by GPC/MALLS (sample concentration 3 mg ml^{-1}).[3] The BSA–dextran complex of molar ratio 2:1 in a solution of concentration 0.5 mg ml^{-1} was investigated under conditions of low-speed sedimentation equilibrium in a Model E analytical ultracentrifuge. The sedimentation coefficient of this same complex (sample concentration 2 mg ml^{-1}) was determined using a Beckman XL-A analytical ultracentrifuge.

3 Results and Discussion

Figure 1 shows the weight-average molecular weight M_w of various BSA–dextran complexes as determined by GPC/MALLS. We see that, of the various samples investigated, only the complex with a composition of 33 mol% dextran has a molecular weight substantially greater than that for the native protein. Table 1 compares the value of M_w obtained from GPC/MALLS for this 2:1 molar ratio complex with that from low-speed sedimentation equilibrium. Also shown for comparison are M_w values for the heat-treated pure BSA (same heating conditions). The results indicate that, although some aggregation of the BSA itself occurs during the

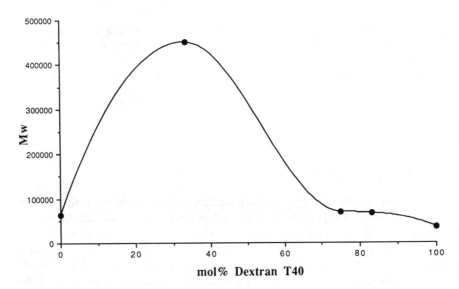

Figure 1 *Weight-average molecular weight M_w of complexes obtained by dry-heating of mixtures of BSA and dextran T40 at various molar ratios*

Table 1 *Molecular weight, sedimentation coefficients, and frictional coefficients*

Sample	M_w (daltons) GPC/MALLS	M_w (daltons) Model E	s_{20} (S)	$s_{20,w}$(S)	f/f_0
BSA native	62 700 ± 5 000	66 700[a]	3.93 ± 0.1	4.04 ± 0.1	1.3 ± 0.1
BSA heat-treated (3 weeks)	158 000 ± 10 000	130 000 ± 10 000	5.43 ± 0.1	5.59 ± 0.1	1.3 ± 0.1
BSA/T40 complex 2:1 ratio	450 000 ± 20 000	330 000 ± 20 000	4.85 ± 0.1	4.98 ± 0.1	3.4 ± 0.1

[a] Value from: K. E. van Holde, 'Physical Biochemistry', Prentice-Hall, 1971

dry-heating process, this does not account for the much larger average molecular weight of the complex. The general trends of behaviour detected by GPC/MALLS and ultracentrifugation are the same. The slight discrepancy in the numerical values can be accounted for by noting that the light-scattering results may be affected to some, extent by additional aggregated material co-existing with the protein–polysaccharide complex, whereas such material might have moved to the cell base in ultracentrifugation measurements and would therefore not have contributed to the weight-average molecular weight.

Figure 2 compares chromatographic elution profiles for the BSA–dextran complex (2:1 molar ratio) with those for the native BSA and the heat-treated BSA. We note the different position of the complex peak (lower elution volume) from that for the protein alone, both from light-scattering analysis and differential refractive index analysis. The peak shape shows that the complex is not a single species but is polydisperse. Figure 3 shows the movement of the sedimenting boundary for the BSA–dextran complex (2:1 molar ratio). The presence of a single sedimenting boundary suggests that the polydispersity is more of a quasi-continuous character than a paucidisperse one.

Sedimentation coefficients of complex, native BSA and heated BSA are also given in Table 1. From the molecular weight and the $s_{20,w}$ value, we can estimate the frictional ratio f/f_0. Assuming that the $s_{20,w}$ value measured is not too far from the infinite dilution value, the large value of f/f_0 derived for the complex is suggestive of either a highly asymmetric or a highly hydrated entity.

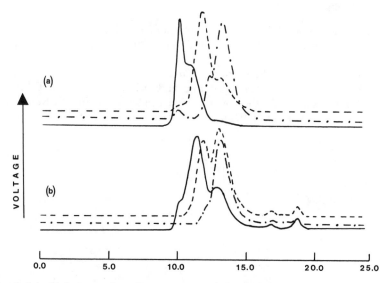

Figure 2 (a) Light-scattering chromatograms (obtained from 90° detector) and (b) Differential refractive index chromatograms: - · · · -, native BSA; - - -, heat treated BSA; —, BSA/dextran T40 (2:1) complex

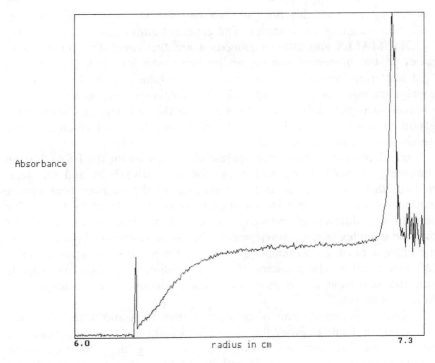

Figure 3 *Sedimenting boundary of BSA/dextran T40* (2:1) *complex*

References

1. E. Dickinson and V. B. Galazka, in 'Gums and Stabilisers in the Food Industry', ed. G. O. Phillips, D. J. Wedlock, and P. A. Williams, Oxford University Press, 1992, Vol. 6, p.351.
2. E. Dickinson and M. G. Semenova, *Colloids Surf.*, 1992, **64**, 299.
3. P. J. Wyatt, in 'Laser Light Scattering in Biochemistry', ed. S. E. Harding, D. B. Satelle, and V. A. Bloomfield, Special Publication No. 99, Royal Society of Chemistry, Cambridge, 1992, p. 35.

A Statistical Model of the Adsorption of Protein–Polysaccharide Complexes

By Eric Dickinson and Stephen R. Euston

PROCTER DEPARTMENT OF FOOD SCIENCE, UNIVERSITY OF LEEDS, LEEDS LS2 9JT, UK

1 Introduction

Recent experimental results[1-4] have shown that enhanced emulsion stability can be obtained if a hybrid molecule, formed by covalent linkage of polysaccharide molecules to globular protein molecules, is used as the stabilizing entity. In this work, we have applied Monte-Carlo computer simulation to investigate the molecular factors that govern adsorption of protein–polysaccharide complexes.

2 Simulation Method

A two-dimensional square lattice of sides $L_x \times L_y$ is set up. On this lattice, 200 hybrid molecules are defined. Each hybrid molecule is comprised of one soft particle (containing 9 subunits) and n_p fractal polymers (each containing 5 subunits). The soft particles are deformable, but still particulate in nature, and are used to model the globular proteins.[5,6] The fractal polymers are flexible, disordered molecules and represent polysaccharide chains. The hybrid molecules are allowed to undergo translational and rotational motion on the lattice subject to the restrictions of excluded volume and importance sampling.[7] In addition, random displacements of individual subunits on soft particles and fractal polymers are carried out, so as to alter the conformation of the two species. Full details of the simulation methods can be found elsewhere.[8,9]

Two hard walls are defined at $y = 0$ and $y = L_y + 1$, at which the soft particle subunits are allowed to adsorb with an adsorption energy of $-1.5\,kT$ per segment. Fractal polymer subunits have no energetic preference for the adsorbed state.

A series of simulations was carried out to investigate the effect of varying the number of fractal polymers per hybrid molecule on the adsorption of soft particle subunits. For comparison, a set of simulations in which the soft particles and fractal polymers were not joined was also modelled.

3 Results

Figure 1 shows a snapshot configuration of a simulation of hybrid molecules comprising one soft particle and one fractal polymer. It was observed[9] that, compared to simulations in which the two species were not bound, the coverage of the adsorbent surface by soft particle subunits

Figure 1 *Instantaneous snapshot of a system containing 200 hybrid molcules adsorbing at a hard wall. Each hybrid molecule comprises one soft particle (9 subunits) joined to one fractal polymer (5 subunits). The soft particle subunit adsorption energy is $E_A^S = -1.5\,kT$, and that of the fractal polymer is $E_A^P = 0.0\,kT$. The total subunit density is 25%*

decreases as the number of fractal polymers per hybrid molecule increases. This is because the non-adsorbing fractal polymer acts as a barrier to the adsorption of the soft particles. The adsorbed fraction of soft particle subunits becomes negligible when five bound fractal polymers are incorporated into the hybrid molecule.

Although binding of non-adsorbing fractal polymer reduces the adsorbed fraction at the hard walls, this does not necessarily lead to an adsorbed layer with a lower capacity for steric stabilization. Figure 2 shows the normalized total subunit density a distance y from the hard walls for simulations where $n_p = 0$, 1, 2, 3, 4, or 5. For hybrid molecules containing 1, 2, or even 3 fractal polymers, although the adsorbed fraction is reduced, the distance which the whole adsorbed layer extends from the surface is increased. Whilst it is not possible to give a quantitative assessment of the relative steric stabilizing ability of adsorbed layers formed by the various hybrid molecules, some general observations can be made. It is probable that, with hybrid molecules containing a high ratio of soft particles to fractal polymers (1:1 or 1:2), the thicker adsorbed layer would lead to an enhanced stability. This would be due to the fractal polymer molecules bound to the adsorbed soft particles. It is equally probable that hybrid

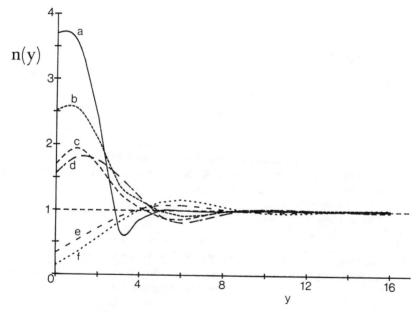

Figure 2 *Subunit density profiles near the adsorbent walls for systems containing hybrid molecules comprised of one soft particle (9 subunits) and n_p fractal polymers (each 5 subunits). The normalized subunit density n(y) is plotted against the distance y from the surface: (a) $n_p = 0$, (b) $n_p = 1$, (c) $n_p = 2$, (d) $n_p = 3$, (e) $n_p = 4$, (f) $n_p = 5$. The density profiles are normalized by dividing the subunit site fraction by the mean bulk density away from the walls. The data are averaged over the two walls*

molecules with a low ratio of particle to polymer (1:4 or 1:5) would form a patchy adsorbed layer with a reduced stabilizing capacity, since the incompletely covered surface would be susceptible to bridging flocculation.[10]

4 Conclusions

The trends observed in these simulations show some similar features to results from recent experiments on the emulsion stabilizing properties of covalent Maillard-type complexes of globular proteins with dextrans of varied molecular weight.[1-3] An individual protein molecule may bind to one or more polysaccharide molecules depending on reaction conditions and macromolecular composition. This produces hybrid molecules of varying hydrophobic/hydrophilic character. For each protein/polysaccharide combination, there is a critical value of the molar protein:polysaccharide ratio which confers optimum stability. Over-reaction of the two species leads to a loss of functionality, presumably because the hybrid molecule is too hydrophilic for rapid adsorption and unfolding at the emulsion droplet surface. Optimum functionality may occur when, on average, there is enough bound polysaccharide to enhance stability, but there are still enough accessible protein hydrophobic groups on the hybrid to allow strong binding to the surface.

Acknowledgement

E. Dickinson acknowledges financial support for S. R. Euston from the Agricultural and Food Research Council.

References

1. E. Dickinson and V. B. Galazka, *Food Hydrocolloids*, 1991, **5**, 281.
2. E. Dickinson and V. B. Galazka, in 'Gums and Stabilisers for the Food Industry', ed. G. O. Phillips, D. J. Wedlock, and P. A. Williams, Oxford University Press, 1992, Vol. 6, p. 351.
3. E. Dickinson and M. G. Semenova, *Colloids Surf.*, 1992, **64**, 299.
4. A. Kato, Y. Sasaki, R. Furuta, and K. Kabayashi, *Agric. Biol. Chem.*, 1990, **54**, 107.
5. J. A. de Feijter and J. Benjamins, *J. Colloid Interface Sci.*, 1982, **90**, 289.
6. E. Dickinson and S. R. Euston, *J. Chem. Soc., Faraday Trans.*, 1990, **86**, 805.
7. M. P. Allen and D. J. Tildesley, 'Computer Simulation of Liquids', Clarendon, Oxford, 1987.
8. E. Dickinson and S. R. Euston, *J. Colloid Interface Sci.*, 1992, **152**, 562.
9. E. Dickinson and S. R. Euston, *Food Hydrocolloids*, 1992, **6**, 345.
10. E. Dickinson and L. Eriksson, *Adv. Colloid Interface Sci.*, 1991, **34**, 1.

On the Structure of Casein Micelles

By Hein J. M. van Dijk[1]

CAMPINA MELKUNIE B. V., PO BOX 222, 3440 AE WOERDEN, THE
NETHERLANDS

We have examined the partition of calcium (Ca), magnesium (Mg), and
phosphate (P) between the casein micelles and various aqueous phases.[1]
After addition of $> 5\%$ NaCl, we have $(Ca_c + Mg_c)/P_{ic} = 2.0$, and, after
addition of NaOH, notable changes in the partition were found to occur at
approximately $P_{ic}/P_o = 2.0$ (subscripts: c = colloidal, i = inorganic, o = organic colloidal). We have also examined the disintegration of the casein
micelles after addition of NaOH. Above pH 9, skim milk slowly becomes
as transparent as whey. The speed of this process is increased if the protein
amino groups have been succinylated or it is carried out at a higher pH,
and is decreased due to addition of NaCl or amidation of carboxyl groups.
Even low concentrations of added NaCl (*i.e.* 10–20 mM) strongly retard
the disintegration but hardly influence the reaggregation.[2]

The following model is proposed for the casein micelles, and especially
for the native (-like) micellar calcium phosphate (nMCP).

Proposition 1

*Most nMCP bridges in raw milk contain 2 phosphoserine residues (P_o),
4 P_{ic} ions, and 8 divalent cations.*

As the value of $(Ca_c + Mg_c)/P_{ic}$ can be lowered from ~2.3 to only 2.0
by addition of NaCl, while roughly half of the Ca_c can be exchanged by
the dissolved Ca in the same period,[3] it appears that $(Ca + Mg)/P_{ic}$ in
nMCP may be 2.0. In addition, our results suggest that $P_{ic}/(P_o$ associated
with nMCP) is also 2.0. Finally, it is most likely that every nMCP ion
cluster interconnects two P_o groups.[2,4]

Proposition 2

*Casein molecules of the same kind are bound together in pairs by a number
of nMCP bridges (nMCP dimers).*

After addition of 6 M urea[5] or NaOH (pH 10)[2] to milk, the casein
molecules associated with the nMCP are in units of, say, 2–20 molecules.

[1] Present affiliation: Tastemaker, PO Box 414, 3770 AK Barneveld, The Netherlands

All P_o groups in these units are associated with the nMCP.[2,6] Since most of these P_o groups are present in clusters (which are not all identical), and since below pH 6.7 in most sub-micelles it is predominantly the α_{s1}-casein which associates with the nMCP,[2] it is barely conceivable that the P_o groups find each other at random. Steric hindrance will not allow this in most cases.

Both before and after the formation of nMCP, most caseins are in the form of 'sub-micelles' of diameter between about 15 and 20 nm.[7] Hydrophobic bonding is most likely involved in their formation.[7]

Proposition 3

Mainly on account of the fact that nMCP and casein-(Ca + Mg) considerably reduce the charge on the sub-micelles, they aggregate into 'casein micelles'.

Since the amount of nMCP grows with increasing pH,[1] and since the nMCP-links between casein molecules have a relatively long lifetime,[2,6] the above results suggest that the casein sub-micelles are not linked together by such 'permanent' nMCP bridges. As the casein molecules will carry on average a net charge of about -17 at pH 6.7, and as 17 Ca, 1 Mg, 5.6 PO_4^{3-}, and 2.4 PO_4^{2-} ions are on average associated per casein molecule,[4,6] these ions compensate for *ca.* 90% of the net charge of the casein molecules in the casein micelle. The above results[2] suggest that this is the main cause for the aggregation of the sub-micelles.

References

1. H.J.M. van Dijk, *Neth. Milk Dairy J.*, 1991, **45**, 241.
2. H.J.M. van Dijk, *Neth. Milk Dairy J.*, 1992, **46**, 101.
3. K. Yamauchi, Y. Yoneda, Y. Koga, and T. Tsugo, *Agric. Biol. Chem.*, 1969, **33**, 907.
4. H.J.M. van Dijk, *Neth. Milk Dairy J.*, 1990, **44**, 65.
5. T. Aoki, Y. Kako, and T. Imamura, *J. Dairy Res.*, 1986, **53**, 53.
6. H.J.M. van Dijk, *Neth. Milk Dairy J.*, 1990, **44**, 111.
7. D. G. Schmidt, *Neth Milk Dairy J.*, 1980, **34**, 42.

Primary and Secondary Structures of Caseins

By C. Holt

HANNAH RESEARCH INSTITUTE, AYR KA6 5HL, UK

1 Introduction

Caseins comprise one of the most evolutionarily divergent families of mammalian proteins. For example, straightforward pairwise comparisons of homologous primary structures reveal degrees of sequence identity between ruminant and non-ruminant proteins typically in the range 40–50%. Moreover, the regions of identity are largely confined to the signal peptide and phosphorylation site sequences. Using standard scoring methods to introduce insertion/deletion points (hereafter called indels) improves the apparent degree of conservation only marginally, suggesting that the primary structures are not, in general, subject to the same functional constraints that filter the mutational drift in genomic sequences coding for globular proteins. It would appear, therefore, that the biological functions of the caseins are not critically dependent on the nature of the amino acid residues over large parts of their sequences.

Such a conclusion is supported by what little is known of the secondary and tertiary structures of the caseins. None has been crystallized, and so we must rely on the meagre information that can be gleaned from solution methods such as circular dichroism and infrared spectroscopy. Both of these methods, together with nuclear magnetic resonance spectroscopy and hydrodynamic measurements, indicate that individual caseins in solution adopt an open conformation with a high degree of configurational freedom of the side chains and little or no α-helical conformations.[1] For all the caseins, much of the polypeptide chain appears to be in an extended or disordered conformation in free solution. This lack of a tendency to form typical globular structures in spite of their predominantly hydrophobic character is what marks out the caseins as a unique group of proteins. Understanding the relation between their primary, secondary, and solution structures will undoubtedly shed light on their uses as gel-forming proteins and emulsifying agents in the food industry. The information would also be useful in helping to direct any attempts at engineering the caseins for altered food functionalities.

In this paper, a preliminary account is given of multiple sequence alignment methods, applied to members of the β-casein group of homologous proteins, rather than the simple pairwise comparisons previously employed. The conformational information that is accessible from the shape of the amide I band in the infrared spectrum of whole (bovine) casein and native casein micelles in ^2H_2O buffers is also assessed.

2 Primary Structures of β-Caseins

Sequence searches and comparisons were made of the OWL composite database maintained at the SERC Daresbury Laboratory in the UK, using the associated suite of programs, principally PDQ and DELPHOS for database queries and NEWSWEEP, ALIGN, and CLUSTAL V for sequence comparisons and alignments. It was found that 490 entries contained the sequence SSEE and 72 entries contained the sequence PQNI, which are the longest conserved sequences found in the mature β-casein polypeptide chain, but a combined search query served to identify only and all the β-caseins that could be found on the database by any other means. The minimum signal peptide sequence LACLVA identified only and all the β-casein precursor sequences in the database. Apart from other caseins, this highly conserved signal peptide sequence was found to show similarity to the egg phospho-proteins known as the vitellogenins, precursors of the phosvitins, possibly indicating some common relationships in secondary processing of the polypeptide chains as both families have the SSEE phosphorylation recognition site. When a comparison of whole sequences was made, the most similar non-casein sequences were those of proline and glutamine-rich seed storage proteins such as the wheat gliadins, maize zeins, and barley hordeins, but it is possibly hazardous to deduce from this any evidence of a recent evolutionary divergence of mammals from the higher plants. Other proteins such as the tooth amelogenins, bone osteopontins, and salivary proline-rich phosphoproteins were also identified in the searches, but at a lower degree of similarity, in spite of a shared mammalian origin and the common factor that they are all expressed in mineralizing tissues.

The precursor sequences of β-caseins from cow, sheep, man, rat, mouse, and rabbit were subjected to a more detailed study of similarity using two different scoring matrices. The identity matrix scores one for every conserved residue and zero for all non-identities, but the alignments were found to be very sensitive to the penalty scores chosen for indels. The Dayhof matrix, based on the probabilities of point mutations being accepted in a period of 250 My (the 250 PAM matrix), was also used and was found to be less sensitive to the introduction of indels near default values for penalty scores. Nevertheless, the alignments produced by this matrix were not acceptable because the main centres of phosphorylation, which are judged highly conserved in any pairwise comparison of β-casein sequences, were not perfectly aligned for all 6 sequences. The reason is that conservation of seryl residues does not score highly in the 250 PAM matrix, but five seryl residues in the N-terminal region of bovine β-casein are known to be phosphorylated in the mature protein and to be important for its biological function. The problem was overcome by modifying the 250 PAM matrix such that seryl residues were assigned the same scores as cysteinyl residues which are much more highly conserved in most proteins. An alternative strategy was also followed in which the sequences were edited to replace seryl or threonyl residues in potential sites of phosphorylation by cysteinyl residues, but this approach was not favoured as it introduced a subjective bias at an early stage of the alignment process. Using fixed and floating gap penalties of 10 with either the 250 PAM or modified 250 PAM matrices gave multiple sequence alignments in the C-terminal region that were virtually the same, indicating that the alignments were fairly robust. Reducing the gap penalties to 8 or less produced a noticeable divergence, and gap penalties of 9 provided a guide for some minor manual editing of the positions of indels to give the alignments shown in Figure 1. It can be seen that a considerable fraction of the sequences show a degree of functional conservation, such as substitutions of similarly hydrophobic residues at equivalent positions and alignments of residues such as prolyl and seryl with generally low propensities to form regular secondary structures.

3 Infrared Spectroscopy of Casein

Whole casein was dissolved in ^2H_2O (hereafter called D_2O), freeze-dried, and redissolved in D_2O containing 100 mM KCl to a final concentration of 65 g l^{-1}. The solution gave a pH meter reading of 5.54 and a spectrum was recorded in a Mattson Galaxy FTIR spectrometer using an out-of-compartment ZnSe horizontal attenuated total reflection cell at 4 cm^{-1} resolution with a deuterated tri-glycine sulfate detector. The spectrum was divided by an air spectrum and converted to absorbance. The absorbance of the buffer was subtracted, negative features were removed, and the spectrum was normalized. A small addition (40–80 μl per 5 ml of solution) of 500 mM NaOD was made, the pH meter reading noted, and the spectrum re-recorded. In this way a total of 17 spectra were acquired covering the apparent pH range 5.54–8.61. Inspection of the spectra revealed at least two pH-dependent regions with small but consistent changes in the strong amide I band at 1760 cm^{-1} and marked changes in amplitude of a weak band at about 975 cm^{-1}. Both features were analysed quantitatively.

A band at about 975 cm^{-1} is expected from stretching modes of the phosphate moiety of phosphorylated residues, and support for this assignment was found in the variation of the area of the band with pH. The entire titration curve could be fitted by a single site ionization model with apparent pK=6.53, or slightly better by a two site model with 38% of groups having a pK of 7.09 and the remainder with a pK of 6.26. Thus the

```
Cow  MKVLILACLVALALARELEELNVPGEIVESLSSSEESITRIN-KKIEKFQSEEQQQTEDE
Shp  MKVLILACLVALALAREQEELNVVGETVESLSSSEESITHIN-KKIEKFQSEEQQQTEDE
Mou  MKVFILACLVALALARET-TFTVSSET-DSI-SSEESVEHINEQKLQKVNLMGQLQAEDV
Rat  MKVFILACLVALALAREKDAFTVSSET-GSI-SSEESVEHINE-KLQKVKLMGQVQSEDV
Rab  MKVLILACLVALALAREKEQLSVPTEAVGSVSSSEEITHINK-QKLETIKHVEQLLREEK
Man  MKVLILACLVALALARE---------TIESLSSSEESITEYK-QKVEKVKHEDQQQGEDE
     ***+************          +  *+ **** +    +   *++ ++  *++ *+


Cow  LQDKIHPFAQTQSLVYPFPGPIPNS---LPQNIPPLTQTPVV--VPPFLQPEVMGVSKVK
Shp  LQDKIHPFAQAQSLVYPFTGPIPNS---LPQNILPLTQTPVV--VPPFLQPEIMGVPKVK
Mou  LQAKVHSSIQSQPQAFPYAQAQTISCNPVPQNIQPIAQPPVVPSLGPVISPELESFLKAK
Rat  LQNKFHSGIQSEPKAIPYAQTISCSP--IPQNIQPIAQPPVVPTDGPIISPELESFLKAK
Rab  LQDKILPFIQS---LFPFAERIPYPT--LPQNILNLAQLDML---LPLLQPEIMEDPKAK
Man  HQDKIYPSFQPQPLIYPFVEPIPYGF--LPQNILPLAQPAVV---LPVPQPEIMEVPKAK
     * *+ + +*+  + *++        +****  ++*+ ++   *++ **+   *+*


Cow  EAMAPKHKEMPFPKYPVE-PFTESQSLTLTDVENLHLPLPLLQSWMHQPHQPLPPTVMFP
Shp  ETMVPKHKEMPFPKYPVE-PFTESQSLTLTDVEKLHLPLPLVQSWMKQPPQPLPPTVMFP
Mou  ATILPKHHKQMPLLNSETVLRLINSQIPSLASLANLHLPQSLVQ-LLAQVVQAFPQTH-LV
Rat  ATVLPKHKQMPFLNSETVLRLFNSQIPSLD-LANLHLPQSPAQ-LQAQIVQAFPQTPAVV
Rab  ETIIPKHKLMPFLKSPKTVPFVDSQILNLREMKNQHLLLPQLLPFMHQVFQPFPQTPIPY
Man  DTVYTKGRVMPVLKSP-TIPFFDPQIPKVTDLENLQLPLPLLQPLMQQVPQPIPQTLALP
     ++++*  +  **+++   + ++*  ++    + ++  *+++  ++ ++  *+  *++*  *   +


Cow  P-QSVLSLSQSKVLPVPQKAVPYPQRDMPIQAFLLYQEPVLGPVRGPFPII---------
Shp  P-QSVLSLSQPKVLPVPQKAVP--QRDMPIQAFLLYQEPVLGPVRGPFPIL---------
Mou  SSQTQLSLPQSKVLYFLQQVAPFLPQDMSVQDLLQYLELL-NPTVQ-FPATPQ-----HS
Rat  SSQPQLSHPQSKSQYLVQQLAPLFQQGMPVQDLLQYLDLLLNPTLQ-FLATQQL----HS
Rab  P-QALLSLPQSKFMPIVPQQVVPYPQRDMPIQALQLFQELLF-PTHQGYPVVQPIAPVN--
Man  P-QPLWSVPQPKVLPIPQQVVPYPQRAVPVQALLLNQELLLNPTHQIYPVTQPLAPVHNP
     + *  ++* +*+* + ++ +++*   ++++*+++++++++  *    +++


Cow  --V
Shp  --V
Mou  VSV
Rat  TSV
Rab  --V
Man  ISV
     *
```

Figure 1 *Multiple sequence alignments of all the pre-β-casein sequences found on the OWL composite database. Alignments were made using the CLUSTAL V program[2] and scored using a 250 PAM matrix modified such that seryl residues were scored as for cysteinyl residues. Fixed gap and floating gap penalties of −10 were used and a small amount of manual editing was employed to move a few small indels by one or two residue positions. Single letter codes are A = Ala, C = Cys, D = Asp, E = Glu, F = Phe, G = Gly, H = His, I = Ile, K = Lys, L = Leu, M = Met, N = Asn, P = Pro, Q = Gln, R = Arg, S = Ser, T = Thr, V = Val, W = Trp, Y = Tyr. Species abbreviations: Shp = sheep, Mou = mouse, Rab = rabbit. *Identity, + Similarity*

band can be assigned to the degenerate stretching mode of the di-anionic form of phosphoseryl (and, possibily, phosphothreonyl) residues. These results are in the same range as the calculated pK values derived from ^{31}P-nmr spectroscopy of individual caseins.[3]

The amide I band shows small but progressive changes of shape with pH which are revealed in greater detail by second derivative curves, as illustrated in Figure 2. The amide I band is fitted to the standard protein model[4] with components at 1633 and 1675 cm^{-1} (extended β-structure), 1654 cm^{-1} (α-helix), 1665 cm^{-1} (β-turns), and 1645 cm^{-1} (disordered),

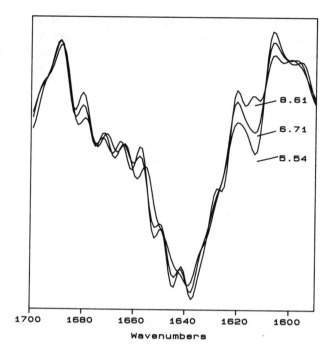

Figure 2 *Second derivative spectra of whole casein in the amide I region showing the consistent pattern of variation of peak shape with pH* (5.54, 6.71, and 8.61)

together with an allowance for an aromatic breathing mode at $1612\ cm^{-1}$. Gaussian band shapes of width $30\ cm^{-1}$ for the 1675 band and $20\ cm^{-1}$ for the rest are assigned, and a linear baseline is assumed for the trough between the amide I and II bands. A rough fitting was performed by eye, by adjustment of peak amplitudes and with the centre frequencies fixed, other parameters being optimized with a non-linear curve fitting routine. The rather featureless amide I band was over-fitted by this complicated model, even when the centre frequencies were held constant. However, by always using the same starting model of band components, consistent trends in the fractions of secondary structure can be obtained, which, even if the absolute values are in error, probably correctly reveal the directions of change in secondary structure with pH. These changes are found not to be large. Over the range of pH studied, the calculated fraction of β-strands fell by *ca.* 8% from 68% to just over 60% as the pH increases, with smaller and approximately equal increases in the other calculated elements of secondary structure. The pH dependent decrease in extended β-structure is closely correlated with the area of the band at $975\ cm^{-1}$ and the shape of the curve is fitted by an ionization model with the same pK as was used in the single site model of casein phosphate group dissociation. It should not be concluded that the phosphate centres of caseins can adopt a β-strand conformation, as the ionization of phosphate groups may exert a 'through-space' effect on other regions of the primary structure. It is also not certain that the detected conformational changes are those of orthodox β-strands, as opposed to extended chain conformations transiently adopted by predominantly flexible proteins. This latter view would be more consistent with what is generally known of casein structure.

References

1. C. Holt and L. Sawyer, *Protein Engineering*, 1988, **2**, 251; **3**, 273.
2. D. G. Higgins, A. J. Bleasby, and R. Fuchs, CABIOS, 1991.
3. R. W. Sleigh, A. G. Mackinlay, and J. M. Pope, *Biochim. Biophys. Acta*, 1983, **742**, 175.
3. D. M. Byler and H. Susi, *Biopolymers*, 1986, **25**, 469.

Structure, Rheology, and Fracture Properties

Time Dependent Fracture Behaviour of Food

By Ton van Vliet, Hannemieke Luyten, and Pieter Walstra

DEPARTMENT OF FOOD SCIENCE, WAGENINGEN AGRICULTURAL
UNIVERSITY, PO BOX 8129, 6700 EV WAGENINGEN, THE NETHERLANDS

1 Introduction

During eating, it is necessary for solid and solid-like food to fracture so as
to facilitate swallowing. Also, during usage of intermediate products for
food preparation, or in the preparation of meals, fracture is often required.
Therefore, fracture-like properties of foods and intermediate products are
often determined as part of quality control, but usually in an empirical or
semi-empirical way. More basic studies are still rather scanty, although
interest is growing.

Fracture properties of engineering materials have been studied much
more profoundly. Theory developed for the fracture of engineering materials mainly concerns situations in which the fracture properties are essentially independent of the speed of deformation.[1,2] The basic assumption is
that all, or nearly all, the deformation energy is stored in the material and
that energy dissipation only occurs due to fracturing of the specimen
(linear elastic fracture mechanics) or due to fracture as well as a limited
amount of irreversible work just around the crack tip (elastic plastic
fracture mechanics). However, experimental evidence shows that for many
food materials the fracture properties do greatly depend on the speed of
deformation of the product.[3-11] The observed stress at fracture and the
energy needed for fracture often increase with increasing deformation rate
while the strain at fracture may decrease,[3,6,7,10] increase,[4-6,8,11] or be
independent[10,12] of the deformation rate. As will be discussed below,
several mechanisms may cause this time dependency.

Fracture is considered to occur when all the bonds between the structural elements in a certain macroscopic plane break, resulting in a
breakdown of the structure of the material over length scales much larger
than the structural elements, and ultimately a falling apart of the material.[13] In the theory of fracture mechanics the starting point is that the
geometric structure of all materials is inhomogeneous. Many materials
contain small (mostly invisible) cracks while the inhomogeneities (defects)

in other materials effectively behave like tiny cracks.[14] The size of these inherent defects probably varies amongst food products. Values reported in the literature are 0.1–0.3 mm in Gouda cheese[11] and *ca.* 0.2 mm in potato starch gels consisting of swollen starch granules.[13] Now the question is whether or not these (tiny) cracks will grow during the application of a stress on the material. Two requirements for growth of a crack to occur may be distinguished.

(i) The stress at the tip of a crack has to be higher than the cohesion or adhesion stresses between the structural elements. If this is the case, a crack will start to grow (fracture initiation).

(ii) The amount of energy released due to crack growth has to be larger than the energy required for further growth of the crack. If this is the case, and the requirement (i) is also met, the crack will propagate spontaneously (fracture propagation).[1,13]

If only stresses are considered in describing the fracturing of materials, one cannot explain the effect of deformation speed on the fracture behaviour. To that end an energy consideration is needed. Below a discussion of the fracture parameters relevant to the subject of this paper is given; this is followed by a consideration of the energy balance.

2 Fracture Parameters

For describing the large deformation and fracture behaviour of a material, several parameters are needed.[1,13,15] An ideal stress–strain curve for a homogeneous linear elastic material is given in Figure 1. Besides the fracture stress and strain, one can calculate from such a graph an elastic modulus (*i.e.* stress σ divided by strain ε for $\varepsilon \to 0$, called 'stiffness' in the field of fracture mechanics) and the energy required to induce fracture ($\int \sigma d\varepsilon$, called 'fracture toughness' in the field of fracture mechanics, and measured in $J\,m^{-3}$). If the experiments are done in uniaxial compression or tension, one has to take into account in calculating the stress (= applied force/area on which the force act) that the area on which the compressive or tension force acts changes during the experiment. As a measure of strain, the natural or Hencky strain is to be preferred[16,17] which is calculated as

$$\varepsilon = |\ln\{(h_0 + \Delta h)/h_0\}| = |\ln(h_t/h_0)|, \tag{1}$$

where Δh is the change in length over the original length h_0.

For most foods the relation between the stress and the strain is much more complicated than that given in Figure 1. Some examples are given in Figures 2–4 below. As can be seen immediately by comparing these plots, there is no direct relation between the various fracture parameters or between the fracture parameters and the elastic modulus.

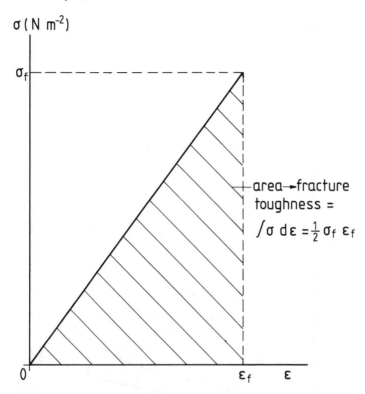

Figure 1 *Stress* σ *versus strain* ε *for an ideal* (*linear*) *elastic material. The material fractures where the curve ends*

Another factor that has to be taken into account in relation to the time dependency of the fracture properties is the difference between the moment that a crack starts to grow (fracture initiation) and the moment that it starts to advance independently of any further external supply of energy (fracture propagation).[13] The latter is normally taken as the fracture point. However, certainly when crack growth proceeds slowly (see below), there is in fact not a single fracture point, but a certain time during which fracture occurs, which depends strongly on the type of material[13] and on the fracture conditions (*e.g.* deformation rate).

An extensive description of various methods to determine large deformation and fracture behaviour of different foods has been published recently by van Vliet[17] and Luyten *et al.*[15]

3 Energy Balances Involved in Fracture

As mentioned above, one has to consider the overall energy balance for a better understanding of the time dependency or the fracture behaviour. In order to deform a material, a certain amount of energy has to be supplied

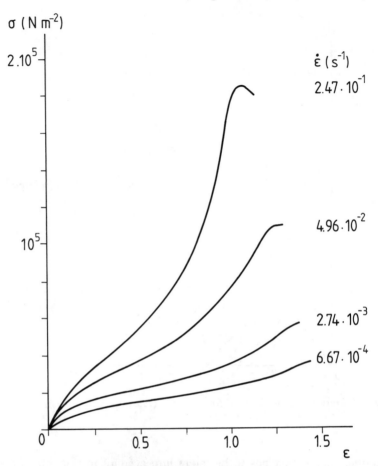

Figure 2 *Stress σ versus strain ε in uniaxial compression for a 2-week-old Gouda cheese at various initial strain rates ε̇ (indicated). (Results from ref. 9.)*

to it. The deformation energy that is put into the material can be stored (elastically), used for fracture, or dissipated in some way other than by fracturing. We can distinguish two processes causing energy dissipation.[10] The first one only acts in visco-elastic materials and is due to viscous flow of the material (or of the stress-carrying network of a composite material). The second process involves energy dissipation due to friction processes between different components of the material caused by inhomogeneous deformation of the material at large strains.[10,11] In principle, it will occur in all materials that are heterogeneous on a larger than molecular scale.

In general, the net energy balance can be written as

$$W = W' + W''_m + W''_c + W_f \tag{2}$$

where W is the amount of energy supplied to the material, W' is the part of the deformation energy that is elastically stored, W''_m is the part

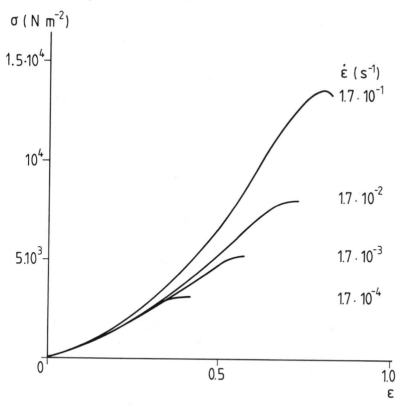

Figure 3 *Stress σ versus strain ε in uniaxial compression for 9-month-old Gouda cheese at various initial strain rates ε̇ (indicated). (Results from ref. 9.)*

dissipated due to flow of the network due to visco-elasticity of the material, W''_c is the part dissipated due to friction processes between different components of the material, and W_f is the fracture energy. Initially W_f is zero. Fast growth of cracks can occur if the value of W' is so high that the energy that can be released due to stress relaxation in the material in the vicinity of the formed crack exceeds W_f.[1,2,13] In general, the amount of energy dissipated due to a certain process depends on the rate of deformation; this causes the fracture properties to be time dependent. In a visco-elastic material, W' and W_f also depend on the deformation rate. Because the time dependency of the different energy parameters depends on the specific properties of the material, it is difficult to give general results on the time dependency of the fracture properties of real foods. However, some general trends can be given according to the prevailing mechanism of energy dissipation.

An extensive discussion on the energetic aspects of the fracture of food has been given before.[13] The effects of the two energy dissipation processes on the fracture properties are discussed further in the following section.

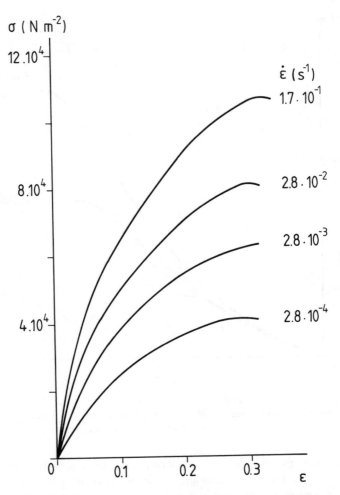

Figure 4 *Stress σ versus strain ε in uniaxial compression for* 10 wt% *potato starch gels at various initial strain rates* ἐ (*indicated*). (*Results from ref. 8.*)

4 Some Results on Time Dependency

In this section, some experimental results on the small- and large-strain behaviour—including fracture—of some foods are given as examples. The reported large-strain properties were determined by uniaxial compression or tension at $20 \pm 0.5\,°C$ at various strain rates using an Overload Dynamics or Zwick traction–compression apparatus.[13] The rheological properties at small shear strain presented in this and the following section were determined using a Deer PDR 81 Rheometer or a Bohlin VOR Rheometer, by applying a sinusoidal varying shear stress or shear strain, respectively, to the test piece and measuring the resulting strain or stress, respectively. For elastic materials the stress and strain are in phase; for visco-elastic materials the strain is out of phase with the stress by an angle δ. We determine the storage modulus G', which is a measure of the energy stored in the material during a periodic application of a stress, the loss modulus G'', which is a measure of the energy dissipated in the material, and $\tan\delta = G''/G'$. A higher value of $\tan\delta$ implies that the material behaves in a relatively more viscous and less elastic manner. It was checked that the experiments were done in the linear region.[18,19] There are several reports on time-dependent fracture properties. Some experimental results are shown in Figures 2–4. Both young and old Gouda cheeses behave visco-elastically at small deformations. In each case, $\tan\delta$ was found to vary from *ca.* 0.4 at low frequencies (5×10^{-3} rad s^{-1}) to *ca.* 0.3 at higher frequencies (5×10^{-2} rad s^{-1}).[9] The starch gels behaved purely elastically at small deformations;[8] the values of Young's modulus determined in uniaxial tension or compression, were independent of the deformation speed.

As can be seen, the fracture strain can increase or decrease with increasing strain-rate, or it may be independent of it. These results indicate that at least two different mechanisms, which lead to time dependency, are acting in food. The fracture stress and the energy supplied to the material increase with strain-rate for all cases shown. Next the results presented will be discussed in the context of the two mechanisms for energy dissipation discussed in section 3.

5 Effects due to Visco-elasticity

Visco-elastic behaviour of a material implies that the reaction of such a material to a stress consists partly of a viscous component and partly of an elastic component. In order to obtain information on the effect of visco-elastic behaviour on the large deformation and fracture behaviour, some model studies were performed with PVA–borax gels. Gels were prepared by dissolving 3 wt% PVA (acetate content 1.5%, 78 000 daltons), adding 2.5 wt% disodium tetraborate, and heating for 2 days at 70 °C. Around room temperature these gels behave predominantly elastically under fast deformations ($\tan\delta$ low at high oscillation frequencies) and predominantly liquid-like at slow deformations ($\tan\delta$ high at low frequencies, *i.e.* $G'' > G'$).[20] Fast deformations produce fracture, while at slow deformations the materials exhibit viscous flow and do not fracture. The value of $\tan\delta$ can be varied by varying the gel temperature. The mechanical behaviour was determined in shear at small strains in dynamic experiments at varying frequencies and temperatures, and also at large strains (including fracture) at varying shear-rate $\dot{\gamma}$ and at the same temperatures as in the small-strain experiments. The observed fracture strain γ_f at a certain deformation speed is plotted as a function of $\tan\delta$ determined over a similar experimental time-scale but at much smaller strains (and stresses). For the comparison of the time-scale, use was made of the empirical, so-called Cox–Merz rule which states that the time-scale of a large deformation may be compared to the time-scale at small deformation by equalizing ω and $1/\dot{\gamma}$.[19] The results are shown in Figure 5. As can be seen, there is a good relation between both parameters which is essentially independent of the measuring temperature in the range investigated.

A similar approximate relation between $\tan \delta$ and fracture strain may also be deduced from Figure 6, where γ_f and $\tan \delta$ are plotted as a function of the time scale of the deformation for skim milk gels formed by rennet action or by acidification in the cold, followed by quiescent heating. For the rennet skim gels, $\tan \delta$ is clearly larger over longer time scales (reciprocal of the oscillation frequency ω in $\mathrm{rad\,s^{-1}}$), while γ_f increases if the rate of deformation (reciprocal of time prior to fracture) is decreased. For the acid induced skim milk gels, both are independent of the rate of deformation. At high deformation speeds γ_f is clearly larger for rennet induced skim milk gels than for acid induced skim milk gels; this is caused by a difference in the spatial structure of these gels.[21]

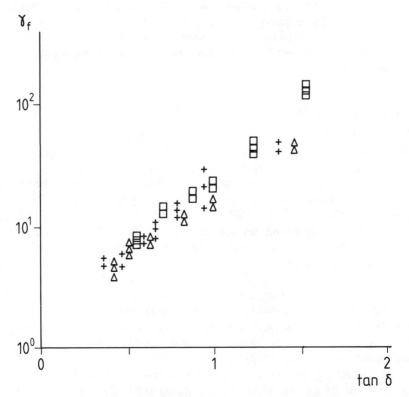

Figure 5 *Fracture strain γ_f of PVA–Borax gels, as determined in shear experiments by applying large stresses as a function of $\tan \delta$, determined at small shear strains over the same time-scale. The PVA concentration is 3 wt%; the disodium tetraborate concentration is 2.5 wt%. Measuring temperature: $\triangle$, 5 °C; +, 10 °C; $\square$, 20 °C*

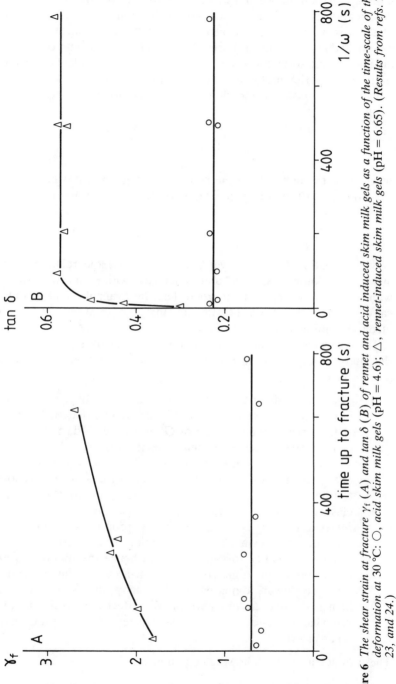

Figure 6 *The shear strain at fracture γ_f (A) and tan δ (B) of rennet and acid induced skim milk gels as a function of the time-scale of the deformation at 30 °C: $\bigcirc$, acid skim milk gels (pH = 4.6); $\triangle$, rennet-induced skim milk gels (pH = 6.65). (Results from refs. 7, 23, and 24.)*

The results presented above clearly show that there is, at least, an indirect relation between the dependence of the fracture strain on the deformation rate and the visco-elastic behaviour of a material. If the ratio of the viscous and elastic contributions to the mechanical behaviour of a material does not depend on the extent of deformation, the energy dissipation due to viscous flow may, in principle, be calculated from G'' or from an experimentally determined apparent viscosity (η^*). Assuming homogeneous flow during deformation, the following equation can be written,

$$W_m'' = \eta^* \cdot \dot{\varepsilon}_v^2 \cdot t \tag{3}$$

where $\dot{\varepsilon}_v$ is the flow-rate of the material. Using $t \approx \varepsilon_f/\dot{\varepsilon}$, where $\dot{\varepsilon} \equiv d\varepsilon/dt$ is the strain-rate of the test piece, equation (3) may be rewritten as:

$$W_m'' = \eta^* \cdot \dot{\varepsilon}_v^2 \cdot \varepsilon_f/\dot{\varepsilon} \tag{4}$$

Solving equation (4) is not yet possible because the relation between $\dot{\varepsilon}_v$ and $\dot{\varepsilon}$ is not known, while η^* and ε_f decrease with $\dot{\varepsilon}$, and η^* also decreases with ε. For these relations no good data are available, and so only trends can be given. Equation (4) implies that the total energy for fracture increases with increasing strain-rate.

The following general statements on the effect of a lower strain-rate on the energy components involved in the fracture of a visco-elastic material may be given.

(i) The energy dissipation due to viscous flow, W_m'', will be relatively more important compared to W', presuming that such a material is characterized by a decrease of G' and G'' with decreasing frequency ($1/t$), where G' decreases much steeper than G''.

(ii) The elastically stored energy, W', will be relatively lower due to the energy dissipation during the deformation and absolutely lower due to the lower G'. Moreover, the transport to the crack tip of elastically stored energy, which comes available due to stress relaxation during crack growth,[13] may proceed less efficiently.

(iii) The energy necessary for the formation of new surfaces, W_f, will be somewhat smaller (because G', and hence the number of effective elastic bonds per unit cross-sectional area, is smaller) and will be relatively lower compared to W_m'' but probably will be somewhat higher relative to the part of W' effectively available for crack propagation, because of a less efficient energy transport during crack growth.

These trends have the following consequences.

(a) The total amount of energy necessary to deform the material to a certain strain decreases with decreasing strain-rate, and so does the value of W needed for fracture. This implies that $\sigma(\varepsilon)$ and also σ_f decrease with decreasing $\dot{\varepsilon}$.

(b) The material has to be deformed further to reach a value of W' high enough for fracture propagation to occur. This implies that ε_f increases with decreasing $\dot{\varepsilon}$. Another way of saying the same thing in fact [see equation (3)] is that, during slow deformation, the material has more time to flow and therefore ε_f is larger.

(c) At very low $\dot{\varepsilon}$, W' stays too low with no fracture propagation and thus no 'fracture' occurs. The material flows, or may first yield and then flow.[13]

Using the above reasoning, the results given in Figure 2 for two-week-old Gouda cheese may in principle be explained. However, the results given in Figure 3 for 9-month-old Gouda cheese cannot be explained in this way, since the extent of visco-elasticity is the same at small deformations. A possible explanation in the second case is that the viscosity is much higher and therefore the flow-rates are much lower; but this is not valid because the stresses involved during deformation are higher to the same extent. Therefore, another energy dissipation mechanism must be active in the old cheese when large deformations are applied; probably this also applies in the young cheese, but to a lesser extent.

6 Effects due to Friction between Structural Elements Caused by Inhomogeneous Deformations

Most foods are composite materials. They consist of various structural elements with different mechanical properties. This means that the material will deform inhomogeneously, certainly at the larger deformations needed for fracture to occur. Such an inhomogeneous deformation involves energy dissipation due to friction between the structural elements, e.g. between starch granules in starch gels, between the dispersed particles and the network in filled gels, and between the liquid and the particles in particle gels.

Energy dissipation due to friction will depend on such factors as the structure of the material, the mechanical properties of the components, and the way the material is deformed. It also depends on the local rate of deformation and hence on the crack speed. In general, we have

$$W''_c \propto V_{crack}^a \tag{5}$$

where V_{crack} is the rate at which the crack advances, and the parameter a is presumed to be constant over a certain range of V_{crack}. Visual observation of the rate by which a spontaneously proceeding crack grows[22] has shown that

$$V_{crack} \propto \dot{\varepsilon}^b \tag{6}$$

where *b* is a constant smaller than 1. So the energy dissipation due to friction will tend to increase with the deformation rate of the test piece. In principle, this energy dissipation may mean that the amount of elastically stored energy W' stays low and also that the transport of energy to the crack tip, which becomes available due to stress relaxation around the growing crack, proceeds inefficiently. This will clearly retard the speed of crack propagation, and in certain cases may lead to quite low crack speeds. The latter factor may lead to the fracture strain increasing with increasing rate of deformation of a material if the crack speed becomes so slow that a measurable deformation of the whole test-piece may occur during the time necessary for the crack to proceed through the test-piece. This can indeed occur, as will be illustrated by the following calculation.

Dickinson and Goulding[4] observed an increase in the strain at fracture ε_f with increasing $\dot{\varepsilon}$ for Cheddar, Cheshire, and Leicester cheeses. Their results for Leicester cheese are given in Table 1. As the measuring method, they used uniaxial compression of cylindrical test-pieces with a height H_0 of 3 cm and a diameter of 2.9 cm. From the Hencky strain-rate given, the deformation rate $V_{comp.}$ of the test-piece can be calculated:

$$V_{comp.} = \dot{\varepsilon} H_0 \qquad (7)$$

The height H_t of the test-piece at the moment that it fractures can be calculated from the strain at fracture. If one assumes that, during the compression at the lowest compression speed, no further compression of the test-piece occurs during the time necessary for the crack to proceed through it, and that effects due to visco-elasticity are absent, the extra compression ΔH for the other specimens during this time is equal to $-H_t$ at fracture minus $-H_t$ at fracture for the lowest compression speed. Such a difference can be explained if the fracture time is equal to $\Delta H/V_{comp.}$.

Table 1 *Fracture strain ε_f (Hencky) of Leicester cheese for various strain rates $\dot{\varepsilon}$ (data from ref. 4) and parameters deduced from them in order to calculate the required crack speeds necessary to explain the observed increase in ε_f with increasing $\dot{\varepsilon}$. H_t is height of test-piece at fracture; t is calculated time needed for the crack to proceed through half of the test piece; and V_{crack} is the calculated crack speed*

$\dot{\varepsilon}$ (s^{-1})	ε_f (–)	H_t (cm)	t (s) = $\Delta H/V_{comp.}$	V_{crack} (cm s^{-1})
0.0028	0.18	2.50		
0.0095	0.20	2.46	1.4	1
0.030	0.20	2.45	0.55	2.6
0.059	0.23	2.39	0.62	2.3
0.14	0.25	2.33	0.40	3.6
0.28	0.30	2.22	0.33	4.4
0.46	0.35	2.11	0.28	5.1

This is the case if the crack speed is equal to about half the diameter of the test piece divided by the fracture time. As can be seen (Table 1), the crack speed necessary to explain the observed increase in ε_f with increasing $\dot{\varepsilon}$ is about $1-5\,\mathrm{cm\,s^{-1}}$. Crack speeds observed visually or by making photographs at regular short intervals are $\geqslant 0.02\,\mathrm{cm\,s^{-1}}$ for young Gouda cheese and $10-100\,\mathrm{cm\,s^{-1}}$ for old Gouda cheese, and so the calculated speeds seem reasonable. As may be expected, some increase of V_{crack} with increasing $\dot{\varepsilon}$ is calculated.

An explanation like that discussed for the Leicester cheese would also yield time-dependent fracture properties for starch gels. For the starch gels shown in Figure 4, the total energy required for fracture has been determined as a function of the strain-rate in tensile experiments[11] (Figure 7). As can be seen, W was independent of $\dot{\varepsilon}$ at low $\dot{\varepsilon}$ and increased with increasing $\dot{\varepsilon}$ at high $\dot{\varepsilon}$. If it is assumed that this increase is due to increasing energy dissipation due to friction, the constant a in equation (5) can be calculated to be $ca.$ 0.3 for this system.[11] For the same systems the time between the start of fracture and its completion has also been determined as a function of $\dot{\varepsilon}$ (see Figure 8).[22] From this relationship, the crack speed, which was estimated as the width of the test-piece divided by the fracture

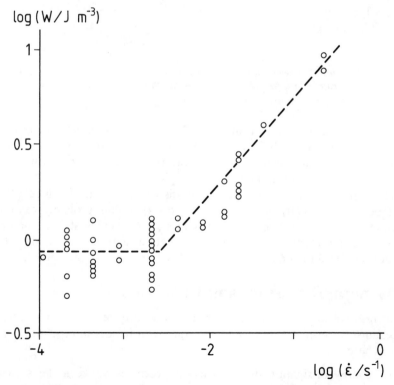

Figure 7 *The total energy required for fracture,* W, *as a function of the strain rate* $\dot{\varepsilon}$ *for 10 wt% potato starch gels in uniaxial tension. (Results from ref. 8.)*

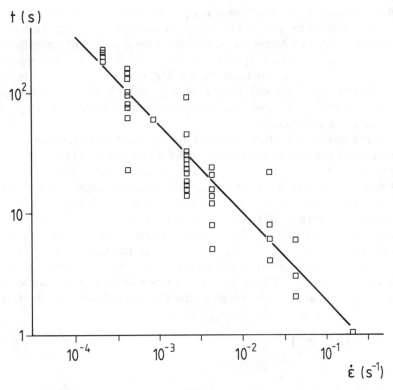

Figure 8 *The time* t *between the start of fracture and its completion as a function of the strain rate* ε̇ *for* 10 wt % *potato starch gels in uniaxial tension*[22]

time, can be calculated as a function of $\dot{\varepsilon}$. It results in a value of the index b in equation (6) of 0.7. The observed value of b implies that V_{crack} increases less than linearly with $\dot{\varepsilon}$ and therefore that the test-piece is deformed more during crack growth at higher $\dot{\varepsilon}$.

The high values found for the energy dissipation in the starch gels (those that behave elastically at small deformations) is thus probably caused by frictional processes between the swollen starch granules at large deformations. During fracturing, we observe that several layers of the starch granules around the crack tip are moved with respect to each other.[22]

7 General Discussion and Conclusions

The results given above show that there are at least two different mechanisms causing time-dependent large deformation and fracture behaviour. These are:

(i) energy dissipation due to friction processes between the structural elements caused by inhomogeneous deformation of many materials at large strains (this can cause ε_f to increase with increasing $\dot{\varepsilon}$);

(ii) energy dissipation due to viscous flow of the whole (or a significant part of) the visco-elastic test-piece, where the flow may also start after an initial yielding of the material (this causes ε_f to decrease with increasing $\dot{\varepsilon}$).

Both these mechanisms normally imply that σ_f and W increase with increasing $\dot{\varepsilon}$. Only if the fracture strain becomes very large at low $\dot{\varepsilon}$, which is a rather exceptional case, may W increase with decreasing $\dot{\varepsilon}$. The first mechanism will always act in composite materials if the rheological properties of the components are not the same, which is mostly the case. This implies that it must also act in the young Gouda cheese discussed in section 5. The crack speeds observed for these cheeses are low (0.02–$0.5\,\mathrm{cm\,s^{-1}}$), and so the friction mechanism probably contributes significantly to the fracture behaviour of young Gouda cheeses. If this were not the case, the effect of visco-elasticity on the fracture strain would be much larger than that observed (Figure 2).

The fact that, for old Gouda cheese (Figure 3), no effect of $\dot{\varepsilon}$ on ε_f is observed is probably due to both mechanisms balancing each other out in this case. The effect of visco-elasticity is probably rather similar for both the young and the old Gouda cheese, because at small deformations $\tan\delta$ is about the same. However, as can be seen in Figure 2, young Gouda cheese yields at fairly small strains causing flow to become more extensive before the test-piece finally fractures (it may not fracture at all at very low $\dot{\varepsilon}$). In the old cheese, no clear yielding occurs. Moreover, due to its lower water content, energy dissipation due to friction probably is more important. That no increase of ε_f with $\dot{\varepsilon}$ is observed will be due to the crack speeds being rather high (around 10–$100\,\mathrm{cm\,s^{-1}}$), so that the time between crack initiation and crack completion is quite short.

Acknowledgement

The authors are indebted to Professor A. G. Atkins, Dr. G. Jeronimidis, and Dr. J. F. V. Vincent of the Biomechanics Group of the University of Reading, UK for stimulating and valuable discussions on fracture mechanics.

References

1. A. G. Atkins and Y.-M. Mai, 'Elastic and Plastic Fracture', Ellis Horwood, Chichester, 1985.
2. H. L. Ewalds and J. R. Wanhill, 'Fracture Mechanics', Delftse Uitgevers Maatschappij, Delft, 1984.
3. J. Culioli and P. Sherman, *J. Texture Stud.*, 1976, **7**, 353.
4. E. Dickinson and I. C. Goulding, *J. Texture Stud.*, 1980, **11**, 51.
5. S. B. Ross-Murphy and S. Todd, *Polymer*, 1983, **24**, 481.
6. H. McEvoy, S. B. Ross-Murphy, and A. H. Clark, *Polymer*, 1985, **26**, 1483.
7. P. Zoon, T. van Vliet, and P. Walstra, *Neth. Milk Dairy J.*, 1989, **43**, 35.

8. H. Luyten and T. van Vliet, in 'Gums and Stabilisers for the Food Industry', ed. G. O. Phillips, D. J. Wedlock, and P. A. Williams, IRL Press, Oxford, 1990, Vol. 5, p. 117.
9. H. Luyten, T. van Vliet, and P. Walstra, *Neth. Milk Dairy J.*, 1991, **45**, 33.
10. H. Luyten, T. van Vliet, and P. Walstra, *Neth. Milk Dairy J.*, 1991, **45**, 55.
11. H. Luyten, M. G. Ramaker, and T. van Vliet, in 'Gums and Stabilisers for the Food Industry', ed. G. O. Phillips, D. J. Wedlock, and P. A. Williams, IRL Press, Oxford, 1992, Vol. 6, p. 101.
12. L. K. Creamer and N. F. Olson, *J. Food Sci.*, 1982, **47**, 631.
13. T. van Vliet, H. Luyten, and P. Walstra, in 'Food Polymers, Gels and Colloids', ed. E. Dickinson, Special Publication No. 82, Royal Society of Chemistry, Cambridge, 1991, p. 392.
14. J. E. Gordon, 'The New Science of Strong Materials', Penguin Books, Middlesex, 1968, Chap. 4.
15. H. Luyten, T. van Vliet, and P. Walstra, *J. Texture Stud.*, in press.
16. M. Peleg, *J. Texture Stud.*, 1984, **15**, 317.
17. T. van Vliet, in 'Rheological and Fracture Properties of Cheese', IDF Bulletin 268, Brussels, 1991, p. 16.
18. P. Zoon, T. van Vliet, and P. Walstra, *Neth. Milk Dairy J.*, 1988, **42**, 249.
19. J. D. Ferry, 'Visco-elastic Properties of Polymers', Wiley, New York/London, 1980.
20. H. Beltman, 'Verdikken en Geleren', Ph.D. Thesis, Wageningen Agricultural University, The Netherlands, 1975.
21. L. G. B. Bremer, B. B. Bijsterbosch, R. Schrijvers, T. van Vliet and P. Walstra, *Colloids Surf.*, 1990, **51**, 98.
22. H. Luyten, M. G. Ramaker, and T. van Vliet, unpublished results.
23. P. Zoon, T. van Vliet, and P. Walstra, *Neth. Milk Dairy J.*, 1988, **42**, 271.
24. S. P. F. M. Roefs and T. van Vliet, *Colloids Surf.*, 1990, **50**, 161.

Mechanical and Fracture Properties of Fruit and Vegetables

By Julian F. V. Vincent

CENTRE FOR BIOMIMETICS, UNIVERSITY OF READING, EARLEY GATE, READING RG6 2AT, UK

1 Introduction

Composites are commonly defined as materials composed of two distinct phases, usually stiff fibres or particles in a more or less amorphous and relatively compliant matrix.[1] Such a model probably suffices for meat (a markedly fibrous composite) and cheeses (which can be modelled as particle-filled composites). But most foods are considerably more complex structurally than this (even neglecting the strange rheological properties that many of them have). For instance, plant foods (vegetables and fruits) are constructed primarily of cells and fibres, whose walls not only are made of a fibrous composite material (cell-wall cellulose in a matrix of assorted polysaccharides and proteins), but which are also arranged in a series of morphological hierarchies (Figure 1) giving a sort of 'supercomposite'.

Plants are worth investigating for other reasons, which will become apparent, but which can be summarized here as follows:

(i) they are composites of composites;
(ii) since they are essentially the product of genetic activity, their textural properties can be manipulated using modern technology (this has a large number of implications, not all of which are touched upon here);
(iii) they can suggest ideas for the packaging of foods;
(iv) their very complexity makes them intrinsically interesting (for instance as examples of complex, yet safe, materials to be modelled in aerospace applications).

2 The Hierarchy

The hierarchy of organization of plants progresses in various size steps, each successive step being at least two orders of magnitude larger, as summarized in Figure 1. This interval corresponds to quite a large step

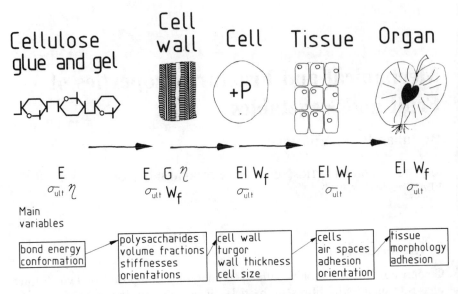

Figure 1 *The hierarchy of organization in plants, culminating in a fruit (it could as well be a stem or leaf). There is a size difference of two or three orders of magnitude between each step in the hierarchy. The bottom row of boxes shows the various factors at each level of the hierarchy which constitutes the first factor in the list of factors at the next highest level. The variables are: E = Young modulus; σ_{ult} = strength; η = viscosity; G = shear modulus; ω_f = work of fracture; EI = flexural rigidity*

size—other hierarchecal biological materials such as collagen and keratin have size jumps of only one order of magnitude.

Cellulose

Cellulose is a much studied polysaccharide; it contains beta-1,4 linked glucose. Its ribbon conformation makes it a fibrous material of very high theoretical modulus (250 GPa) although measured moduli are necessarily lower (80 GPa). Some mechanical properties of cellulose are shown in Table 1.

Table 1 *Mechanical properties of cellulose*

	Young Modulus (GPa)	Tensile Strength (GPa)	Ultimate Strain
Theoretical	250	25	
Dry fibre (ramie)	80	0.9	2.3

In the cell wall, the fibres of cellulose are embedded in a mixture (matrix) of proteins and polysaccharides which are more or less highly hydrated.[2] Much has been written about plant cell walls. It has to be realized that we must understand how the mechanical properties of plant tissues are generated and controlled at this level of organization in order to be able to apply rational genetic control of these properties. This in turn can lead to all sorts of advantages in areas such as general texture, processability, harvesting, *etc.* At present, though, we know relatively little. There is, as far as I know, no model for plant cell walls based on a composite model of cellulose fibres in a matrix.

The Cell Wall

This is a multilayered composite, which is frequently anisotropic due to controlled orientation of the cellulose, resulting in a further drop in the modulus to *ca.* 1 GPa.[3] The walls of living, non-lignified cells are extended by an internal (turgor) pressure of the order of 1 MPa (10 atmospheres). This has two main mechanical effects: it makes the cell stiff, and it stretches the cell wall, so providing strain energy which can propagate a crack in the cell wall, thus embrittling the cell. This aspect of the fracture mechanics of plant tissues is probably basic to an understanding of 'crispness'.

As a material fractures, new surface is created on either side of the crack. This surface has an intrinsic energy, analogous to surface tension, associated with it, which is provided by strain energy stored in the material.[4] As the cell gets larger, the amount of available strain energy increases as the cube of the cell diameter, whereas the strain energy required to propagate a crack increases only as the square of diameter. There is therefore a strong virtue in smallness. In some systems the cells are so large and weak that they are liable to burst if they attain their maximum turgor. An example is the apple, where the stiffness and strength of apple tissue which has been soaked in distilled water are dramatically reduced due to burst cells (Figure 2). Apple cells are about 100 μm in diameter with thin cell walls (about 2 μm), and have a turgor pressure of about 1 MPa. The strain energy stored in such a cell is about 2 mJ, which is sufficient to propagate a crack around two-thirds of the cell if the work of fracture of the cell wall is 10^3 J m^{-2}.[5] The crack still has to be started, perhaps at a flaw or hole such as might be generated by plasmadesmata extending between cells, or it may be that the tensile strength of the cell wall has to be surpassed.

Tissues

The cells are assembled into tissues. In the parenchyma (the flesh) the cells are more or less uniform, and they are stuck together either by mutual pressure as a result of being confined within a limiting epidermis (basically,

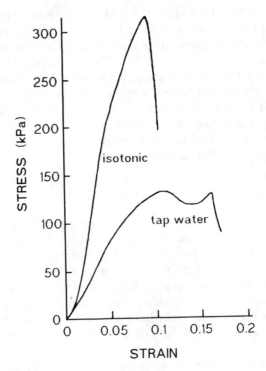

Figure 2 *Change in stiffness and strength of apple parenchyma depending on the solution (tap water or isotonic mannitol) in which the cells have been bathed. Tap water has caused cells to swell and about half of them have broken*

therefore, a turgor mechanism), or alternatively by adhesive pectins. This effective adhesion gives the tissue a shear stiffness which is in some way proportional to the degree of adhesion. Tensile stiffness of the tissue, being less dependent upon morphology, is hardly affected by turgor, though obviously the cells must still adhere to each other.

The influence of turgor on structural stiffness (*e.g.* bending) is shown when a plant wilts. The effect of reducing lateral stickiness is not so easily shown, but it occurs in apples which have over-ripened. During this process the pectins are broken down and become soluble, freeing the cells from each other (they can be rubbed off the parenchyma when they produce a feeling rather like that of sand; individual cells are easily seen, being *ca.* 0.1 mm in diameter) and eliminating the dependence of shear stiffness on density. This effect is most readily observed in early-cropping apples, such as Yellow Transparent and James Greave, which tend to be of lower density overall, indicating that the cells are less well stuck together with larger air spaces between.[6] We have recently observed the opposite phenomenon—of tissues becoming stiffer even in the absence of externally available distilled water. This could be due to the repartitioning of water within the cell and the cell wall, so increasing turgor pressure; or to the migration of Ca^{2+} into the cell wall, cross-linking the matrix polysaccharides, and therefore stiffening them; or to the action of cross-linking enzymes resident in the cell wall. The stimulus for this response (at least the one which we have used) is the application of low uniaxial compressive loads within the elastic range. The strain involved is less than 0.01; the stress of the order of 10^{-4} Pa.

The combination of high turgor and good adhesion leads to lower fracture energy of the tissue, *i.e.* 'crispness'. This is due to: better shear transfer from cell to cell (strain energy can be lost as ductility in the adhesive layer if adhesion is weak); fewer effective mechanical interfaces present which can stop the crack by blunting it or deflecting it; and, of course, a higher proportion of the total strain energy required for fracture available in the prestrained cell walls themselves.

Fracture

The fracture energy of plant parenchyma can be measured in a number of ways.

Standard tensile tests are difficult to perform since the test piece needs to be large enough to mount in a test rig and stiff enough to be gripped (in clamps) or glued (to metal tabs, which then can be gripped). Tensile tests are then possible on specimens which have been notched in the central portion so that the fracture occurs away from the clamps (which might otherwise have a weakening effect). It is fairly simple with such tests to vary turgor and to determine its effect on fracture. Bending tests on notched beams are also relatively easy, though are not much use when turgor is reduced since all that happens is that the beam bends further but the crack does not propagate. We have developed some tests which are particularly useful for small pieces of plant tissue which may be inhomogeneous.

The wedge fracture test[6,7] gives the same morphology as a tensile test on a notched specimen, but the strain is supplied by forcing the two 'ears' of a specimen apart with a wedge of included angle of 10–30 degrees (Figure 3). The wedge does not cut, though the tip should have a small radius of the order of 0.1 mm. The ideal is to get the force–displacement trace showing a plateau, indicating that the rate of application of strain energy exactly balances its rate of use as fracture energy. This can be achieved by varying the effective width of the specimen, by literally tuning the amount of tissue available for storing strain energy to the amount of tissue to be cleaved. When this equilibrium can be reached, the calculation of toughness follows from

$$W_f = F/w \tag{1}$$

where W_f is the work of fracture, F is the plateau force on the wedge, and

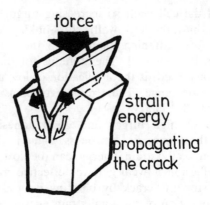

Figure 3 *The use of a wedge to impart strain energy to a specimen and thus cause propagation of a crack*

w is the width of the specimen. It is less simple if equilibrium is not reached, when the work done has to be estimated from the area under the force–deflection curve. This wedge test can be performed on samples down to 2 mm across.

Another test involves cutting on an instrumented microtome (Figure 4).[8,9] The geometry is the same as for the equilibrium case with the wedge. Both these techniques can be adapted to other moderately soft foods such as cheese or certain types of meat.

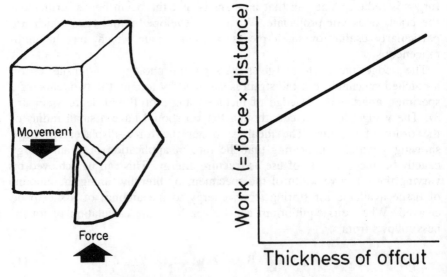

Figure 4 *The use of a knife, cutting thin slices of material, to estimate a minimum fracture energy, equivalent to a brittle fracture*

A more recently developed test uses a pair of scissors in a standard materials test machine.[9] Since the geometry of the scissors is changing while they cut, the data have to be captured by computer and corrections calculated. This can be done 'on line', enabling fracture data to be gathered very simply and directly for small pieces of material which can be very heterogeneous. The method was developed for studying the fracture of grass, but it has been applied (in a rather less developed form) to a variety of leaves, and to meat and insects (considered as food for bats). These tests yield similar results for a variety of plant tissues so long as they are turgid.

The correspondence between results with the microtome (which, under ideal conditions, totally directs the path of the crack) and the wedge (which allows the crack to propagate freely and thus to wander from side to side and around obstacles, which would lead to higher values for the fracture energy) indicates that turgid non-fibrous parenchyma fractures in a brittle manner. Flaccid tissue has a higher work of fracture, showing partly that other mechanisms are operating, which can absorb this energy (*e.g.* sliding between cells), and that the contribution of the prestrained cell wall to the global supply of strain energy has been reduced.

The Organ

The assembly of parenchyma into organs is illustrated by the fracture properties of apples. The parenchyma (or flesh) is structured with (amongst other things) radially arranged air spaces.[10] These make the fracture brittle in the radial direction but relatively tough in the tangential direction since the air spaces stop and redirect cracks which run into them (Figure 5).[11]

You can easily detect this anisotropy during biting using your incisors as a pair of opposed wedges.[7,11] Take a fresh, crisp, apple such as a Granny Smith. Cut two pieces of the same size, approximately cubes, from the cheek of the apple. Remove the skin and remnants of the core. Bite, slowly, into each piece with your incisors such that the crack travels radially or tangentially. With the radial crack, your teeth should penetrate a short way (about 1/3 the total thickness of the cube) and the apple should fracture freely with a sharp noise. Tangentially, the crack uses much more energy and you will probably have to bite all the way through and never generate a free-running crack. This anisotropy is also shown in compressive and tensile tests where both modulus and failure mode are affected.

There is some confusion in the literature as to the definition of stiffness and the failure mode(s) of apple tissue and what it means. It is obviously important with all plant materials to be aware of anisotropy and to investigate it as a matter of course. In tests to measure the shear stiffness of apple parenchyma, it has been found impossible to get a correlation between density and stiffness of the tissue if the section is taken across the

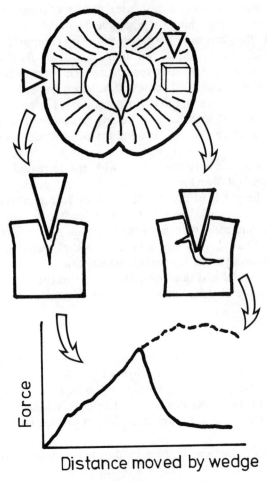

Figure 5 *The orientation of spaces or interfaces in a material can influence the direction and energy of cracking*

radial air spaces. When anisotropy is taken into account and samples are taken with the air spaces parallel to their log axis, the correlation between stiffness and density is excellent.[10]

At the organ level of organization, other tissues take part in the structure. In the apple, these other tissues are represented mainly by the skin and the core. There is also some vascular tissue in the parenchyma, but this seems not to make much difference to the overall mechanics. However, properties and interactions of the other tissues are important.

It is not easy to measure properties of the skin, primarily because it is very difficult to define the skin. If one accepts a working definition of 'the thinnest layer that can reasonably be separated from the outside of the fruit', one is left with a multilayered material, the outer surface of which is relatively brittle (in a fracture test it can be seen to be breaking more readily than the inner layers), and an inner surface which is derived from small compacted cells. One way around this is to take different thicknesses of 'skin' which include varying amounts of parenchyma. This can give some feeling for the properties of the outer layer.

Apple skin is not remarkably tough.[6] A small imperfection can weaken it quite severely, a characteristic known as 'notch sensitivity'. Casual observation suggests that, whilst it is not difficult to start a fracture in apple skin, it is more difficult to propagate it. This is in contrast to the skin of tomato, grape, plum, *etc.* which is considerably more brittle, but it is much more difficult to start a crack. There is no way to express this analytically at present—this has to remain an intriguing but unquantifiable characteristic. The reason for such a difference in the properties of skin is obscure.

The core of the apple is relatively diffuse in that it forms bulges around the seed cavities. But some simple tests can show how the core, skin, and parenchyma of the apple interact mechanically. The experimental variables are: removal of the core; removal of the skin; and decoupling of skin and core by cutting around the insertion of the core at the stem and flower ends of the fruit. If the apple is now compressed with the core vertically extending between the platens (Figure 6), the core will be placed in compression and the skin in tension. Under this loading the skin seems to be important in stopping the apple splitting (Figure 7). If the core is oriented horizontally, compression of the whole apple will put the core into tension (Figures 6 and 8). The apple cannot then deform so freely, and the skin breaks at a lower overall compression (Table 2) and bruising is more extensive, even when the skin has been removed (Table 3). The experiments suggest that the forces are transmitted to the core mainly by the skin. So the integrated structure is rather like a single turgor-driven cell with the skin playing the same role as the cell wall.

3 Some Lessons

One desirable outcome of studies on the 'texture' of plant tissues is to understand more completely the generation of these properties at the most basic, biochemical/genetic level. The mechanical properties can be modelled (using a combination of finite element analysis and composite theory) to a degree sufficient to identify the contribution of the various cell wall components (cellulose fibres and other matrix-forming polysaccharides). This enables the mechanical properties to be redefined as chemical properties, and hence the genetic control of mechanical properties can follow using standard genetic techniques.

However, it may be more stimulating to take a more abstract approach and to regard the apple, or other fruit, as nature's version of 'animal food'. The rationale is that these fruits have been evolved by the plant as attractive for animals to eat in order that the plant's seed can be

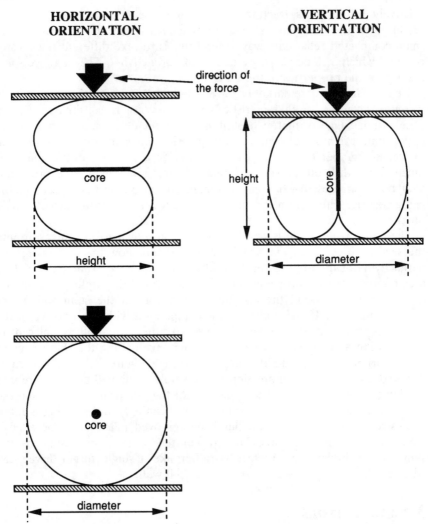

Figure 6 *Orientations of an apple being compressed: only if the core is parallel to the platens will it be stressed*

distributed and fertilized more effectively. Hence, the natural selection of the fruit has been performed by the animal itself and the fruit may fairly be considered to represent some sort of 'ideal' or 'optimal' way of presenting food. This, in turn, leads to the idea of using 'fruit' as a paradigm for food packaging.

Consider the concept of storing nutritious liquid (soup, for instance) in a large number of small containers. This would form a very safe storage system, more or less leak-proof, since the disruption of any one container would allow the loss of only a very small amount of the total liquid. The

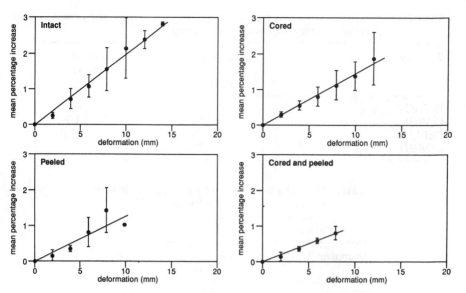

Figure 7 *Percentage increase in diameter of the Golden Delicious apple as a result of vertical compression (core lying along the direction of force). The fruit was compressed in steps of 2 mm in an Instron and the diameter measured (n = 4). The point where the regression line stops indicates the approximate deformation at which the apple first splits*

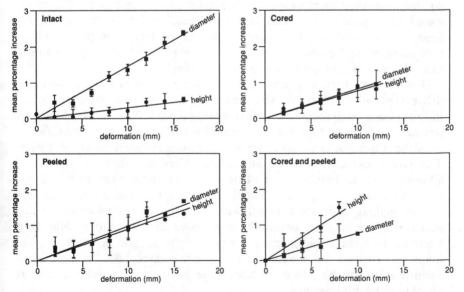

Figure 8 *Percentage increase in height and diameter of Golden Delicious apple as a result of horizontal compression (core at right angles to direction of force). The fruit was compressed in steps of 2 mm in an Instron and the dimensions measured (n = 4)*

Table 2 *Distance of deformation required to produce the first tensile failure in fruit compressed in vertical and horizontal orientations*

Treatment	Compression at tensile failure (mm)	
	Vertical orientation	Horizontal orientation
INTACT	13.90	16.55
CORED	13.95	13.08
PEELED	10.20	14.60
CORED + PEELED	9.75	10.18

Table 3 *Mean bruise volume shown at a percentage of the total volume of the fruit in Golden Delicious apples compressed horizontally by 10 mm (n = 3)*

Treatment	Mean bruise volume	s.d.
INTACT	34.75	2.73
CORED	15.72	5.32
PEELED	31.99	5.84
CORED + PEELED	9.52	1.11

containers could be manufactured from, or contain, sufficient solids for the complete package to represent a balanced source of nutrition. The reality would be a packaging system based on 'cells' which could contain any food, be safe and drip-free, sterile, capable of long storage at ambient temperatures, and edible in small amounts as required. An apple with one side a beef steak, the other side treacle pudding!

How to make such a system? One way would be to produce a somewhat dehydrated gel which is coarsely emulsified into another phase containing the necessary components for micro-encapsulation. The microspheres are separated off, glued together, and bounded by a tensile membrane (which could be a form of rice paper, for instance) to produce a block of 'tissue'. This could, of course, be of any shape. They are then prestressed by allowing them to imbibe water. This will make them more brittle. There are virtues in the design of the apple to be a sphere since this is the best way of making a pressure vessel with the minimum amount of surface, which is more or less uniformly prestressed and therefore safer from fracture. However, packaging problems might suggest a more bar-like shape as evolved by manufacturers of chocolates and biscuits. Extra sophistication can be added by having the microcapsular walls capable of dissolution by gut enzymes.

Foods could be tailored to individual nutritional, or even medical, needs by mixing in capsules containing preferred flavours or necessary drugs. It might even be possible to mix 'n' match a food item from capsules at the point of sale, enabling close tailoring of the food to the preferences of the individual customer. We now have an engineered highly hydrated but drip-free food with a pleasing crisp texture and of infinitely variable nutritional character. Such a food would be less liable to bacterial or fungal infection, and so could be relatively easily stored; it could be carried around easily, even when partly eaten, since it is essentially leak-proof. It could therefore be used, at least at first, by people 'on the move' such as members of the armed forces, spacemen, *etc.*

In more immediately practical terms, a better understanding of the factors controlling mechanical and fracture properties of fruit will help in developing better varieties, but only if the measurements are made using the techniques developed by materials science.

References

1. B. Harris, in 'The Mechanical Properties of Biological Materials', ed. J. F. V. Vincent and J. D. Currey, Cambridge University Press, 1980, p. 37.
2. D. A. Rees, 'Polysaccharide Shapes', Chapman and Hall, London, 1977.
3. M. C. Probine and R. D. Preston, *J. Exp. Bot.,* 1962, **13**, 111.
4. J. E. Gordon, 'The Science of Structures and Materials', Freeman, NY, 1987.
5. G. Jeronimidis, in 'The Mechanical Properties of Biological Materials', ed. J. F. V. Vincent and J. D. Currey, Cambridge University Press, 1980, p. 169.
6. J. F. V. Vincent, *Adv. Bot. Res.,* 1990, **17**, 235.
7. J. F. V. Vincent, G. Jeronimidis, A. A. Khan, and H. Luyten, *J. Text. Stud.,* 1991, **22**, 45.
8. A. G. Atkins and J. F. V. Vincent, *J. Mater. Sci. Lett.,* 1984, **3**, 310.
9. J. F. V. Vincent, in 'Biomechanics—Materials, a Practical Approach,' ed. J. F. V. Vincent, IRL Press, Oxford, 1992, p. 206.
10. J. F. V. Vincent, *J. Sci. Food Agric.,* 1989, **47**, 446.
11. A. A. Khan, 'Mechanical and Fracture Properties of Fruit and Vegetables', Ph.D. Thesis, University of Reading, 1988.
12. R. E. Pitt and H. L. Chen, *Trans. ASAE,* 1983, **26**, 1275.

Mouthfeel of Foods

By D. W. de Bruijne, H. A. C. M. Hendrickx, L. Alderliesten, and J. de Looff

UNILEVER RESEARCH LABORATORY, PO BOX 114, 3130 AC VLAARDINGEN, THE NETHERLANDS

1 Introduction

Foods are materials with a consistency at room temperature that has been designed to perform optimally in meal preparation operations such as cutting, slicing, spreading, or mixing. During eating and mastication the food loses its initial consistency. What is felt during that process, and what is characteristic of many foods, is precisely this decay of consistency. The first stage of the mastication process, however, is often a cutting process: the food is comminuted by the action of the teeth into particles with size typically a few millimetres. At this stage of the mastication process, the food is still rather close to its original consistency. What is 'felt' in this first stage appears to be related directly to various rheological parameters. Extensive research in this field, by Kokini et al.,[1-3] has resulted in correlations between such parameters and various aspects of mouthfeel.

For many foods, mastication does not end with the comminution process. The food particles will usually soften further as a consequence of their rise in temperature in the oral cavity and moisture uptake from the saliva. When this decay in consistency has proceeded far enough, the material may experience the level of stresses in the saliva flow in the oral cavity. When this occurs, the food particles will be broken down to a much smaller size, which is determined by the hydrodynamics of 'flowing' saliva. As this occurs, the food will rapidly form a homogeneous mix with the saliva in the mouth; the mouthfeel of the product will at this stage be 'smooth'. If the consistency does not decay to a sufficient degree, such that break-up in the saliva flow cannot take place, the masticated food will remain thicker than the saliva. The consumer or panellist, tasting that food, will then complain about product defects such as 'graininess', 'stickiness', or 'waxiness'. The last description applies to fat continuous foods only.

The reduction of the size of the food particles during the mastication process is also highly relevant to the flavour release of that food. Flavour release is in this context to be understood as the percentage of the flavour substance in the food that finally reaches the flavour-sensitive organs located in the nose. This evidently is a mass transfer process in which the diffusion of the flavour out of the food particle can be the rate determining step. This is particularly the case if the particle remains large. Assuming for the moment that these food particles are approximately spherical particles of radius R, one may estimate the time required for release of the flavour content by means of penetration theory.[4] The time t is given by

$$t \approx \frac{R^2}{D} \tag{1}$$

Here D is the diffusion coefficient; its numerical value depends only slightly on the nature of the flavour components and mainly on the nature of the food. A good estimate for dilute aqueous foods is $10^{-9}\,\mathrm{m^2\,s^{-1}}$, while the diffusion coefficient in fat continuous foods is about 50 times lower. In order to be perceived during the mastication process, flavours have to be released from the food particle within a few seconds. This implies that the comminution and break-up processes in the oral cavity should at least result in particle sizes (radii) of 70 μm for aqueous foods in order to release all flavours properly. Because of the lower diffusion coefficient in oils, smaller particle sizes are preferred for fat continuous foods. Since small food particles release their flavour components faster than larger particles, one may also conclude that the flavour release profile, *i.e.* the change in the intensity of the perceived flavour in the course of the mastication process, will depend on the size distribution of the particles obtained in the oral break-up process. The present study aims at a better understanding of this break-up process.

2 Break up

Two foods, being at first sight equivalent and fulfilling similar functions during meal preparation or eating, may still show a completely different break-up behaviour during mastication. This can be established by masticating samples of various products for a certain standardized time and analysing the resulting samples for their drop-size distribution. Examples of these expectorated products are shown in Figure 1. The products 1–4 are spreads of different consistency at body temperature. After mastication, there are still rather big product lumps, of a size of about 1 mm, in the thickest product (4), while the break-up proceeds to entities of more than ten times smaller size for the thinnest of the four products (1).

It is evident that break-up of food in the saliva flow will in general be determined by a balance between the forces tending to deform the food particle and the forces resisting that deformation. The first type are the hydrodynamic forces exerted by the saliva flow. The forces resisting the deformation are determined by the interfacial tension and by the rheolo-

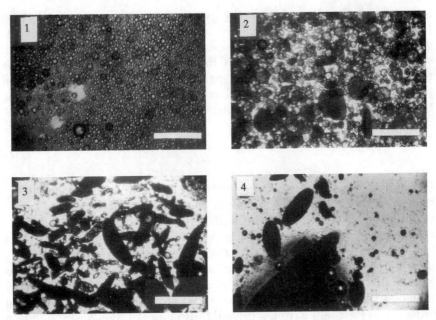

Figure 1 *Expectorated spread samples with increasing consistency at body temp-erature (1 > 2 > 3 > 4). The bar length represents 1 mm in each case*

gical properties of the food particles. The complete set of forces relevant for the break up process thus comprises:

  (i)   the rheological properties of that food material,
  (ii) the stresses exerted by the saliva flow,
  (iii) the interfacial tension between the food material and the saliva.

This last property is relevant for both non-aqueous and fat-continuous foods. The rheological properties of the food and its interfacial tension with saliva can be measured directly and relatively simply. This is not the case for the mean stress level in saliva as it 'flows' during mastication. The relation beween the ultimate drop size and the above parameters (rheolo-gical properties, interfacial tension, and hydrodynamic stresses exerted on the drops) is, however, precisely known[5,6] for Newtonian liquids, with which they are not miscible. The results of these studies[5,6] are used here to determine the in-mouth stresses in flowing saliva.

Break-up of Newtonian Fluids

The case that is the most easy to understand is that of break-up in purely elongational flow. In this flow each element of volume is being stretched without rotation of the direction of stretching. In a more general type of flow, the direction of stretching is not fixed but rotates. The rate of rotation of the axis of stretching and the rate of stretching are equal in the case of 'simple shear' flow.

In elongational flow, the drops deform and the stess σ_c acting on each drop is approximately equal to the stress in the continuous phase, *i.e.*

$$\sigma_c \approx \eta_c \, \dot{\varepsilon}_c \tag{2}$$

Here, η_c and $\dot{\varepsilon}_c$ are the viscosity and the rate of deformation of the continuous phase. The surface tension tends to preserve the spherical shape of the drop. It can be shown, by dimensional arguments, that this effect can be accounted for by means of a 'Young's' modulus E of the drop, which is related to the interfacial tension γ and the radius of the drop by

$$E \approx \frac{2\gamma}{R} \tag{3}$$

For the degree of deformation of the drop (ε_d) one may then write:

$$\varepsilon_d \approx \frac{\sigma_c}{E} \approx \frac{\dot{\varepsilon}_c \, \eta_c \, R}{2\gamma} \tag{4}$$

When the drop elongation exceeds a certain value, the drop will break into smaller drops. The deformation at which breakage occurs will be independent of the size of the drops. This implies that the largest drops that survive a certain elongational flow field are characterized by a constant value of the right hand side of equation (4). This right hand side equals, apart from a factor of four, the so-called 'capillary number', which is usually defined as

$$\Omega = \frac{\eta_c \, \dot{\varepsilon}_c \, d}{\gamma} \tag{5}$$

A very detailed analysis of break-up of Newtonian liquids has been performed by Grace and others.[5,6] In these studies a four-roll mill was used to subject the drops to all kinds of flow between purely elongational flow and simple shear. The break-up in simple shear was studied in a Couette apparatus. The results of those studies are reproduced here in Figure 2. It appears from this figure that the critical capillary number is not fully constant in elongational flow ($\alpha = 1$), but it increases somewhat if the drop viscosity becomes considerably smaller than the viscosity of the continuous phase. The more difficult break-up of low viscosity drops is due to the fact that the external stresses have little 'grip' on them.

In simple shear flow, break-up is considerably more difficult than it is in elongational flow; the critical capillary numbers are higher, in particular when the drop viscosity is larger than the viscosity of the continuous phase:

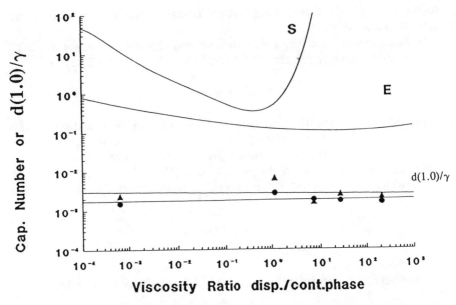

Figure 2 *Critical capillary number for break-up of drops of Newtonian liquids in steady elongational (E) and shear (S) flow. The two sets of points (▲, ●), and the best-fit lines through them, represent break-up results in the oral cavity for two individuals of a number of 'calibration' oils. To plot these data, the external stress and saliva viscosity are taken as 1 Pa and 50 mPa s, respectively*

The difference has to do with the time required to deform the drops. On dimensional grounds, one may deduce that this drop 'relaxation' time is

$$t_d \sim \frac{\eta_d R}{\gamma} \tag{6}$$

where η_d is the viscosity of the drop. The proportionality factor in equation (6) is *ca.* 10, and a more detailed analysis would show that the relaxation time also depends somewhat on the continuous phase viscosity. A more general type of flow than purely elongational flow can be considered, as mentioned earlier, as a combination of elongation and rotation: each element of volume in such a flow is deformed as it rotates. The rates of deformation ($\dot{\varepsilon}_c$) and rotation (ω) are just equal in simple shear flow. When the rotation is faster than the rate of elongation of the drops ($\omega t_d > 1$), break-up becomes increasingly difficult, since the axis of elongation then already rotates to a new position before the drop actually elongates. Using equation (6), one may conclude from this, in agreement with Figure 2, that break-up in shear flow becomes increasingly more difficult when the drop viscosity becomes larger than the continuous phase viscosity. This fundamental difference between elongation and shear flow is used here to establish the type of flow in the oral cavity.

Break-up of Newtonian Liquids in the Mouth

The break-up process in the oral cavity was studied by masticating small samples of oils of different viscosities. After mastication these samples were expectorated and analysed for their drop-size distribution. The weight of the oil samples was about 0.5 g, which means that the mastication process takes place in an excess of saliva. Although the mastication time was varied between 5 and 60 seconds, it appeared that this variation had little effect on the ultimate drop sizes. To inhibit drop coalescence after mastication, the samples were expectorated in a suitable aqueous detergent solution (Twinco), and gently mixed with that solution. Under a confocal laser light microscope,[7,8] photographs were taken of these diluted samples, and these were analysed by means of a computerized image analysis technique. The result of this is a drop-size distribution of the expectorated samples. Since the prime interest here is with the smallest drops, the number-averaged drop diameter is calculated from this distribution. This quantity is taken as a measure of the degree of break-up.

The viscosities of the oils were measured at 37 °C using a Haake CV100 viscometer. The interfacial tensions with saliva were determined by the Wilhelmy plate technique, also at 37 °C. These interfacial tensions were for all oils about 15 mN m^{-1}; lower interfacial tensions were obtained by adding lecithin to the oils. In the latter case the measurements were done against tap water. Air was included as a dispersed phase to extend the range of viscosities. From these data, one cannot directly calculate the capillary numbers because the value of the stress exerted by the saliva flow is not known. These stresses were, however, about equal in all tests, since mastication was in an excess of saliva. For different individuals, these stresses are not equal. Two individuals did this type of test; there appears to be a systematic difference of about 50% in the size d of the oil drops in their saliva.

One may calculate the value of d/γ for the different tests. This parameter represents the capillary number for a flow in which the stress exerted on the drops is 1 Pa. Plotting this parameter against the ratio of the viscosities of the oil and saliva, one obtains results as shown in Figure 2. It appears that our results are in parallel with results obtained from the literature for elongational flow. This implies that the effective flow in the oral cavity is purely elongational. The vertical distance on the graph between the present results, calculated for a saliva stress of 1 Pa, and literature results for elongational flow is approximately a factor of 50. The consequence of this is that we may estimate the actual saliva stress in the mastication process to be *ca.* 50 Pa.

Break-up of Non-Newtonian Liquids

Actual foods are seldom purely Newtonian liquids. Most often their rheological behaviour can be approximated by that of a Bingham liquid. In elongational flow, one may write for such a fluid:

$$\sigma = 2k_b + \eta_b \, \dot{\varepsilon} \qquad\qquad (7)$$

Here $2k_b$ is the yield stress in elongation. It is assumed here that the 'Tresca' yield criterion applies,[9] *i.e.* that the yield stress in elongation is twice that in shear (k_b). The differential viscosity η_b is called the Bingham viscosity.

Soft foods like dressings, jams, sauces, and yoghurts show such a Bingham-like consistency at room temperature. Other foods such as spreads, cheeses, creams, puddings, and some fruits are more solid at room temperature and only become more liquid-like during mastication. This softening is most often a consequence of moisture up-take from saliva. However, melting also plays a role, as is the case for creams, spreads, and some cheeses. In the case of starch-containing puddings and sauces, the amylase activity of human saliva may also be important in the decay of consistency during mastication. The yield stress of such real foods may decrease by many orders of magnitude during the mastication process; they are for this reason not very suitable as model systems in a quantitative study of the oral dispersion process.

A systematic study, as done by Grace[5] and Bentley and Leal[6] for Newtonian fluids, does not exist for Bingham fluids. However, even in the absence of such studies, it is clear that a Bingham fluid will not be broken up in a flow in which the stresses do not exceed the yield stress. This means that break-up in the oral cavity is difficult for foods with a yield stress above 50 Pa. Further, it can be expected that the time t_d required to actually deform the drop will be similar to that for Newtonian drops, provided we use the Bingham viscosity η_b in equation (6). This parameter is, however, of much less importance here, since break-up will not occur at all when the other rheological parameter, k_b, is too large compared with the exerted stresses.

Break-up of Bingham Fluids in the Mouth

The imitation foods chosen for this study were water-in-oil (W/O) emulsions. These can be made perfectly stable against coalescence, also at 37 °C, by using Admul WOL (a polyglycero ester of polycondensed ricenoleic acid, ex Quest) as an emulsifying agent. This emulsifier was dissolved in the soya oil first, whereafter water was added slowly under stirring with a Turrax stirrer. The yield stress of the emulsions was adjusted by varying the percentage of water phase. The rheological behaviour of the emulsions was characterized by means of a so-called 'sweep in shear-rate' on the Haake CV100 viscometer. The rate of shear is increased in such an experiment linearly with time, whereafter the shear-rate is decreased, again linearly with time. Since break-up in the oral cavity actually occurs after the food has been subjected to a significant degree of deformation, the stresses measured at a decreasing rate of shear are most representative of the oral process. The results are given in Figure 3. It appears from these

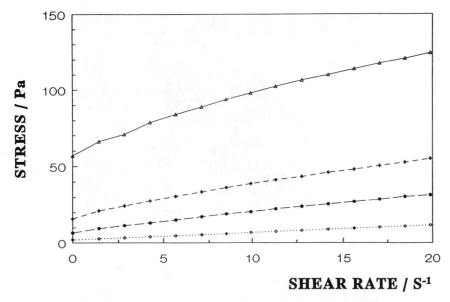

Figure 3 *Rheological characteristics of model emulsions. These emulsions exhibit for each shear-rate higher stresses after an increase in shear than after a decrease. Since foods are deformed considerably when taken in the mouth, the stresses after a decrease in shear-rate are taken as the prime characteristic. Symbols refer to various water volume fractions in the W/O emulsions:* ◇, 60%; ●, 70%; +, 75%; △, 80%

results that a 60% W/O emulsion is still Newtonian, while a 70% W/O emulsion already shows a small yield stress (in shear). The value of this parameter doubles for a 75% emulsion, whereas an 80% emulsion already exhibits a yield stress well above 50 Pa.

These emulsions were masticated for 30 seconds, expectorated, and as before analysed for their average drop size. Two of the resulting photographs are shown in Figure 4: that of the 60% W/O emulsion with a zero yield stress, and that of the 80% emulsion with a yield stress above 50 Pa. Though the largest drops are not noticeably different, it appears that only the 60% emulsion shows quite small drops after mastication. Inside the large drops one may still observe the water drops of the original emulsion. This shows that coalescence of the internal phase does not occur during the dispersion process.

3 Discussion

The mastication experiments with the Newtonian oils have shown that the saliva flow in the oral cavity is elongational. However, if one analyses how the saliva flow is set into motion, one has to conclude that this flow cannot be a perfectly steady flow. It can, at best, be an intermittent flow. Break-up in such a flow can evidently occur only when this flow is

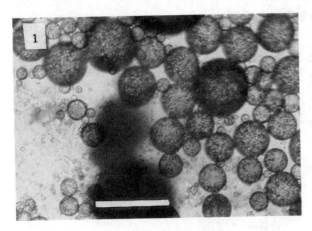

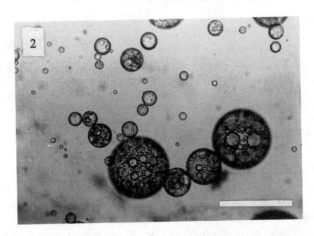

Figure 4 *Micrographs of expectorated samples of model food emulsions:* (1) *sample with yield stress above* 50 Pa; (2) *sample with yield stress below* 50 Pa. *The bar length represents* 1 mm

maintained long enough, *i.e.* longer than the relaxation times of the drops considered. By means of the approximate equation (6), one may estimate this relaxation time for the drops considered. For the most viscous oil, having a viscosity of 6 Pa s and an interfacial tension with saliva of 15 mPa m, one finds an ultimate drop size of about 20 μm. The drop relaxation time thus is about 5×10^{-3} s. The drops were probably 10–100 times bigger at the onset of the dispersion process. This implies that the largest drop relaxation times in this study are probably between 5×10^{-1} and 5×10^{-2} s. Since these drops do break up, one may conclude that the elongational flow in the oral cavity remains steady for at least such periods of time. When one considers, on the other hand, how the jaws and the tongue actually drive the saliva flow, one must conclude that this flow cannot be kept steady for much longer times.

The limited duration of the elongational flow in the oral cavity will be most important for food materials showing visco-elastic behaviour at large degrees of deformation. When the relaxation times of such visco-elastic foods become larger than the duration of the elongational flow, break-up is not expected to take place. Though strong visco-elasticity is a rare property in foods, it can occur in the hydrocolloid area, some such thickeners showing elastic behaviour on small time-scales of *ca.* 0.05 s. The oral dispersion of these thickeners will then be poorer than for purely Newtonian thickeners.

Except under taste panel circumstances, food materials are seldom eaten in the absence of other food components. These other components aid in dispersing the material considered, since they increase the hydrodynamic stresses exerted on the material of interest. Expectorated samples of spreads masticated together with bread show considerably smaller drops than samples masticated in the absence of bread.

4 Conclusions

The present study shows that the saliva flow in the oral cavity is effectively elongational. The flow is intermittent over longer periods of time but steady for shorter times. The duration of the steady periods is probably longer than 0.05 s. This has consequences for visco-elastic food materials. Break-up will become difficult when the relaxation times of such foods exceed this time.

The mean stress level in the saliva flow is about 50 Pa. Foods with a yield stress higher than this cannot be broken up and dispersed in the saliva flow. The mouthfeel is in this case not smooth—but grainy, sticky, or waxy. Under such conditions, flavour will tend not to be released adequately.

References

1. J. L. Kokini, *J. Food Eng.*, 1987, **6**, 51.
2. A. M. Dickie and J. L. Kokini, *J. Food Sci.*, 1983, **48**, 57.
3. J. L. Kokini, *Food Techol.*, 1985, **39** (11), 86.
4. R. B. Bird, W. E. Stewart, and E. N. Lightfoot, 'Transport Phenomena', John Wiley, New York, 1960, p. 668.
5. H. P. Grace, *Chem. Eng. Commun.*, 1983, **14**, 225.
6. B. J. Bentley and L. G. Leal, *J. Fluid Mech.*, 1986, **167**, 219.
7. I. Heertje, P. van der Vlist, J. G. G. Blonk, H. A. C. M. Hendrickx, and G. J. Brakenhoff, *Food Microstructure*, 1987, **6**, 115.
8. 'Confocal Microscopy', ed. T. Wilson, Academic Press, London, 1990.
9. W. Prager and P. G. Hodge, 'Theory of Perfectly Plastic Solids', John Wiley, New York, 1951.

Influence of Soluble Polymers on the Elasticity of Concentrated Dispersions of Deformable Food Microgel Particles

By Ian David Evans and Alexander Lips

UNILEVER RESEARCH, COLWORTH LABORATORY, SHARNBROOK,
BEDFORDSHIRE MK44 1LQ, UK

1 Introduction

Many systems are thickened by close-packed arrays of deformable particles consisting of elastic micro-networks of polymeric material. The most important of these 'microgel' particles[1] in foods are gelatinized starch granules.[2,3] The rheology of these systems is known to be determined by a combination of particle swelling and deformability,[3,4,5] but the understanding has remained largely qualitative until .recently. Elsewhere we have described a quantitative model for the elasticity of microgel dispersions.[6] This was tested with aqueous dispersions of Sephadex particles—neutral, essentially spherical cross-linked dextran moieties—available in a range of grades for use in molecular exclusion chromatography. The elasticity behaviour of these dispersions closely follows the theory,[6] whilst non-retrograded starch dispersions show a richer phenomenology,[7] perhaps related to the presence of solubilized amylose. Effects due to the presence of soluble polymers are examined here using Sephadex dispersions.

2 The Elasticity of Simple Microgel Dispersions

If a pair potential $V(r)$ can be defined, the elasticity of a system of monodisperse particles in an inelastic medium at infinite frequency—or in 'plateau' regions with limited frequency dependence—is given by the equation[9,10]

$$G = Nk_BT + \frac{2\pi N^2}{15} \int_0^\infty g\,(r)\, \frac{d}{dr} \left[r^4 \left(\frac{dV(r)}{dr} \right) \right] dr \qquad (1)$$

Here, N is the particle number density, k_B is the Boltzmann constant, and $g(r)$ is the radial distribution function. At high volume fractions both monodisperse and polydisperse dispersions can be expected to have strong short-range order. This allows a cell model to be used,[11] and $g(r)$ then approximates to a weighted delta function at the nearest neighbour separation R_{sep}, yielding the equation

$$G = Nk_BT + \frac{\phi n}{40\pi R_0^3}\left(4R_{sep}\frac{dV}{dr} + R_{sep}^2\frac{d^2V}{dr^2}\right) \quad (2)$$

Here, the derivatives are evaluated at $r = R_{sep}$, n is the number of nearest neighbours, R_0 is the undeformed radius, and the phase volume is given as $\phi = 4\pi R_0^3 N/3$ for full swelling. The treatment given here is formally equivalent to that of Buscall *et al.*[11] except for their omission of the entropic term, Nk_BT. Several small errors in their derivation also led to an incorrect equation. The entropic term is negligible for our dispersions and is subsequently ignored.

The force law for spherical microgel particles should be similar to that for macroscopic elastic spheres,[6] as analysed by Hertz.[12,13] Combining his results with equation (2) gives equation (3) below relating the macroscopic elasticity G with G_p, the elasticity of the network comprising the particles, σ being the corresponding Poisson ratio:

$$G = \frac{\phi_{CP}nG_p}{5\pi(1-\sigma)}\left[\phi_{red}^{1/3}(1-\phi_{red}^{-1/3})^{1/2} - (8/3)\phi_{red}^{2/3}(1-\phi_{red}^{-1/3})^{3/2}\right] \quad (3)$$

Here, ϕ_{CP} is the phase volume for close-packing and ϕ_{red} is the reduced phase volume, ϕ/ϕ_{CP}, which is given as $\phi_{red} = (R_0/R_{sep})^3$.

At high particle concentrations, the value of $4\pi R_0^3 N/3$ will exceed unity, swelling will become restricted, and theories for deswelling of gel networks will become more appropriate. Flory theory[14] suggests G then scales with $c^{1/3}$, where c is the polymer concentration. Refinements lead to power laws with exponents in the range $1/3-2/3$.[15,16] Analysis of our data yields an exponent of *ca.* 0.6.

3 Materials and Methods

Dextran T2000 ($M_W = 2 \times 10^6$) and Sephadex G200 superfine (G200S) were from Pharmacia. Dispersions of G200S were prepared and examined at 25 °C. Starches were commercial samples, and were used without further purification. Two chemically modified starches were examined, an acetylated adipic acid cross-linked waxy maize starch, I, and a hydroxy-propylated phosphate cross-linked waxy maize starch, II. Gelatinization procedures are described elsewhere.[7] Starches were tested at 60 °C to avoid retrogradation, amylose molecules staying in a metastable, solubilized state at this temperature. Concentrations are expressed on a wt% basis. Swelling behaviour was examined using a dye exclusion method.[3] Rheological measurements were made with a modified Weissenberg Rheogonio-meter,[3,17] type R16, using a cone-and-plate geometry, with radius 37.5 mm and 0.07 radian cone angle. Intrinsic viscosity was determined at 25 °C with a Contraves Low Shear 30 viscometer.

4 Results

The elasticity of G200S microgel dispersions is well described by our model[6] as shown in Figure 1. Similar experiments with gelatinized starch systems led to results which were dependent on the starch type,[7] as shown in Figure 2. The 'notional phase volume' axis plots the value of $4\pi R_0^3 N/3$, whilst the elasticity axis is normalized to the value at unit notional phase volume.

The cross-linked starches behave similarly to Sephadex dispersions, albeit with a lower exponent of *ca.* 0.33 at high concentrations, and a lower close-packing phase volume of *ca.* 0.5, as expected for starch granules. These results are thus consistent with the theoretical model. For the native starches this is evidently not the case. At 60 °C leached amylose molecules, which are absent from the modified starches examined here, are in a metastable, non-associated form producing the low viscosity, inelastic solutions,[3] although at lower temperatures they associate to give elastic networks.[8,18] Resolubilization of these retrograded networks is difficult. It is thus not straightforward to perform 'reconstitution' experiments to test

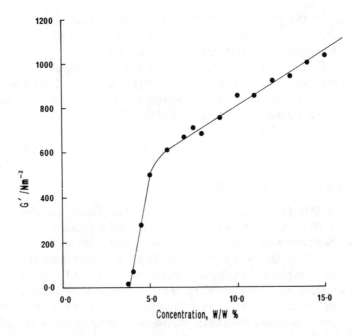

Figure 1 *Storage modulus* G' *at* 25 Hz *and* 25 °C *for Sephadex particles as a function of particle concentration. The solid curve shows the theoretical fit*

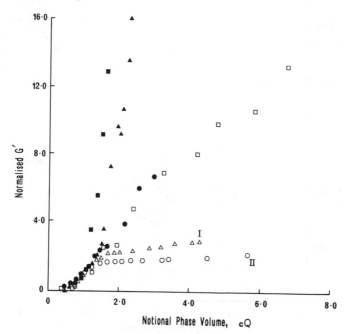

Figure 2 *Concentration dependence of elasticity of gelatinized starch dispersions at 60 °C. Normalized modulus G′ is plotted against notional phase volume (see text): I, II, cross-linked waxy maize starch; ■, wheat starch; ●, tapioca starch; ▲, corn starch; □, potato starch*

the role of soluble material. Instead we examine systems made by mixing G200S with high-molecular-weight dextran T2000 as models for starch behaviour.

Intrinsic viscosity measurements yield a radius of gyration for dextran T2000 via the Flory–Fox equation[19] of $R_g = 31.8$ nm, in good correspondence, via the anticipated $c^{1/2}$ scaling, with the literature value of $R_g = 16$ nm[20] for dextran T500 (5×10^5 daltons). The overlap concentration of 1.5 wt% was obtained from published osmotic data,[21,22] and it compares well with the value of 1.8 wt% obtained on the basis of simple volume-filling considerations.[23]

Dispersions of G200S and T2000 were found to show similar visco-elastic behaviour to G200S dispersions.[6] Thus the dispersions showed only slight frequency dependence and were predominantly elastic in character. The behaviour can thus be summarized in the form of the elasticity modulus at an arbitrary frequency of 0.25 Hz. Dispersions were examined at several fixed weight ratios close to typical values of the amylopectin/amylose ratio in starches. Figures 3 and 4 illustrate the concentration scaling of the elasticity for two such G200S/T2000 ratios. Figure 5 illustrates behaviour at a fixed level of 5 wt% T2000. In each case the data for G200S alone are included together with three sets of curves as discussed below.

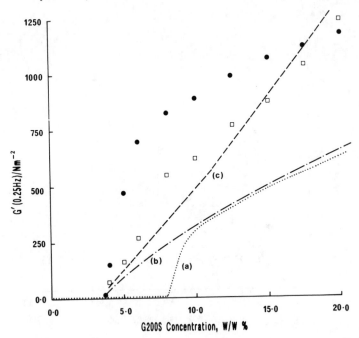

Figure 3 *Storage modulus G′ for Sephadex particles and dextran T2000 (4:1) (□) and particles without dextran (●) as a function of particle concentration. Origins of theoretical curves (a), (b), and (c) are given in the text*

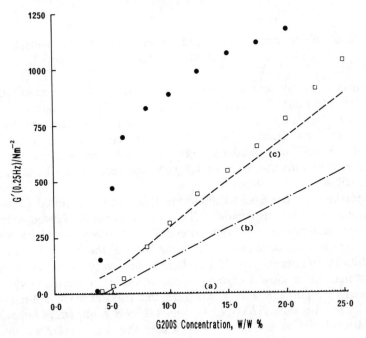

Figure 4 *As for Figure 3 except with 2:1 particle/dextran ratio*

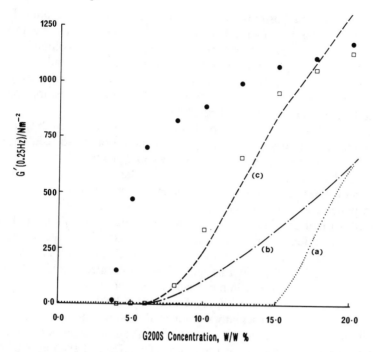

Figure 5 *As for Figure 3, except with constant amount of dextran T2000 present* (5 wt %)

5 Discussion

The data in Figures 3–5 show the marked effect of dextran T2000 on the elasticity of Sephadex G200S dispersions. Significantly, the concentration scaling changes qualitatively towards the nearly linear scaling behaviour observed for potato starch and tapioca starch. Several types of interactions are possible for these systems. The most obvious is that of osmotic competition. Dextran is excluded from the Sephadex due to its high molecular weight, making the T2000 concentration outside the particles greater than the average concentration whilst particle swelling is restricted, reducing phase volume and increasing rigidity. Dextran solutions have negligible elasticity at all relevant concentrations. Consequently the enhanced polymer concentration should have no direct effect on the elasticity. It is fairly straightforward, using the cell model, to calculate the anticipated effects of osmotic competition. We have used published osmotic data[22] to calculate changes in phase volume and the external dextran concentration via an iterative scheme and we have assumed that the elasticity of the material comprising the particles scales with the concentration inside the granules with a $c^{0.61}$ scaling as indicated by previous work.[6] Results of these calculations are shown as curves (a) in Figures 3–5. These curves suggest that osmotic competition alone cannot account for the observed behaviour. Some other form of interaction must be involved, apparently inducing particle interactions at relatively low volume fractions. Thus, an attractive potential between the particles might lead to relatively 'open' flocculated structures. Alternatively, dextran molecules may allow interaction between non-touching granules. Two possibilities are polymer–particle bridging and depletion flocculation.

Dextran T2000 could cause bridging of G200S particles by virtue of specific chemical interactions or by partial penetration of adjacent granules. The former seems unlikely since Sephadex and dextran are chemically so similar, and dextran itself shows no such tendency. Whilst whole dextran T2000 molecules are essentially completely excluded by G200S, some limited penetration by portions of the molecules may be possible leading to 'bridged' structures. This would cause visco-elastic coupling and would increase the effective particle radius somewhat. This seems unlikely, however, to be important here for two reasons. Firstly, the relative sizes of Sephadex particles ($R_0 \approx 50$ μm) and T2000 molecules would only lead to an increase in effective volume fraction of about 0.1%. Secondly, the self-entanglement relaxation times of dextran T2000 are much less than 1 second as indicated by oscillatory studies. Reptation theory[23] suggests that this will also be the case for relaxation of dextran molecules entangled with the intra-particle networks reported here. The effect of such entanglements would thus not be observable in our experiments.

One might anticipate a polysaccharide depletion layer around the particles, as would be the case for non-adsorbing solid particles.[20,24] This would be expected to affect the rheology.[25] Various approaches have been made to modelling such depletion.[26,27] Here we seek also to include the effect of particle deformability. Depletion provides an attractive component to the pair potential, which, for hard spheres, can be calculated simply[28,29] providing one knows the depletion length scale Δ. For deformable particles this calculation is valid until the particles touch. However, after contact, the geometry becomes more complex, and contact flattening allows large reductions in depletion volume. If one assumes that Hertz's results apply, one can calculate the shape of the deformed particles[13] and the associated change in depletion volume and potential. Typical results are presented in Figure 6 for the Hertzian potential V_{Hz}, the depletion potential V_{Dep}, and their sum V_{Tot}. It is assumed that Δ is equal to R_g at low dextran concentrations and scales with $c^{-0.75}$ above the overlap concentration.

The absolute values of the depletion minima calculated for these systems are very large compared to $k_B T$. Consequently touching particles will 'sit' at the energy minima, allowing evaluation of R_{sep} in equation (2). The first derivative of V with respect to r then is zero, and numerical evaluation of the second derivative allows calculation of the system elasticity via equation (2). Two problems remain. Firstly, it is necessary to predict how Δ varies with dextran concentration. Secondly, we need to know the average number of elastically active contacts n as a function of phase volume.

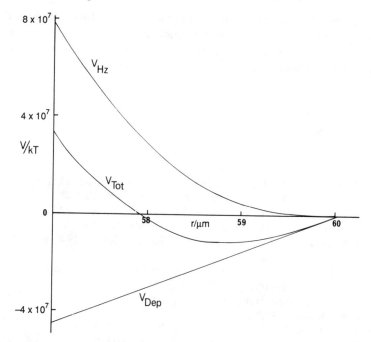

Figure 6 *Calculated interparticle potential* V(r) *for a system of* 15 wt % *Sephadex particles and* 7.5 wt % *dextran T2000*: V_{Dep}, *depletion potential*; V_{Hz}, *Hertzian repulsive potential*; V_{Tot}, *total potential*

Various theories[30,31] and some practical data[32] exist which suggest that, below the overlap concentration, Δ equals R_g, whilst at higher concentrations an inverse power-law dependence of Δ on concentration is expected. The correct form of scaling for any given polymer is at present unclear, expecially for the somewhat branched molecules considered here. We thus consider the likely extremes of scaling, curves (b) in Figures 3–5 reflecting a $c^{-0.75}$ dependence,[30] and curves (c) reflecting a constant value of Δ equal to R_g. The dependence of n on phase volume is difficult to predict, but it presumably follows a percolation-like scaling. At the normal close-packing fraction of *ca.* 0.71, it seems likely that the value of $n \approx 11$ will be attained, as was found[6] is the absence of external polymer. Furthermore, we have never observed measurable elasticity at phase volumes below *ca.* 0.5. So we use the simplest model consistent with these comments, and assume that n varies linearly with phase volume over this range.

Depletion effects do seem inevitable in these systems. Figures 3–5 show that they can account satisfactorily for the observed elasticity/concentration results, and provide a promising working hypothesis, notwithstanding the assumptions and approximations required for our calculations. We believe this is also the case for potato or tapioca starch behaviour. We have not, however, succeeded in simulating the concentration dependence observed with the cereal starches, either experimentally or theoretically. It thus seems likely that some other phenomenon governs the high concentration behaviour of gelatinized cereal starches: possibly it might result from some internal elastic inhomogeneity.[7] Further work is necessary to resolve this issue.

References

1. R. J. Ketz, R. K. Prud'homme, and W. W. Graessley, *Rheol. Acta*, 1988, **27**, 531.
2. E. B. Bagley and D. D. Christianson, *J. Text. Stud.*, 1982, **13**, 115.
3. I. D. Evans and D. R. Haisman, *J. Text. Stud.*, 1979, **10**, 347.
4. T. J. Schoch, *Wallwerstein Research Communication*, 1969, **32**, 149.
5. L. L Navickis and E. B. Bagley, *J. Rheol.*, 1983, **27**, 519.
6. I. D. Evans and A. Lips, *J. Chem. Soc., Faraday Trans.*, 1990, **86**, 3413.
7. I. D. Evans and A. Lips, *J. Text. Stud.*, 1992, in press.
8. W. Banks and C. T. Greenwood, 'Starch and its Components', Edinburgh University Press, Edinburgh, 1975, Chap. 2.
9. G. S. Green, 'Proceedings of the 1st International Congress on Rheology', Plenum, Holland, 1948, p. 1.
10. R. Zwanzig and R. Mountain, *J. Chem. Phys.*, 1965, **43**, 4464.
11. R. Buscall, J. W. Goodwin, M. W. Hawkins, and R. H. Ottewill, *J. Chem. Soc., Faraday Trans. 1*, 1982, **78**, 2889.
12. H. Hertz, 'Gesammelte Werke', Leipzig, 1895.
13. L. D. Landau and E. M. Lifshitz, 'Theory of Elasticity', 2nd edn, Pergamon, Oxford, 1984.
14. L. R. G. Treloar, 'The Physics of Rubber Elasticity', 3rd edn, Clarendon Press, Oxford, 1975.
15. J. Bastide, S. Candau, and L. Leibler, *Macromolecules*, 1981, **14**, 719.
16. H. McEvoy, S. B. Ross-Murphy, and A. H. Clark, *Polymer*, 1985, **26**, 1493.
17. C. Roberts, *Rheol. Acta*, 1971, **10**, 135.
18. M. J. Miles, V. J. Morris, and S. G. Ring, *Carbohydr. Res.*, 1985, **135**, 257.
19. M. Bohdanecky and J. Kovar, 'Viscosity of Polymer Solutions', Elsevier, Amsterdam, 1982, Vol. II.
20. P. D. Patel and W. B. Russel, *J. Colloid Interface Sci.*, 1989, **131**, 192.
21. H. J. Granger, S. H. Laine, and G. A. Laine, *Microcirc. End. Lymph.*, 1985, **2**, 85.
22. E. Edmond, S. Farquhar, J. R. Dunstone, and A. G. Ogston, *Biochem. J.*, 1968, **108**, 775.
23. M. Doi and S. F. Edwards, 'The Theory of Polymer Dynamics', Clarendon Press, Oxford, 1986.
24. B. Vincent, J. Edwards, S. Emmett, and A. Jones, *Colloids Surf.*, 1986, **18**, 261.
25. Th. F. Tadros, *Prog. Colloid Polym. Sci.*, 1990, **83**, 36.
26. G. J. Fleer, J. M. H. M. Scheutjens, and M. A. Cohen Stuart, *Colloids Surf.*, 1988, **31**, 1.
27. P. G. de Gennes, 'Scaling Concepts in Polymer Physics', Cornell University Press, Ithaca, New York, 1979.
28. P. R. Sperry, *J. Colloid Interface Sci.*, 1982, **87**, 375.

29. H. de Hek and A. Vrij, *J. Colloid Interface Sci.*, 1981, **84**, 409.
30. J. F. Joanny, L. Leibler, and P. G. de Gennes, *J. Polym. Sci., Polym. Phys. Ed.*, 1978, **17**, 1073.
31. S. F. Edwards, *Proc. Phys. Soc. London*, 1966, **88**, 265.
32. N. S. Davidson, R. W. Richards, and A. Maconnaichie, *Macromolecules*, 1986, **19**, 434.

Creaming Behaviour of Dispersed Particles in Dilute Xanthan Solutions

By H. Luyten, M. Jonkman, W. Kloek, and T. van Vliet

DEPARTMENT OF FOOD SCIENCE, WAGENINGEN AGRICULTURAL
UNIVERSITY, PO BOX 8129, 6700 EV WAGENINGEN, THE NETHERLANDS

1 Introduction

The polysaccharide xanthan is often used to stabilize dispersed particles against sedimentation or creaming in products like soups, sauces, dressings, etc.[1,2] A possible mechanism for the stabilization is the high viscosity obtained at low xanthan concentrations. We indeed are able to explain quantitatively the observed rate of sedimentation of a single inert particle in a xanthan solution in terms of the viscosity measured at the relevant shear-rates.[3] (Instead of 'solution' the term 'molecular dispersion' would probably give a better description of the situation, but this will not be used here as it could create confusion when considering the creaming behaviour of dispersed emulsion droplets.) In mixtures of xanthan with a galactomannan the correct explanation of the sedimentation of a single inert particle is more difficult to obtain because of the presence of a yield stress and the thixotropic character, but quantitative agreement has been achieved.[4] However, when using xanthan to stabilize an emulsion against creaming, the rate of creaming of the emulsion droplets is, even at low xanthan concentrations, much faster than predicted from the rheological properties. The reason for this is the flocculation of the suspended particles causing enhanced creaming. This phenomenon has been observed by several workers,[5-7] but the exact cause of the flocculation is not yet known. Theoretically, polysaccharides can induce flocculation of suspended particles both by adsorption at the particle surface and by a depletion mechanism.[8]

In the literature, some evidence for adsorption of xanthan at interfaces can be found. Young and Torres[9] found that the surface tension (water –air) of 0.1 to 1 wt% xanthan solutions decreased very slowly in time. Also an adsorption of xanthan of 0.1 to 0.8 $mg\,m^{-2}$ at polystyrene latex particles was found for 0.01 to 0.18 wt% xanthan solutions.[10] This adsorption could induce flocculation of the particles, e.g. by bridging. Bergenståhl[11] showed that xanthan and other polymers are able to stabilize soy bean oil emulsions ($d_{32} = 0.3-0.8\,\mu$m) against flocculation at concentrations much lower than would be necessary to stabilize the emulsion by an increase in viscosity. The amount of xanthan needed, depending on the emulsifier, was found to be between 10^{-4} and 10^{-2} wt %. According to the author,[11] a possible explanation is adsorption of the polymer on the emulsifier layer resulting in the formation of a so-called bilayer. Xanthan itself is not sufficiently surface-active to replace the emulsifier.

Nevertheless, according to most authors, adsorption of xanthan at a particle or droplet surface is unlikely. Xanthan does not affect the interfacial tension of an oil–water interface.[12,13] For some galactomannans a small decrease has been found,[12] but, according to the authors, this was due to residual impurities. Because xanthan is not considered surface-active, a depletion mechanism is often used to explain the flocculation observed.[6,7,13] However, for food-like systems, strong evidence for depletion flocculation has not been given yet. For non-food systems, it has been found[14-17] that calculations of the depth of the secondary minimum in the interaction energy between particles, taking depletion into account, can explain the change in stability against flocculation with increasing concentration, particle size, etc.

In this paper we will give some results on the enhanced creaming of emulsion droplets in low concentration xanthan solutions. Both the effects of the xanthan concentration and of the mean droplet size are studied, together with the interfacial tension at the oil–water interface. Results will be compared with calculated estimates of the depth of the secondary minimum, taking a depletion mechanism into account.

2 Materials and Methods

Solutions of xanthan (a gift from Suiker Unie, The Netherlands) were made by boiling and stirring 0.1 wt % xanthan in demineralized water for 5 to 20 minutes. No effect of heating time on the determined properties of the solution was found. A stock emulsion was made by homogenizing at room temperature a pre-emulsion of 40 vol % n-hexadecane and a 5 wt % solution of Tween 20 in demineralized water at a pressure of 100 bar (small droplets, A/S N. Foss Electric Denmark, type 12705) or at a pressure of 5 bar (larger droplets, Rannie homogenizer type C 168). The emulsions were diluted with demineralized water and 0.1 wt % xanthan solution to concentrations of 0–0.03 wt % xanthan and 4–5 vol % n-hexadecane. Thiomersal (0.01 wt %) was added as a preservative.

The creaming of the emulsions was followed visually at 20 °C in glass tubes of height 31.6 cm and diameter 3.7 cm. The oil content of the different layers was determined by a modified Gerber method, and the mean droplet size by a spectrophotometric method.[18] The xanthan concentration present was calculated as the amount of neutral polysaccharides.[19] The interfacial tension was measured at the n-hexadecane–water interface by a Wilhelmy plate method. The rheological properties of the creamed layer were established using a Bohlin VOR Rheometer equipped with concentric cylinders at a constant shear-rate of 1.8×10^{-3} s^{-1} at 20 °C.

3 Results

The rate of creaming and the height of the creamed layer were found to be strongly dependent on the amount of xanthan added to the emulsion as well as on the size of the emulsion droplets (Figures 1 and 2).

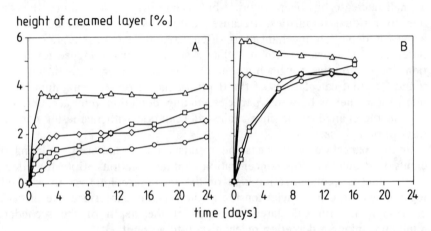

Figure 1 *The height of the creamed layer as a function of time for two different average emulsion droplet sizes. (Volume fraction of n-hexadecane = 0.04.) (A) $d_{32} = 0.22$ μm; xanthan concentration = 0 wt % (□); 0.010 wt % (○); 0.014 wt % (◇); 0.020 wt % (△). (B) $d_{32} = 0.83$ μm; 0 wt % (□); 0.003 wt % (○); 0.006 wt % (◇); 0.015 wt % (△)*

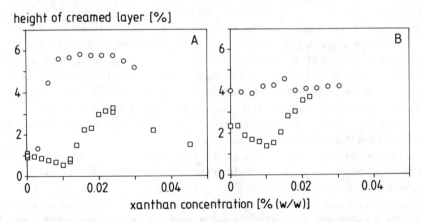

Figure 2 *The height of the creamed layer as a function of the xanthan concentration for emulsions with d_{32} 0.22 μm (□) and 0.83 μm (○) as determined (A) after 24 hours and (B) after 2 weeks. (Same emulsions as in Figure 1.)*

With increasing xanthan concentration, the emulsions with $d_{32} = 0.83$ μm became less stable against creaming over short time-scales. Only at the two highest xanthan concentrations tested, the stability possibly increased somewhat over short times. For the higher xanthan concentrations, the rate of creaming at the start of the experiment was very fast. Later on, the creamed layer became more compact with time. The emulsions without added xanthan, and those with the lowest xanthan concentration tested (0.003 wt %), creamed relatively slowly. After 2 weeks, there was no clear effect of the xanthan concentration on the height of the creamed layer. However, at higher xanthan concentrations, the creamed layer appeared to be more compact and rigid and the oil concentration was higher (*ca.* 50 vol % for 0 and 0.003 wt % xanthan increasing to *ca.* 65 vol % for 0.03 wt % xanthan). In all emulsions with xanthan, flocs were visible. The size of these flocs appeared to increase with increasing xanthan concentration. A very thin oil layer appeared on all emulsions with a xanthan concentration less than 0.021 wt %. However, the mean size of the emulsion droplets was unchanged.

The stability of the emulsions with $d_{32} = 0.22$ μm also depended sensitively on the amount of xanthan added. At low xanthan concentrations, the emulsions appeared to be more stable against creaming than without added xanthan. However, at larger xanthan concentrations, the creaming was enhanced. For xanthan concentrations of 0.012–0.022 wt % the observed initial creaming rate was very fast, but the height of the creamed layer changed very little after 2–3 days. When no xanthan was added, or at xanthan concentrations of 0.002–0.010 wt %, the initial rate of creaming appeared to be slower, but the creaming lasted longer. At high xanthan concentrations, the creamed layer contained a somewhat higher oil content than at lower concentrations (40–45 vol %). The mean size of the emulsion droplets did not change.

The creaming of all emulsions (except for the oil layer with the larger emulsion droplets) was reversible to gentle shaking. The rate of creaming and the appearance of the layers after shaking were not different from those observed before shaking.

The rheological properties of some creamed layers were studied at a constant shear rate of 1.85×10^{-3} s^{-1}. An example of the results observed is shown in Figure 3. The rheological properties of a comparable emulsion without xanthan could not be determined with the equipment used, because the stresses were smaller than the inaccuracy in the stress determination (about 0.02 Pa, indicated as bars in the figure). A clear structure in the creamed layers could clearly be observed when xanthan was added and flocs could be seen during creaming. During a short waiting time after shearing, the structure became more rigid, as can be seen from the higher shear stress and the occurrence of an overshoot during the second and subsequent application of the shear-rate to the same material. A possible explanation could be that, due to shearing, the emulsion became more compact, maybe comparable to the compacting found for the same creamed layer stored for longer times. The observed stresses during slow shearing (0.15–0.2 Pa) were much lower than the stress in the creamed layer due to gravitation. This explains why the structure in this layer can rearrange and the height of the creamed layer becomes less with time.

The xanthan concentration in the aqueous phase of the lower layer was found to be higher than the original added amount of xanthan to the aqueous phase. Some results are shown in Table 1. The large spread in results is probably caused by difficulties in the removal of all the fat from the lower layer.

The interfacial tension of the n-hexadecane–water interface was found to decrease when an excess of Tween 20 was added (Table 2). Xanthan (0.03 wt % added to the water) did not significantly change the interfacial tension. No further changes in the interfacial tension were observed over longer waiting times.

Table 1 *Xanthan concentration in the aqueous phase of the lower emulsion layer after 24 days creaming* ($d_{32} = 0.22$ μm, 4 vol % *n-hexadecane*)

Original concentration wt %	Final concentration wt %
0	0.0002
0.002	0.0038
0.004	0.0044
0.006	0.0062
0.008	0.0082
0.010	0.0106
0.012	0.0134
0.014	0.0168
0.016	not determined
0.018	0.0207
0.020	0.0242
0.022	0.0275
0.024	0.0271

Table 2 *Interfacial tension determined in the n-hexadecane–water interface at 20 °C*

	Interfacial tension/mN m^{-1}	
	Directly	Equilibrium
n-hexadecane–water	47.7	43.3[a]
+ 0.03 % xanthan	46.4	42.1[a]
n-hexadecane–water		
+ Tween 20	7.5	6.9[b]
+ 0.03 % xanthan	8.5	7.0[b]

[a] After 3–5 hours. [b] After 20 minutes.

4 Discussion

The addition of small amounts of the polysaccharide xanthan can cause emulsion droplets to flocculate and to cream rapidly. For the emulsions with the smallest droplet size tested ($d_{32} = 0.22\ \mu m$), this effect was found for xanthan concentrations higher than *ca.* 0.014 wt %. The emulsion with $d_{32} = 0.83\ \mu m$ creamed even faster at a xanthan concentration of 0.003 wt %. At higher concentrations, the stability against creaming may be increased again somewhat. Because flocs were still visible, this increased stability may be due to an increase in the viscosity.

Xanthan causes a high viscosity and the formation of structure in the creamed layer (Figure 3). Oil droplets at a volume fraction of 0.458 will cause the viscosity of the emulsion to be *ca.* 6.3 times as high as the viscosity of the continuous phase. Xanthan solutions of comparable concentrations have, at low strain-rates (the first Newtonian plateau), apparent viscosities of *ca.* 1.6×10^{-2} Pa s (0.01 wt % xanthan) to 5.1×10^{-2} Pa s (0.05 wt % xanthan).[3] We measured, for the creamed layer with 45.8 vol % oil and about 0.03 wt % xanthan in the water phase, a viscosity of *ca.*

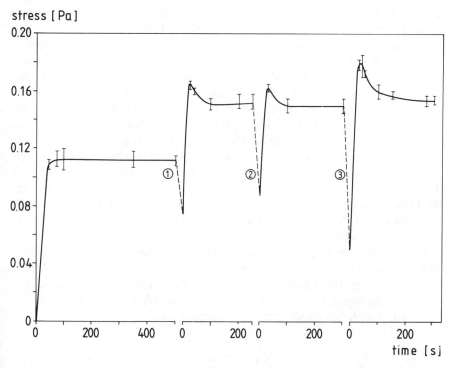

Figure 3 *Stress–time curves determined at a constant strain-rate of $1.85 \times 10^{-3}\ s^{-1}$ of a creamed layer with 45.8 vol % oil obtained from an emulsion with 4 vol % oil, 0.03 wt % xanthan, and a d_{32} of 0.22 μm after 13 days creaming. The waiting time between the experiments was (1) 72 seconds, (2) 56 seconds, and (3) 15 minutes*

50–80 Pa s, *i.e.* more than 100 times higher than expected simply from the oil and xanthan content. From the overshoot, a yield stress of the creamed layer of *ca.* 0.15–0.20 Pa could be estimated. This relatively low value explains why the flocculation was reversible for shaking by hand and why the creamed layer became more compact in time. Nevertheless, this observed weakness in the structure of the creamed layer and probably of the flocs does not provide a decisive answer about the mechanism of flocculation.

Xanthan does not give a significant decrease in the interfacial tension at the n-hexadecane–water interface. This implies that xanthan very probably does not adsorb at that interface. An adsorption on the top of the emulsifier layer, and the formation of a so-called bilayer as proposed by Bergenståhl,[11] cannot be excluded by this experiment. However, as the xanthan concentration in the water phase of the lower layer of the creamed emulsion was found to be higher than the original xanthan concentration in the aqueous phase, this points to a depletion mechanism as the cause of the flocculation. Depletion of xanthan from an interface implies that a layer of the aqueous phase around each emulsion droplet is not available for xanthan. This causes the polymer concentration in the aqueous phase 'far' away from the interface to be somewhat higher. The exact increase depends on the size of the emulsion droplets, their volume fraction, the thickness of the depletion layer, and the extent to which depletion layers are shared between two or more droplets (for instance due to flocculation). When the last factor is not taken into account, we can estimate the maximum change in xanthan concentration in the bulk water phase to be about 20% (depletion layer thickness $\Delta = 150$ nm;[20] $d_{32} = 0.22$ μm; $\phi = 0.04$). The measured changes in the xanthan concentration in the bulk aqueous phase are roughly within this range. So these results do point to a depletion mechanism as the cause of the flocculation, but because of the enormous variation in the results we cannot use this definitively to prove a depletion mechanism. Therefore, below, we compare the results found for the effect of the emulsion droplet size and the xanthan concentration with those based on the calculated interaction energy between the corresponding emulsion droplets.

To calculate the interaction energy between emulsion droplets stabilized with Tween 20 (a non-ionic surfactant), contributions of the van der Waals energy, the depletion energy, and steric repulsion have all to be taken into account. The van der Waals energy, V_A, can be calculated as a function of the shortest distance h between the droplets using[21]

$$V_A = -Aa/12h \qquad (1)$$

with A the Hamaker constant and a the radius of the droplets. For n-hexadecane droplets in water, the value of A is *ca.* 1.33 kT.[22] The depletion energy, V_{depl}, can be calculated as[16,23]

$$V_{depl} = 2\pi a(-P_{osm})(\Delta - h/2)^2(1 + 2\Delta/(3a) + h/(6a)) \qquad (2)$$

where Δ is the thickness of the depletion layer and P_{osm} is the osmotic pressure of the polysaccharide solution. The osmotic pressure can be calculated as[17]

$$P_{osm} = RTc(1/M_n + A_2c + \ldots) \qquad (3)$$

where R is the gas constant, T is the absolute temperature, c is the polysaccharide concentration, M_n is the number-weight molecular weight of the polysaccharide, and A_2 is the second virial coefficient. For this last factor a value of 4.94×10^{-4} cm^3 mol g^{-2} has been found[24] for xanthan with $M_w = 3 \times 10^6$ daltons in 0.1 M NaCl. Although the xanthan we used for our experiments probably had $M_n \approx 1 \times 10^6$, and we also did not add NaCl, we use this value here as it is the only one available. Nevertheless, due to the low xanthan concentrations used, the calculated value for the osmotic pressure is not greatly affected by the value chosen for A_2.

The steric repulsion is caused by the presence of Tween 20 at the n-hexadecane–water interface. The exact magnitude of this energy as a function of the distance between the droplets is difficult to calculate. It will be negligible at distances greater than about twice the length of the emulsifier molecule in the water phase, and it will increase very steeply at shorter distances. Therefore, we assume that the minimum in total interaction energy is at this separation and that it can be estimated as the sum of V_A and V_{depl} at this distance. The length of the part of the polyoxyethylene sorbitan monolaurate molecule (Tween 20) that is in the aqueous phase is probably ca. 5–7 nm, which implies that repulsion energy is important for distances shorter than ca. 10–15 nm. The sum $V_A + V_{depl}$ is calculated for the different xanthan concentrations and emulsion droplet sizes. Results are given in Table 3.

The exact size of the interaction energy is difficult to estimate, especially the osmotic pressure P_{osm} and the depletion layer thickness Δ. Ausserre et al.[20] found a sudden and strong decrease in Δ with increasing concentration above 0.01 wt% xanthan (0.1 M NaCl; $M_n \approx 1.3 \times 10^6$). Such a decrease would be expected to greatly affect the calculated minima. In fact, if we use their data for the depletion layer thickness, the sum of $V_A + V_{depl}$ would decrease at higher xanthan concentrations, implying that the extent of flocculation would become less. This was not observed. For lower xanthan concentrations (≤ 0.010 wt%) they estimate a depletion layer thickness of about 150 nm. We are not sure if it is correct to use their figures for Δ because the molecular weight of the xanthan used by us was lower and we did not add NaCl. Moreover, our emulsion droplets were small with respect to the size of the xanthan molecule and the depletion layer thickness. Therefore, we assume, rather arbitrarily, the depletion layer thickness to be 150 nm and independent of the xanthan concentration. If a layer thickness of 100 or 200 nm is taken, the value for V_{depl}

Table 3 *Estimated minimum in interaction energy (in units of* kT) *for different distances as a function of the xanthan concentration in the aqueous phase and the emulsion droplet diameter. Depletion layer thickness is* 150 nm

$d_{32} = 0.22\ \mu m$

conc. [wt %]	0	0.002	0.004	0.006	0.008	0.010	0.012	0.014	0.016	0.018	0.02
distance [nm]											
10	−1.2	−1.6	−1.9	−2.2	−2.6	−3.0	−3.3	−3.7	−4.1	−4.5	−4.9
15	−0.8	−1.1	−1.5	−1.8	−2.2	−2.5	−2.9	−3.2	−3.6	−4.0	−4.3
20	−0.6	−0.9	−1.2	−1.6	−1.9	−2.2	−2.6	−2.9	−3.3	−3.7	−4.0

$d_{32} = 0.83\ \mu m$

conc. [wt %]	0	0.003	0.006	0.009	0.012	0.015	0.018	0.021	0.024	0.027	0.03
distance [nm]											
10	−4.6	−5.8	−7.1	−8.4	−9.7	−11.1	−12.5	−14.0	−15.5	−17.0	−18.5
15	−3.1	−4.3	−5.5	−6.7	−8.0	−9.4	−10.7	−12.1	−13.6	−15.1	−16.6
20	−2.3	−3.5	−4.6	−5.9	−7.1	−8.4	−9.7	−11.1	−12.5	−13.9	−15.3

would be approximately 60% lower, or 100% higher, respectively. The osmotic pressure is mainly affected by the assumed molecular weight.

For the smallest droplets and the lowest xanthan concentrations, the calculated minima in the pair interaction energy are not deep enough to account for stable flocculation, which we indeed did not observe. When a depletion mechanism is not taken into account, also the minima in energy at larger xanthan concentrations are not deep enough to explain the flocculation observed. For the larger droplets, the calculated minima in interaction energy were deeper. This may be the reason (as well as the larger droplets) for the larger flocs observed and the more rigid appearance of the creamed layer. The increase in floc size with increasing xanthan concentration may also be explained by the calculated increase in the depth of the secondary minima.

5 Conclusions

The enhanced creaming of an emulsion on adding xanthan is due to flocculation of the emulsion caused by the presence of this polysaccharide. The exact mechanism causing the flocculation is difficult to establish. Because the interfacial tension of an n-hexadecane–water interface with Tween 20 present is not affected by the addition of xanthan to the water, an adsorption mechanism is not likely. Moreover, xanthan does not accumulate in the creamed layer as would be expected for adsorption at the interface or on the emulsifier. In contrast, the xanthan concentration in the bulk water phase in an emulsion is higher than in the orignal water phase, which indicates a depletion mechanism.

Estimates of the depth of the secondary minimum in the pair interaction energy between emulsion droplets, taking a depletion layer into account, can explain some of our results. Firstly, the larger floc size and the more rigid appearance of the creamed layer for the larger emulsion droplets is due to the deeper secondary minima under these conditions. Secondly, the relatively stable emulsions obtained for the smaller droplets and lower xanthan concentrations goes along with the correspondingly very small secondary minima. Thirdly, the increase in floc size with increasing xanthan concentration is consistent with an increase in the depletion energy.

Summarizing, we believe that a depletion mechanism is very likely to be the reason for flocculation of emulsions containing xanthan, but we are not yet able to prove this unequivocally.

Acknowledgement

This work was supported by Suiker Unie, Roosendaal, and the Dutch Innovation Oriented Program Carbohydrates (IOP-k).

References

1. V. J. Morris, in 'Gums and Stabilisers for the Food Industry', ed. G. O. Phillips, D. J. Wedlock, and P. A. Williams, IRL Press, Oxford, 1990, Vol. 5, p. 315.
2. C. M. Gordon, in 'Gums and Stabilisers for the Food Industry', ed. G. O. Phillips, D. J. Wedlock, and P. A. Williams, IRL Press, Oxford, 1990, Vol. 5, p. 351.
3. H. Luyten, W. Kloek, and T. van Vliet, in 'Third European Rheology Conference', ed. D. R. Oliver, 1990, p. 323.
4. H. Luyten, T. van Vliet, and W. Kloek, to be published in 'Proceedings of the XIth International Conference on Rheology', Brussels, 1992.
5. E. Dickinson, in 'Gums and Stabilisers for the Food Industry', ed. G. O. Phillips, D. J. Wedlock, and P. A. Williams, IRL Press, Oxford, 1988, Vol. 4, p. 249.
6. P. A. Gunning, D. J. Hibberd, A. M. Howe, and M. M. Robins, *Food Hydrocolloids*, 1988, **2**, 119.
7. Y. Cao, E. Dickinson, and D. J. Wedlock, *Food Hydrocolloids*, 1990, **4**, 185.
8. E. Dickinson, *J. Colloid Interface Sci.*, 1989, **132**, 274.
9. S.-L. Young and J. A. Torres, *Food Hydrocolloids*, 1989, **3**, 365.
10. S. Takigami, P. A. Williams, and G. O. Phillips, in 'Gums and Stabilisers for the Food Industry', ed. G. O. Phillips, D. J. Wedlock, and P. A. Williams, IRL Press, Oxford, 1992, Vol. 6, p. 371.

11. B. Bergenståhl, in 'Gums and Stabilisers for the Food Industry', ed. G. O. Phillips, D. J. Wedlock, and P. A. Williams, IRL Press, Oxford, 1988, Vol. 4, p. 363.
12. A. G. Gaonkar, *Food Hydrocolloids*, 1991, **5**, 329.
13. E. Dickinson and S. R. Euston, in 'Food Polymers, Gels and Colloids', ed. E. Dickinson, Special Publication No. 82, Royal Society of Chemistry, Cambridge, 1991, p. 132.
14. P. R. Sperry, *J. Colloid Interface Sci.*, 1982, **87**, 375.
15. D. H. Napper, in 'The Effect of Polymers on Dispersion Properties', ed. Th. F. Tadros, Academic Press, London, 1982, p. 199.
16. B. Vincent, J. Edwards, S. Emmett, and R. Croot, *Colloids Surf.*, 1988, **31**, 267.
17. A. Lips, I. J. Campbell, and E. G. Pelan, in 'Food Polymers, Gels and Colloids', ed. E. Dickinson, Special Publication No. 82, Royal Society of Chemistry, Cambridge, 1991, p. 1.
18. P. Walstra, H. Oortwijn, and J. J. de Graaf, *Neth. Milk Dairy J.*, 1969, **23**, 33.
19. M. Dubois, K. A. Gilles, J. K. Hamilton, P. A. Rebers, and F. Smith, *Anal. Chem.*, 1951, **28**, 350.
20. D. Ausserre, H. Gervet, and F. Rondelez, *Macromolecules*, 1986, **19**, 85.
21. J. Lyklema, 'Fundamentals of Interface and Colloid Science', Academic Press, London, 1991, Vol. 1.
22. D. B. Hough and L. R. White, *Adv. Colloid Interface Sci.*, 1980, **14**, 3.
23. M. J. Snowden, S. M. Clegg, P. A. Williams, and I. D. Robb, *J. Chem. Soc., Faraday Trans.*, 1991, **87**, 2201.
24. T. Coviello, K. Kajiwara, W. Burchard, M. Dentini, and V. Crescenzi, *Macromolecules*, 1986, **19**, 2826.

NMR Micro-imaging Studies of Water Diffusivity in Saturated Microporous Systems

By B. P. Hills

AFRC INSTITUTE OF FOOD RESEARCH, NORWICH RESEARCH PARK, COLNEY, NORWICH NR4 7UA, UK

1 Introduction

With the development of the technique of Magnetic Resonance Imaging (MRI), it is now possible to observe the time course of water transport in microporous food materials non-invasively. For this reason MRI has great potential in the food industry where water transport is known to affect food quality and safety.[1] In this paper it is shown how MRI measurements of water diffusivity in microporous systems, when combined with water proton relaxation times and pulsed gradient spin-echo measurements can be used to quantify the system microstructure. The relationship between water diffusivity and the Darcy flow permeability is also discussed.

2 The NMR Micro-imaging Diffusion Method

Since NMR image intensity is proportional to the water proton fraction, water diffusivity can be measured from the changes in image intensity as D_2O (or H_2O) diffuses into an H_2O-(or D_2O-)saturated microporous system. Figure 1 shows the time course of the intensity profiles along a cylindrical sample tube as D_2O diffuses into the sample through the right-hand boundary. The intensity is proportional to the D_2O concentration and profiles were obtained as differences from the water proton spin densities. The sample in this case consisted of packed cellulose fibres set in a 10% gelatin gel matrix. By fitting these image profiles with numerical solutions to the diffusion equation, an effective water self-diffusion coefficient (D_∞) can be deduced. With small samples (*ca.* 1 cm in length), diffusion times of 2–3 hours are sufficient for determination of D_∞. However, with such small samples errors can arise from irregularities in the shape of the sample–D_2O boundary (which in the above example should, ideally, be planar). This causes small deviations in the measured and

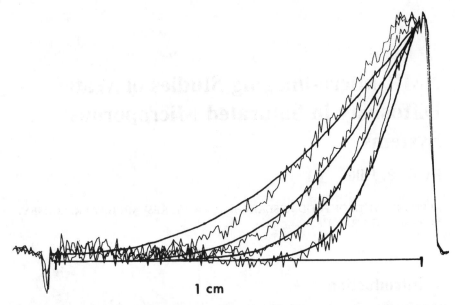

Figure 1 *A comparison of experimental and theoretical diffusion profiles for the influx of D_2O into packed cellulose powder saturated with 10% gelatin gel. The profiles were measured at times 990, 1890, 3690, and 7290 seconds after contact with D_2O. The theoretical profiles were calculated with $D_\infty = 1.2 \times 10^{-5} \, cm^2 \, s^{-1}$*

calculated diffusion profiles which can be seen in Figure 1. Such errors can be alleviated, at the expense of increased diffusion times, by scaling up the sample size by an order of magnitude to *ca.* 10–15 cm, and monitoring the increase in image intensity at a particular point several centimetres from the boundary. Figure 2 shows a typical proton fraction against time curve for water (H_2O) diffusing into a packed suspension of silica gel initially saturated with D_2O. The solid line is the calculated increase for a diffusion coefficient, D_∞, equal to $(1.2 \pm 0.05) \times 10^{-5} \, cm^2 \, s^{-1}$. Table 1 lists values of D_∞ at 295 K for several small (*ca.* 1 cm) and large (*ca.* 15 cm) saturated model microporous systems.

3 Interpretation of D_∞ in Microporous Systems

While values for the water diffusivity (D_∞) are useful in their own right for many applied problems in food processing and spoilage, the water diffusivity data also contain valuable information about the system microstructure. Figure 3 shows an illustrative cross-section through a randomly packed saturated suspension of water-impenetrable particles such as silica. The measured water diffusivity (D_∞) is clearly determined by morphological factors such as the pore-size distribution and the geometry of the 'throats' connecting the pores. As a first step to quantifying the relationship between D_∞ and the microstructure, the complex three-dimensional

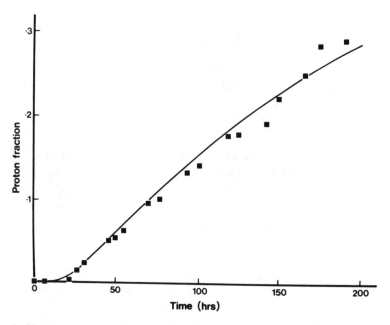

Figure 2 *The time course of the proton fraction at a fixed pixel position as water exchanges with D_2O in a packed suspension of fine silica powder at 295 K. The line shows the theoretical result for a diffusion coefficient D_∞ = 1.2 × 10⁻⁵ cm² s⁻¹*

Table 1 *Values of D_∞ at 295 K for various model systems*

System	Mean Particle Size	$D_\infty/10^{-5}$ cm² s⁻¹
10% Cross-linked gelatin gel	Submicron junction zones	2.0 ± 0.05
Fine silica packed in water	40–60 μm	1.2 ± 0.05
Coarse silica packed in water	130–250 μm	0.88 ± 0.05
Cellulose packed in water	Fibre (100–250 × 15 μm)	1.4 ± 0.05
Fine silica in 10% gelatin gel	40–60 μm	0.75 ± 0.1
Coarse silica in 10% gelatin gel	130–250 μm	0.75 ± 0.1
Starch in 10% gelatin gel	ca. 50 μm	0.6 ± 0.1
Cellulose in 10% gelatin gel	Fibre (100–250 × 15 μm)	1.2 ± 0.1
Parenchyma apple tissue	Cell diam. ca. 150 μm	1.0 ± 0.1
Parenchyma carrot tissue	Cell diam. ca. 100 μm	1.5 ± 0.1

morphology can be treated using a simple one-dimensional diffusion model by representing a pore as a lamellar region of width l_i, associated with the self diffusion coefficient of pure water D_w ($= 2 \times 10^{-5}$ cm² s⁻¹). Each pore is separated from its neighbours by a 'throat' which is modelled as a small interpore space of width δ and associated with an effective water self diffusion coefficient D_t ($\leqslant D_w$). For simplicity, all throats are assumed to be the same, and so it is sufficient to fix the effective throat width δ at

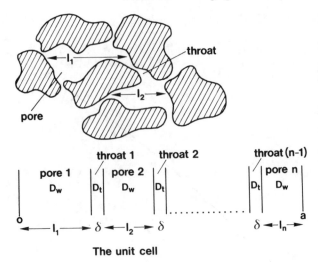

The unit cell

Figure 3 *An illustrative cross-section through a randomly packed particle suspension showing pores and throats. The one-dimensional model corresponding to this complex morphology is also shown. (See text for details.)*

some arbitrary value ($\delta \ll l_i$) and vary the pore connectivity by altering the diffusion coefficient in the throats, D_t. The 'unit cell' shown in Figure 3 is infinitely repeated by assuming reflective boundary conditions at $x = 0$ and a. For this model the effective water diffusivity D_∞ measured by the micro-imaging method is given as[2]

$$D_\infty = \left(\sum_{i=1}^{n} l_i + (n-1)\delta\right)\bigg/\left(\sum_{j=1}^{n} (l_j/D_w) + (n-1)\delta/D_t\right), \qquad (1)$$

where n is the number of pores in the unit cell. The calculation of D_∞ therefore reduces to identifying the number of pores (n) in a representative unit cell, the pore sizes (l_i) and the pore connectivity (D_t). Fortunately, NMR water-proton relaxation time measurements and pulsed field-gradient spin-echo measurements can provide independent information about these parameters.

Figure 4 shows the measured distribution in water-proton transverse relaxation times for a packed aqueous suspension of fine silica. This originates from the diffusion of water in the pores (and throats) to the silica surface where it rapidly relaxes and can be analysed with the model shown in Figure 3 by introducing the bulk water relaxation rate γ_w ($\sim 0.5\ \text{s}^{-1}$) for water inside the pores and an intrinsic relaxation rate γ_t in the throats, and numerically solving the Bloch–Torrey equations[3] for the time evolution of the transverse magnetization density within the unit cell. In this way the effective unit cell and pore sizes can be deduced. For example, a simple unit cell consisting of three pores ($n = 3$) such that

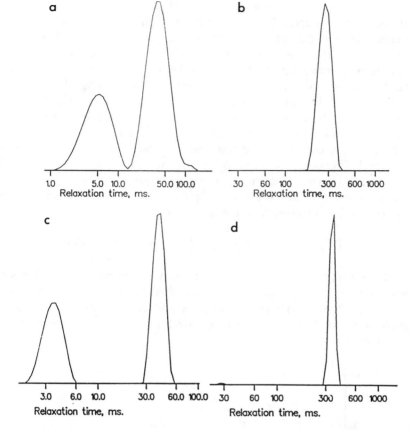

Figure 4 *A comparison of the experimental and theoretical water-proton relaxation behaviour for a packed, saturated, aqueous suspension of fine silica: (a) the experimental transverse relaxation time distribution; (b) the experimental longitudinal relaxation time distribution measured at 100 MHz; (c) the theoretical transverse relaxation time distribution (see text for parameter values); (d) the theoretical longitudinal relaxation time distribution for $\gamma_t = 30 \text{ s}^{-1}$ and a pore connectivity $D_t = 2 \times 10^{-6} \text{ cm}^2 \text{ s}^{-1}$. Note the different time axes*

$2l_1 = l_2 = 4.4 \, \mu\text{m}$, $l_3 = 11.7 \, \mu\text{m}$, $\delta = 0.87 \, \mu\text{m}$, $\gamma_t = 3400 \text{ s}^{-1}$, $\gamma_w = 0.5 \text{ s}^{-1}$, and $D_w = 2 \times 10^{-5} \text{ cm}^2 \text{ s}^{-1}$ is sufficient to reproduce the major features of the observed 'bimodal' transverse relaxation (see Figure 4). The fitting is insensitive to the pore connectivity D_t, since transverse magnetization relaxes so rapidly in the throats that it essentially isolates the pores.

Unlike the transverse relaxation, the longitudinal relaxation is a single exponential and is shifted to much longer relaxation times (Figure 4). This observation cannot, however, be explained merely by reducing the throat relaxation rate γ_t since at small values of γ_t the calculated relaxation distribution also depends critically on the pore connectivity D_t. For

$D_t \leq 10^{-6} \text{ cm}^2 \text{ s}^{-1}$, the calculated longitudinal relaxation is bi-exponential with percentage populations as shown in Figure 5 for $\gamma_t = 30 \text{ s}^{-1}$. The observed single exponential T_1 data are compatible with $\gamma_t = 30 \text{ s}^{-1}$ and a pore connectivity D_t greater than *ca.* $10^{-6} \text{ cm}^2 \text{ s}^{-1}$. The value of D_t can be found by substituting the measured value of D_∞ into equation (1) along with the values of n, l_i, and δ determined by fitting the transverse relaxation data. For fine silica this give $D_t = 2.3 \times 10^{-6} \text{ cm}^2 \text{ s}^{-1}$ which is consistent with the observed single exponential T_1. This interpretation can be further checked using data from NMR pulsed field-gradient spin-echo (PGSE) measurements based on the standard Stejskal–Tanner pulse sequence.[4] The observed echo amplitude $S(q, \Delta, \tau)$ is a function of the wave vector q corresponding to the pulsed gradient area, the gradient pulse separation Δ, and the 90–180° pulse spacing τ. An effective PGSE diffusion coefficient $D(q, \Delta, \tau)$ can also be defined as

$$D(q, \Delta, \tau) = - \ln (S(q)/S(q = 0))/q^2(\Delta - 1/3\delta) \qquad (2)$$

Figure 6 shows the Δ-dependence of $D(q, \Delta, \tau)$ measured for a fine silica suspension in gelatin, together with the theoretical curve obtained using the computer model in Figure 3 with the unit cell parameters listed in the legend to Figure 6. A very similar approach has also recently been applied to cellular tissue.[5]

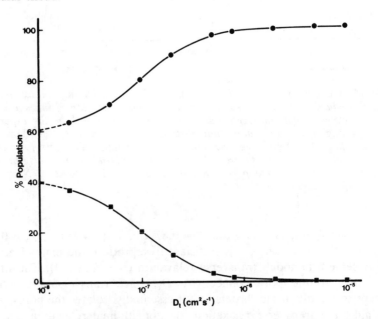

Figure 5 *The calculated effect of varying the pore connectivity* (D_t) *on the water-proton longitudinal relaxation in packed, saturated fine silica for* $\gamma_t = 30 \text{ s}^{-1}$. *For* $D_t < 10^{-6} \text{ cm}^2 \text{ s}^{-1}$, *the relaxation is seen to be a double exponential. The observed single exponential longitudinal relaxation in Figure 4 corresponds to* $D_t \geq 10^{-6} \text{ cm}^2 \text{ s}^{-1}$

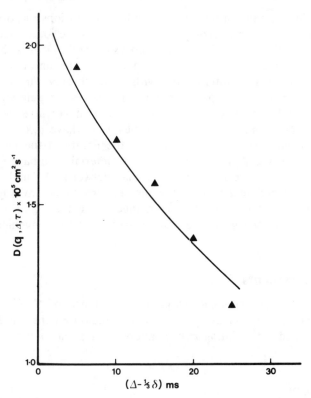

Figure 6 *A comparison of the experimental* (▲) *and theoretical (solid line)* Δ*-dependence of* D(q, Δ, τ) *for a packed suspension of fine silica in 10% gelatin gel for a wave vector* q *of* $0.3745 \times 10^3 \ \text{cm}^{-1}$, $\delta = 0.5$ ms, $\tau = 15.5$ ms. *Parameters used in the unit cell model are:* D_w $= 2.3 \times 10^{-5} \ \text{cm}^2 \text{s}^{-1}$, $D_t = 2.3 \times 10^{-6} \ \text{cm}^2 \text{s}^{-1}$, $\gamma_w = 30 \ \text{s}^{-1}$, $\gamma_t = 2200 \ \text{s}^{-1}$, $2l_1 = l_2 = 14.13 \ \mu m$

4 Relationship of D_∞ to other Macroscopic Transport Coefficients

Fluid flow through porous media under an applied pressure gradient (∇p) is characterized by the Darcy flow permeability K. The flux $\mathbf{J}$ is given by

$$\mathbf{J} = - (K/\mu) \ \nabla p, \tag{3}$$

where ∇p is the pressure gradient and μ is the fluid shear viscosity. The force–flux relationships of non-equilibrium thermodynamics can be used to show that, when both diffusion and flow are limited by the throats, the Darcy coefficient K is linearly proportional to D_∞ such that $K = \lambda^2(D_\infty/D_w)$ were λ is a length parameter related to the system microstructure and D_w is the self-diffusion coefficient of pure water. Similarly it can be shown that, for the diffusion and conductivity of ions in porous media, we have $(D_\infty)_i/(D_0)_i = (\sigma_i/(\sigma_0)_i)\ \phi^{-1}$, where σ_i is the ionic conductivity in the porous medium, ϕ is the medium porosity, and $(D_\infty)_i$ is the diffusion coefficient of the ions in the porous medium. If water and ions experience the same restrictive interaction with the throats, it follows that $(D_\infty/D_0)_{water} = (\sigma_i/(\sigma_0)_i)\ \phi^{-1}$. This is an interesting relationship since Roberts and Schwartz have previously shown[6] that, for a related family of porous systems having a range of porosities ϕ, we have $\sigma_i/(\sigma_0)_i = \phi^m$ with m ranging from 1.8 to 2.0 depending on the nature of the medium. This implies that, for the same family of porous materials, we have $(D_\infty/D_0) \propto \phi^{m-1}$. Whether this relationship, and that between D_∞ and K, can be usefully established in porous materials remains to be investigated. The micro-imaging method for D_∞ combined with flow and conductivity measurements on the same material provides a very convenient way for doing this.

Acknowledgements

The author wishes to acknowledge the assistance of S. C. Smart and V. M. Quantin in undertaking the diffusion measurements, and Dr. P. S. Belton for many stimulating conversations during the preparation of this work.

References

1. C. C. Seow, 'Food Preservation by Moisture Control', Elsevier Applied Science, Barking, 1988.
2. J. Crank, 'The Mathematics of Diffusion', 2nd edn, Clarendon, Oxford, 1985.
3. H. C. Torrey, *Phys. Rev.*, 1956, **104**, 563.
4. E. O. Stejskal and J. E. Tanner, *J. Chem. Phys.*, 1965, **42**, 288.
5. B. P. Hills and J. E. M. Snaar, *Molec. Phys.*, 1992, **76**, 979.
6. J. N. Roberts and L. M. Schwartz, *Phys. Rev. B.*, 1985, **31**, 5990.
7. J. Bear, 'Dynamics of Fluids in Porous Media', American Elsevier Publishing, New York, 1972.

Crystallization in Simple Paraffins and Monoacid Saturated Triacylglycerols Dispersed in Water

By John Coupland, Eric Dickinson, D. Julian McClements, Malcolm Povey, and Christine de Rancourt de Mimmerand

PROCTER DEPARTMENT OF FOOD SCIENCE, UNIVERSITY OF LEEDS, LEEDS LS2 9JT, UK

1 Introduction

The degree of crystallization of fat droplets in a food oil-in-water emulsion is an important factor affecting the stability of the colloidal system and the distribution of solute molecules between dispersed and continuous phases. Solid fat content is relevant to the texture of food colloids such as ice-cream and whipped cream, and also to the stability of food emulsions in shear flow.[1,2] Emulsion stability is an important consideration in the food industry, ultimately having a bearing on both shelf-life and microbiological safety.

We have been studying crystallization and melting behaviour in emulsions of types n-alkane-in-water and triacylglycerol-in-water. The oil phases n-hexadecane and n-octadecane have been selected for their relative chemical simplicity and the absence of polymorphism. The triacylglycerols, glycerol trilaurate and glycerol trioleate, have been chosen as examples of fats that can be obtained in a relatively pure form, and which differ primarily in their melting point. The presence of one carbon double bond in the fatty acid chains of glycerol trioleate depresses its freezing point compared with glycerol trilaurate, ensuring that it remains liquid over the entire range of temperature employed in our experiments.

Droplet size, the nature of the surfactant, the temperature history of the sample, and the nature of the oil phase—all these influence crystallization in emulsions, which has been conventionally characterized using calorimetric methods, pulsed NMR, and densitometry.[3] We and others have also recently detected changes of state of the droplets in dispersions of n-alkanes in water using ultrasonic techniques.[4-10] The approach basically relies on the change in ultrasonic velocity as the oil changes state from liquid to solid.[11]

The utility of the ultrasound technique has been demonstrated by us in investigations of crystallization and melting behaviour in (i) emulsions made from pure n-hexadecane,[4-6] and (ii) emulsions made from a mixed hydrocarbon oil comprising n-hexadecane and n-octadecane.[7,12] In this paper, we present new data for emulsions made from the two triacylglycerols, glycerol trilaurate and glycerol trioleate. Our data indicate that, in addition to the twin influences of particle size and temperature history, the dispersed phase mass transport and the droplet crystallization are also dependent on the chemical nature of the emulsifier. In particular, differences have been found between emulsions stabilized by a low-molecular-weight non-ionic surfactant and emulsions stabilized by protein.

2 Materials and Methods

The preparation of the n-alkane emulsions has been described previously.[12] The triacylglycerols were obtained from Unilever Research, Colworth Laboratory. The glycerol trilaurate was very pure (approximately 96% pure); the glycerol trioleate was of a much less pure grade (approximately 70% pure). The triacylglycerol oil-in-water emulsions were made in the same manner as the n-alkane emulsions, but the oil concentration was reduced from 20 wt% in the n-alkane emulsions to 5 wt% because emulsions containing a higher proportion of oil were found to be unstable. A proportionate reduction in the Tween 20 (polyoxyethylene monolaurate) concentration to 0.5 wt% was also made.

The droplet-size distribution was measured before and after each set of ultrasonic experiments to see if it was effected by the experimental procedure. The size distibutions were measured using a Malvern Mastersizer S2.01.

The velocity of ultrasound was measured using a low powered pulse echo technique at a frequency of 1.25 ± 0.15 MHz. The equipment and method are described elsewhere.[5] Some of this work employed a different ultrasonic cell[13] which enabled us to carry out velocity and attenuation measurements as a function of frequency. Temperature was controlled to ± 0.1 °C by immersing the ultrasonic cell in a Grant W38 thermostatically controlled water bath. A minimum of 30 min was required for thermal equilibration to be achieved whenever the temperature was changed.

3 Results and Discussion

In earlier work[5] we demonstrated that a 20 vol% n-hexadecane-in-water emulsion with mean particle size $d_{32} = 0.36$ μm exhibits supercooling of approximately 15 K (Figure 1). The sharpness of the freezing transition and its dependence on particle size are consistent with either surface nucleation or homogeneous nucleation.[14] On heating the emulsion containing fully solidified drops from 0 °C to a temperature 8 °C below the bulk melting point, a portion of the emulsion dispersed phase (approximately 7%) is

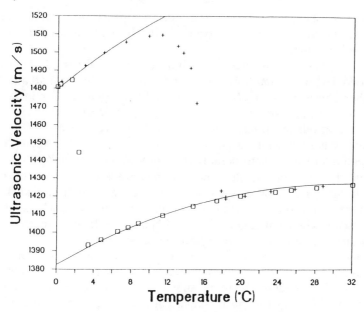

Figure 1 *Variation of ultrasonic velocity with temperature for a* 20 vol% *n-hexa-decane-in-water emulsion* ($d_{32} = 0.36 \mu m$): $\square$, *cooling*; +, *heating. The curves are predictions from ultrasonic scattering theory*

found to melt. Cycling the temperature between 7 °C and 12 °C produces no hysteresis in the melting curve of Figure 1. We conclude from this that a portion of every drop melts. If any droplet had melted completely, then it would not be expected to crystallize again, since it would be above its supercooled freezing point (we know that liquid droplets are stable for weeks at these elevated temperatures). Hence, if this were the case, temperature cycling would gradually cause more and more droplets to melt, and the velocity would gradually drop to the value expected for a wholly liquid system. As this does not happen, it seems logical to assume that it is the surface regions of the droplets which are pre-melting since they are in intimate contact with surfactant: in the Tween 20 stabilized system, the lauric acid chains reside predominantly in the oil phase. Additionally, some of the hydrocarbon oil could also be solubilized in surfactant micelles in the aqueous phase. Such oil would be expected also to melt at temperature lower than that of oil-in-emulsion droplets. However, the quantities of such solubilized oil can only be small because otherwise we would not see the excellent fit shown in Figure 1 between our calculation of ultrasonic velocity and the measured value. These calculations are based on the assumption that the droplets are either wholly liquid or wholly solid and that all the oil is dispersed as droplets and is not solubilized in micelles. We have shown that the type of emulsifier affects the amount of pre-melting; the effect is far less pronounced when sodium caseinate is used.[12] Adsorbed casein does not penetrate so far in the bulk oil phase; nor does the protein form micelles in the aqueous phase capable of solubilizing n-hexadecane.

We have also demonstrated[12] that inter-mixing of dispersed phase occurs when two Tween 20 stabilized emulsions, one consisting of n-alkane A and the other consisting of n-alkane B, are added together. For instance, A might be n-hexadecane and B might be n-octadecane. It seems highly probable that oils are transferred through the aqueous phase by Tween 20 micelles. When sodium caseinate is used in place of Tween 20, this inter-mixing process is considerably slowed down or even halted.

In all the experiments described so far, we see no evidence of any Ostwald ripening. Within experimental uncertainty, particle size distributions were found to be the same at the start and finish of each experiment. In some cases the combined cooling and heating cycle was repeated two or three times over a period of weeks on the same emulsion.

We have now confirmed that the general phenomena described above for n-alkane-in-water emulsions also occur in triacylglycerol emulsions. The experimental procedures were arranged to be the same for the triacylglycerol emulsions as for the n-alkane emulsions, apart from the adoption of a lower oil content, which was necessary due to the tendency of the triacylglycerol emulsions to flocculate and gel on cooling.[14] In Figure 2 we present results for a 5 wt % trilaurin oil-in-water emulsion. The graph demonstrates the same general features that we see with the n-alkane

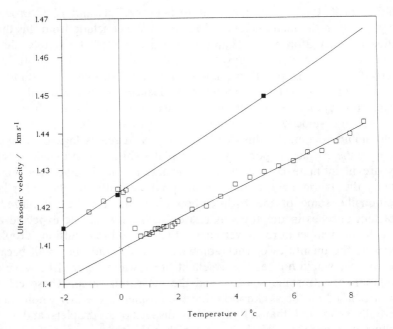

Figure 2 *Supercooling and crystallization in a* 5 vol % *trilaurin oil-in-water emulsion* ($d_{32} = 0.36 \ \mu$m). □, *cooling*; ■, *heating*

emulsions—a large supercooling and extremely rapid crystallization consistent with homogeneous nucleation. When a 20 wt % trilaurin oil-in-water emulsion is mixed in 50:50 proportions with a triolein emulsion, we observe that the freezing point of the trilaurin emulsion becomes further depressed, and that the freezing takes place over a wider temperature range (Figure 3), consistent with possible dilution of the trilaurin droplets with triolein, which had presumably been previously solubilized in Tween 20 micelles and then transported across the aqueous phase. The driving force for this process is the entropy of mixing which is far greater than the driving force for Ostwald ripening in a pure oil emulsion.[12] At higher concentrations of trilaurin oil, we observe partial coalescence of the crystallized droplets, resulting in the formation of a flocculated emulsion gel. The ultrasound is more highly attenuated in the presence of the gel network, and the velocity of sound measurements are therefore much less accurate.

Finally, we have yet to confirm whether the slow heterogeneous crystallization of supercooled liquid droplets in the presence of solid droplets, which occurs in Tween-stabilized n-alkane emulsions during isothermal storage,[7] also occurs in the case of triacylglycerols. This phenomenon implies that crystalline material is able to pass into or through the stabilizing film of non-ionic surfactant and hence somehow nucleate fat crystallization during a droplet encounter. The type of surfactant is likely

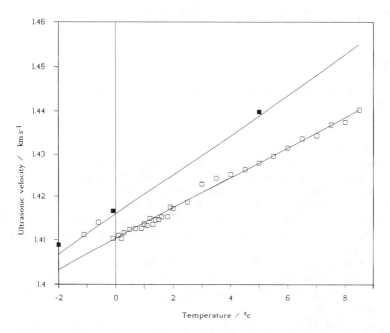

Figure 3 *Supercooling and crystallization in a* 5 vol% 50:50 *mixture of* 5 vol% *trilaurin-in-water and* 5 vol% *triolein-in-water emulsions* ($d_{32} = 0.36 \mu m$): □, *cooling*; ■, *heating*

to be important in controlling how close droplets are able to approach and to what extent any protruding crystals are able to penetrate another droplet surface. The precise mechanism by which this occurs may shed some light on the processes whereby real food oil-in-water emulsions become destabilized during crystallization. In the case of the triacylglycerols, it will also be necessary to account for the influence of crystal polymorphism, crystal growth habit, and the impact of surfactants and small-molecule impurities (such as the mono and di-acylglycerols) on the crystal growth mechanism and the crystal defect structure.

4 Conclusion

We conclude that fine triacylglycerol emulsions exhibit the same sort of behaviour—homogeneous nucleation and micelle promoted inter-mixing in the presence of non-ionic surfactant Tween 20—that we reported previously[12] for n-alkane emulsions. It seems reasonable to assume also that another mechanism observed by us in n-alkane emulsions occurs also in dilute triacylglycerol emulsions—that is, slow heterogeneous nucleation during isothermal storage of a mixture of solid and supercooled liquid droplets of the same oil. Superimposed on this behaviour with *pure* triacylglycerols is the partial coalescence of crystallized triacylglycerol oil droplets leading to gelation; we have observed this at oil concentrations above 5 vol% in trilaurin oil-in-water emulsions.

References

1. E. Dickinson and G. Stainsby, 'Colloids in Food', Applied Science, London, 1982.
2. P. Walstra and R. Jenness, 'Dairy Chemistry and Physics', Wiley, New York, 1984.
3. D. Waddington, in 'Analysis of Oils and Fats', ed. R. J. Hamilton and J. B. Rossel, Elsevier Applied Science, London, 1986, p. 341
4. E. Dickinson, M. I. Goller, D. J. McClements, J. Peasgood, and M. J. W. Povey, *J. Chem. Soc., Faraday Trans.*, 1990, **86**, 1147.
5. M. J. W. Povey, in 'Advances in Food Emulsions and Foams', ed. E. Dickinson and G. Stainsby, Elsevier Applied Science, London, 1988, p. 285.
6. E. Dickinson, D. J. McClements and M. J. W. Povey, *J. Colloid Interface Sci.*, 1991, **142**, 103.
7. D.J. McClements, M. J. W. Povey, and E. Dickinson, in 'Ultrasonics International 91 Conference Proceedings', Butterworths, London, 1991, p. 107.
8. D. J. McClements, E. Dickinson, and M. J. W. Povey, *Chem. Phys. Lett.*, 1990, **172**, 449.
9. C. Javanaud, N. R. Gladwell, S. J. Gouldby, D. J. Hibberd, A. Thomas, and M. M. Robins, *Ultrasonics*, 1991, **29**, 331.
10. E. Betsanis, M. Jury, D.J. McClements, and M. J. W. Povey, *Ultrasonics*, 1990, **28**, 266.

11. D. J. McClements and M. J. W. Povey, *Int. J. Food Sci. Technol.*, 1988, 23, 159.
12. E. Dickinson, M. I. Goller, D. J. McClements, and M. J. W. Povey, in 'Food Polymers, Gels and Colloids', ed. E. Dickinson, Special Publication No. 82, Royal Society of Chemistry, Cambridge, 1991, p. 171.
13. D. J. McClements, *J. Acoust. Soc. Am.*, 1992, 91, 849.
14. J. Coupland, BSc. project thesis, University of Leeds, 1991.

Effects of Plasticizers on the Fracture Behaviour of Wheat Starch

By A. R. Kirby, R. Parker, and A. C. Smith

AFRC INSTITUTE OF FOOD RESEARCH, NORWICH LABORATORY, NORWICH
RESEARCH PARK, COLNEY, NORWICH NR4 7UA, UK

1 Introduction

Starch is a major component of many foods such as pasta, breakfast cereals, biscuits, and various snack products. The sensory texture of starch-based products has often been correlated to water activity or water content.[1-3]

The mechanical properties of starch-based materials have been studied as a function of water content by a number of authors.[1-6] The textural attributes of crispness, crunchiness, and hardness involve large strain processes. Ward[7] has remarked that transitions in failure behaviour, such as the brittle–ductile transitions in synthetic polymers, are not necessarily related to transitions in the small-strain mechanical response, e.g. the glass transition. In an earlier study[8,9] the flexural modulus of starch was found to fall rapidly at a water content corresponding to T_g, and the addition of glucose was found to displace T_g to lower water contents. The earlier work included a study of the large strain deformation behaviour of samples and the failure processes operating at different water contents based on complementary electron microscopy of the fracture surfaces.[8,9] This present paper considers the large strain deformation of wheat starch plasticized by water and glycerol or xylitol in combination with optical microscopic studies of fracture surfaces.

2 Experimental

Sample Preparation

Samples were prepared by extruding narrow sheets using a Baker Perkins MPF 50D co-rotating twin-screw extruder fitted with a slit die as described elsewhere.[8,9] Specimens were cut from the extruded sheets to conform to the British Standards for three-point bend testing of plastics.[10,11]

Samples of different water contents were prepared in two ways: (i) by drying samples in the laboratory for periods of 5–50 h followed by wrapping them in polyethylene film, or (ii) by conditioning the samples in controlled humidity enclosures using standard inorganic salt solutions over a period of 30 to 35 days. The water content of the xylitol-plasticized samples was determined gravimetrically, whereas a Karl–Fischer titration had to be used for the water content determination of the glycerol-plasticized samples. Water contents are quoted on a wet weight basis throughout.

Mechanical properties

The force–deflection response of each sample was obtained using a three-point bend test in an Instron 1122 testing machine in accordance with the British Standards[10,11] at a crosshead speed of 5 mm min^{-1}.

3 Results

The various samples exhibited different types of stress–strain behaviour; we call these 'fracture', 'bend', and 'tear'. The plots are linear at low strains (< 0.015). Fracture is marked by a rapid decrease in the stress in a time period of less than 30 ms (the minimum time between logged forces). Tearing, a slow fracturing process, is characterized by a broad maximum in the stress followed by a rapid decrease in stress with failure occurring at a strain larger than that for fracture. The bending mode is typified by a curvature of the stress–strain plot towards the strain axis with no fracture.

The failure mode was found to change from the fracture processes ('fracture' and 'tear') to remaining intact ('bend') at a distinct water content. Whereas for starch plus glycerol all the samples below 12.3 wt% water failed, for starch plus xylitol only all the samples below 10.8 wt% water failed (Figure 1). Tearing occurs at compositions close to these critical water contents.

For the starch plus glycerol samples at the lowest water content (6.5 wt%), a V-shaped wedge of material from the fracture zone was found to fragment into small pieces. In Figure 2 we see the two remaining pieces fitted together in their original configuration. In the contact area, the recombined pieces fit together perfectly indicating a brittle fracture. The highest water content samples which fracture (9.9 wt%) showed different surface features. The pieces no longer fitted together (Figure 3) due to yielding in the area where the fracture was initiated.

4 Conclusion

In this study well-defined fracture was found to become less common in favour of tearing with increasing water content, and failure no longer takes

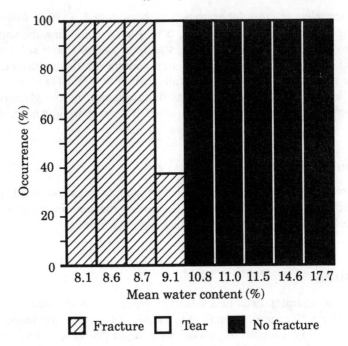

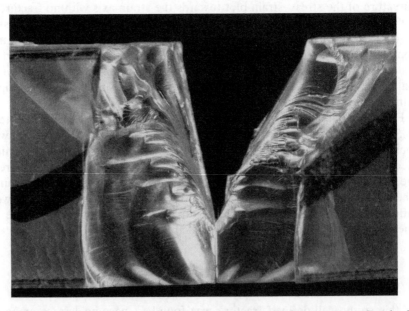

Figure 1 *The occurrence of different types of stress–strain behaviour at different water contents in starch plus xylitol samples*

Figure 2 *The V-shaped wedge left by fragmentation in the fracture zone. Starch plus glycerol sample fractured at a water content of 6.5 wt %*

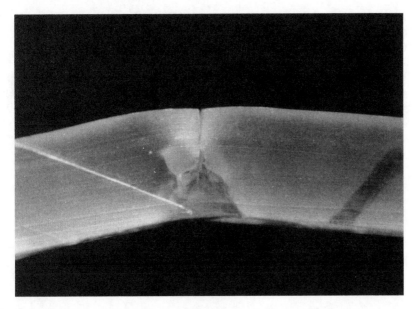

Figure 3 *The pieces of a starch plus glycerol sample fractured at a water content of 9.9 wt % showing plastic deformation*

place at water contents above about 10 wt % for the starch with polyol systems. The evidence from microscopy indicates that the mechanism of fracture changes with increasing water content up to 10 wt %, and some permanent deformation occurs in the samples fractured at this water content. There are, however, no corresponding changes in the failure stress and failure strain with water content.[12]

References

1. E. Katz and T. P. Labuza, *J. Food Sci.*, 1981, **46**, 403.
2. F. Sauvageot and G. Blond, *J. Text. Stud.*, 1991, **22**, 423.
3. G. E. Attenburrow, R. M. Goodband, L. J. Taylor, and P. J. Lillford, *J. Cereal Sci.*, 1989, **9**, 61.
4. R. E. Wetton and R. D. L. Marsh, in 'Rheology of Food, Pharmaceutical and Biological Materials', ed. R. E. Carter, Elsevier Applied Science, London, 1990, p. 231.
5. M. T. Kalichevsky, E. M. Jaroszkiewicz, S. Ablett, J. M. V. Blanshard, and P. J. Lillford, *Carbohydr. Polym.*, 1992, **18**, 77.
6. R. J. Hutchinson, S. A. Mantle, and A. C. Smith, *J. Mater. Sci.*, 1989, **24**, 3249.
7. I. M. Ward, 'Mechanical Properties of Solid Polymers', 2nd edn, Wiley, New York, 1982.
8. A.-L. Ollett, R. Parker, and A. C. Smith, *J. Mater. Sci.*, 1991, **26**, 1351.

9. A.-L. Ollett, R. Parker, and A. C. Smith, in 'Food Polymers, Gels and Colloids', ed. E. Dickinson, Special Publication No. 82, Royal Society of Chemistry, Cambridge, 1990, p. 537.
10. BS 2782 1977: Part 10: method 1005.
11. BS 2782 1978: Part 3: method 335A.
12. A. R. Kirby, S. A. Clark, R. Parker, and A. C. Smith, *J. Mater. Sci.*, submitted.

Rheology of the Gel Protein Fraction of Wheat Flour

By G. Oliver and P. E. Pritchard

FLOUR MILLING AND BAKING RESEARCH ASSOCIATION, CHORLEYWOOD, HERTS WD3 5SH, UK

1 Introduction

Work at FMBRA has shown that the mixing requirements of wheat varieties in the Chorleywood Bread Process (CBP) are related to their high-molecular-weight (HMW) glutenin subunit proteins. A wheat variety may contain up to five different subunits coded by the three chromosomes 1A, 1B, and 1D. Recently a number of varieties have been introduced in the UK with so called 'extra-strong' characteristics. Despite having a normal HMW glutenin composition, the varieties have a high work-input requirement ($> 11\,\mathrm{W\,h\,kg^{-1}}$) in the CBP. The work presented here describes one technique used to elucidate the cause of this phenomenon.

It is possible to isolate from flour a fraction, which is rich in HMW glutenins, called gel protein.[1] The molecular interactions between HMW glutenin subunits in gel protein are related to the interactions that exist in wheat grains. The interactions have to change for the formation of a visco-elastic dough. This fraction has been extracted from the varieties Urban (normal) and Sicco (extra-strong). The varieties contain the same HMW glutenin subunits $(1, 7 + 9, 5 + 10)$, but they differ in their processing requirements. Biochemical and rheological analyses were performed on the extracted gel protein.

2 Experimental

Gel Protein

All flours were tested for gel protein content according to the following procedure. Flour (10 g) was de-fatted with 25 ml petroleum ether (b.p. 40–60 °C) for 1 h, filtered, and dried. De-fatted flour (5 g) was stirred with 90 ml of 1.5% sodium dodecyl sulfate (SDS) for 10 min at 10 °C and then centrifuged at 63 000 g for 40 min. The gel protein occurring as a transparent layer in between the opaque starch-rich sediment and the liquid supernatant was then weighed.

The response of the gel protein to mixing was assessed as follows. Flour (50 g) was mixed with 0.9 g sodium chloride and water. Aliquots (20 g) were taken at intervals, frozen immediately after removal, freeze-dried, and ground so as to pass through a 250 μm sieve. The dried material was de-fatted and gel protein was determined as described above.

Rheological Testing

Gel protein was tested immediately after isolation on a Bohlin VOR rheometer operating in oscillatory mode. The operating conditions are as shown in Table 1.

3 Results and Discussion

Effect of Mixing Time on the Gel Protein

The gel protein content of Sicco flour prior to mixing was found to be greater than that of Urban flour. The two varieties also displayed marked differences in their breakdown behaviour during mixing (Figure 1). Urban

Table 1 *Experimental conditions in rheology measurements with Bohlin Rheometer*

Measuring system	Concentric cylinder (*bob diameter* 14 mm)
Torsion bar	17.88 g cm
Frequency sweep	0.1 to 20 Hz
Amplitude	100 %
Temperature	25 °C

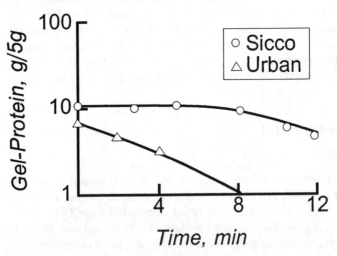

Figure 1 *Change in gel protein extracted from flour during mixing for two different wheat varieties*

decayed almost linearly with mixing time, but Sicco showed little loss in gel protein during the early stages of mixing. After about 8 minutes, however, the loss of gel protein in Sicco increased to a rate similar to that for Urban.

Oscillatory Testing of the Gel Protein

Rheological testing of gel proteins extracted from the flours before mixing showed Sicco to have a larger modulus (Figure 2) and viscosity (Figure 3) than Urban.

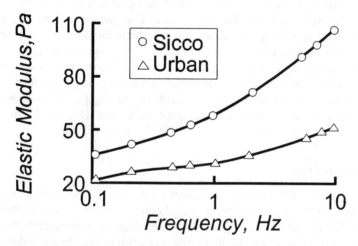

Figure 2 *Rheology of gel protein extracted* **before** *mixing. Modulus G' is plotted against frequency*

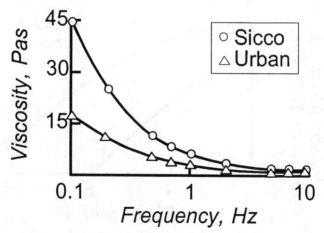

Figure 3 *Rheology of gel protein extracted* **before** *mixing. Shear viscosity is plotted against frequency*

The higher elastic modulus G' found in Sicco suggests that the cross-linking in the gel protein of this variety is more extensive. This may result in the gel protein being stronger and more resistant to breakdown. The higher viscosity is possibly an indication that the glutenin aggregates are larger in Sicco than in Urban.

The decay in rheological properties of the varieties during mixing shows similar behaviour to their rates of breakdown. For comparison, the elastic modulus (Figure 4) and viscosity (Figure 5) data recorded at a frequency of 1 Hz are shown.

Urban loses its modulus and viscosity at a constant rate during mixing. Sicco, however, has a lag phase of several minutes during which there is only a small decrease in modulus; after this period the decay in modulus occurs at a similar rate to that for Urban. The viscosity of the gel protein extracted during mixing decreases linearly with mixing time. On mixing, the glutenin aggregates gradually decrease in size until they reach a threshold size at which they become soluble. Above the threshold, decreasing molecular size will cause a decrease in viscosity of gel protein, but the total quantity remains unchanged. Therefore, the lag phase observed when measuring the decay in either gel protein weight (Figure 1) or elastic modulus (Figure 4) is not observed for the viscosity (Figure 5).

4 Conclusion

The gel protein of Sicco, an 'extra-strong' wheat variety, has a lag phase during mixing before it begins to breakdown. Urban, which contains the same HMW glutenin subunits as Sicco, depolymerizes linearly with mixing time. For full dough development the gel protein has to be substantially

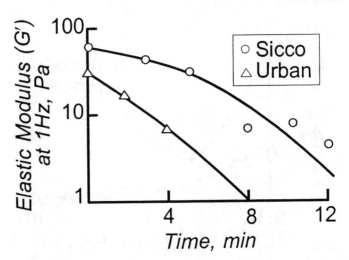

Figure 4 *Change in rheology of gel protein extracted **during** mixing. Modulus G' is plotted against mixing time*

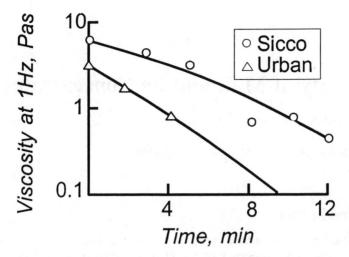

Figure 5 *Change in rheology of gel protein extracted **during** mixing. Shear viscosity is plotted against mixing time*

broken down during mixing so that the different interactions characteristic of a visco-elastic dough can form. For this to occur, Sicco requires a higher level of work input during mixing than Urban. Rheological tests have indicated that, in comparison to Urban, the gel protein in Sicco consists of larger aggregates that are more extensively cross-linked. The extra work input required for Sicco appears necessary in order for the aggregates to become broken down to a similar size to those found in Urban. These techniques are currently being used to evaluate the baking quality of various UK and EC wheats.

References

1. A. Graveland, P. Bongers, and P. Bosveld, *J. Sci. Food Agric.*, 1979, **30**, 71.

Viscosity of Milk and its Concentrates

By David S. Horne

HANNAH RESEARCH INSTITUTE, AYR KA6 5HL, UK

1 Introduction

As well as being influenced by the size, shape, and number of dispersed particles, the flow properties of a colloidal dispersion also depend on the nature and strength of the interparticle forces, the properties of the continuous phase, and the stress applied to the dispersion. Understanding flow behaviour is important for both food product development, where viscosity is a major determinant of texture, and for manufacturing process control, where such knowledge dictates the choice of pumps, heat exchangers, and mixing systems.

Previous studies of the viscosity behaviour of milk and its concentrates have tended to emphasize these important technological aspects.[1-6] Recent advances in the study of the rheological response of concentrated colloidal systems have utilized polystyrene latex suspensions as model systems.[7-9] Our interest is in testing the applicability of such theories in a system of major practical interest. In this paper on the rheology of milk, we seek information on the interaction forces between the constituent colloidal particles, the casein micelles. In this brief report, only evaporated skim milks will be considered, with viscosity measured as a function of volume fraction and steady shear-rate.

2 Materials and Methods

Bulk whole milk from the morning milking of the Hannah Research Institute herd was skimmed by centrifugation and then concentrated by evaporation at 35 °C using a laboratory rotary evaporator. The micro-Kjeldahl method was used to determine the degree of concentration by measuring the crude protein levels in the original and evaporated milks. The maximum degree of concentration achieved was $6.22 \times N$, where N is the original concentration. Intermediate levels between this maximum and the original were obtained by diluting the higher concentration with water, or alternatively by evaporation of a further non-concentrated sample. The data presented were accumulated over a period of a week with fresh concentrates prepared daily.

All viscosity measurements were made at 20 °C in a Bohlin VOR rheometer fitted with one of two geometries: single-gap concentric cylinders and double-gap concentric cylinders, with interchangeable torsion bars to cover the range of sensitivity required. Flow curves (shear stress τ *versus* shear-rate $\dot{\gamma}$) were obtained by scanning shear-rate, each data point being the average stress measured during 30 s of shear. Shear stress and viscosity were reproducible for both increasing and decreasing shear-rate sweeps, with no discernible hysteresis behaviour.

3 Results and Discussion

All flow curves obtained using the Bohlin rheometer showed a decreasing viscosity with increasing shear-rate. This was as true for distilled water or lactose solution as it was for concentrated milk samples, and, in the case of the low viscosity solutions, was presumably associated with overcoming residual inertia in the air-bearing. All curves therefore tended to a limiting shear stress value at low shear-rates. In the case of Newtonian liquids this was of the order of a few mPa. Low concentration milks followed Newtonian behaviour, but, as concentration was increased, the samples became markedly more shear thinning, though in all cases a region was reached at high shear-rates where stress rose proportionately with the rate of shear.

To confer consistency of interpretation on all our data, the flow curves were analysed according to the Bingham equation

$$\tau = \tau_B + \eta_{PL}\dot{\gamma} \tag{1}$$

over the linear high-shear portion of the τ *versus* $\dot{\gamma}$ curve between shear-rates of 1.33 and 291 s^{-1}. The quantity τ_B is the extrapolated or apparent Bingham yield stress and the slope of the straight line defines the apparent or plastic viscosity, η_{PL}. Good linear fits were obtained in all cases, with the correlation coefficients exceeding 0.999. For Newtonian liquids, with zero yield stress, this equation reduces to a direct proportionality between shear stress and shear-rate, the slope defining the viscosity of the suspension.

For low concentrations ($<1.5 \times N$), the plastic viscosities were fitted to a quadratic function in the concentration multiplier. When plotted against volume fraction and normalized to the infinite dilution viscosity, this set of data should define a quadratic function, the coefficient of whose linear term has the Einstein value, 2.5. With this assumption, and knowing the casein concentration of the original milk, the micellar voluminosity was calculated to be 3.85 ml g^{-1} from the coefficient of the linear term in the raw data fit. In this same fit, the intercept value (η_0) of 1.16 mPa s was close to that of a 4 wt % lactose solution (1.15 mPa s) measured under the same conditions. When replotted as relative viscosity (η_0/η_{PL}) against volume fraction, the viscosity data at all concentrations gave a good fit to a

Krieger–Dougherty[10] type equation with ϕ_m, the maximum packing fraction, equal to 0.66 ± 0.02.

The Bingham yield stress is shown as a function of the micellar volume fraction in Figure 1, where it is clear that the yield stress increases steeply with concentration above $\phi \sim 0.3$. The yield stresses we obtain are similar in magnitude to those recorded by Buscall *et al.*[8] for weakly coagulated polystyrene latex suspensions. Treating our results from a scaling viewpoint, we find, from the double log plot of τ_B *versus* ϕ (inset to Figure 1), that the Bingham yield stress scales with volume fraction to the power 4.1. This value for the exponent is also close to the value of 3.85 obtained for the weakly flocculated latex suspension.[8]

The presence of a yield stress can be understood in terms of a three-dimensional structure which has sufficient strength to prevent flow when the applied stress is less than τ_0. For $\tau > \tau_0$, the structure may collapse suddenly to produce flow units which are not subsequently affected by shear. This is ideal Bingham behaviour and so we have $\tau_0 = \tau_B$. More usually, the breakdown is progressive and the flow units become smaller, with the result that the differential viscosity, defined as the slope of the flow curve, decreases with increasing shear until an apparently linear region is reached, as encountered in these studies of concentrated milk.

This rheological behaviour is therefore implying that our concentrated milks are behaving as weakly flocculated suspensions. In such circumstances, it is possible to relate the Bingham yield stress to the interparticle

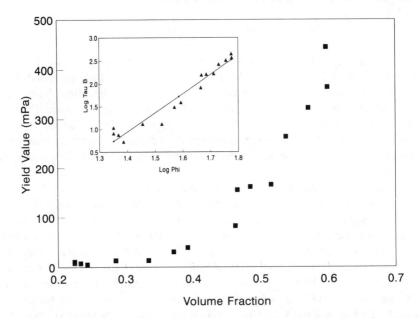

Figure 1 *Bingham yield stresses, extrapolated from flow curves, as a function of volume fraction of evaporated skim milk at 20 °C. Inset is a log–log plot of same data*

interaction, by equating this energy to the amount of energy required to separate the flocs into single units,[7] *i.e.*

$$\tau_B = N E_{sep} \qquad (2)$$

where N is the total number of contacts between particles in flocs, and E_{sep} is the energy required to break each contact. The total number of contacts N may be related to the volume fraction ϕ and the average number of contacts per particle (the co-ordination number), n, by

$$N = \frac{1}{2} \left\{ \frac{3\phi n}{4\pi a^3} \right\} \qquad (3)$$

where a is the particle radius. Combining the latter two equations gives

$$\tau_B = \frac{3\phi n E_{sep}}{8\pi a^3} \qquad (4)$$

Thus, E_{sep} can be calculated from τ_B, provided that a value can be assigned to the co-ordination number, and the micellar radius is known. For our calculations, a value of 8 has been assumed for n, and the micellar radius taken as 100 nm. A further assumption in this calculation of E_{sep} from τ_B is that, above the yield point, all contacts are broken. This assumption is probably justified in this instance where reversible breakdown is observed under high shear.

Having obtained E_{sep} as a function of volume fraction, albeit that the values are small compared to kT and imply only weak attraction, it is interesting to carry the model further by calculating the dependence of this energy on the surface-to-surface separation between the micelles in the concentrated milks. This distance h scales as

$$h \sim 2a[(\phi_n/\phi)^{1/3} - 1] \qquad (5)$$

where ϕ_n is a close-packing fraction characteristic of the type of array, *e.g.* 0.68 for body-centred cubic, or 0.64 for random arrangements. For our calculations a body-centred cubic arrangement was selected, giving rise to the plot of E_{sep} against h shown in Figure 2. The energy required to bring about total separation of the flocs, E_{sep}, decreases rapidly with increasing surface-to-surface separation, indicating a sharp decrease in inter-micellar attraction forces. This decrease is exponential (inset Figure 2) with a characteristic decay length of some 16 nm.

4 Conclusions

Rheological data derived from measurements on concentrated milks can be interpreted using theories hitherto applied only to model systems to reveal

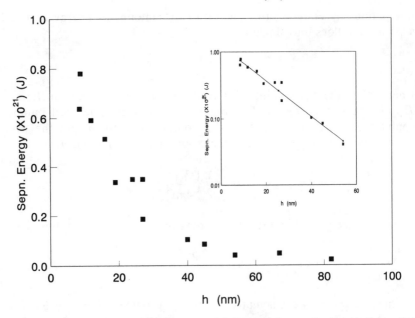

Figure 2 *Variation of* E_{sep}, *calculated from experimental* τ_B *values, with intermicellar surface-to-surface distance* h. *Inset is a semi-log plot of the same data*

the nature of intermicellar interactions in these systems. The appearance of a yield stress in measurements with highly concentrated milks shows the presence of weak flocculation under these circumstances, the strength of attraction between micelles being calculable from these measured yield stresses.

Acknowledgements

The author wishes to thank Ms. JoAnn Smith for her skilled technical assistance. This research was funded by the Scottish Office Agriculture and Fisheries Department.

References

1. B. L. Stepp and D. E. Smith, *Milchwissenschaft*, 1991, **46**, 484.
2. M. Hallstrom and P. Dejmek, *Milchwissenschaft*, 1988, **43**, 31.
3. M. Hallstrom and P. Dejmek, *Milchwissenschaft*, 1988, **43**, 95.
4. K. L. Langley and D. M. Temple, *J. Dairy Res.*, 1985, **52**, 223.
5. T. H. M. Snoeren, A. J. Damman, and H. J. Klok, *Neth. Milk Dairy J.*, 1982, **36**, 305.
6. T. H. M. Snoeren, J. A. Brinkhuis, A. J. Damman, and H. J. Klok, *Neth. Milk Dairy J.*, 1984, **38**, 41.
7. Th. F. Tadros and A. Hopkinson, *Faraday Discuss. Chem. Soc.*, 1991, **90**, 41.

8. R. Buscall, I. J. McGowan, and C. A. Mumme-Young, *Faraday Discuss. Chem. Soc.,* 1991, **90**, 115.
9. C. J. de Kruif, E. M. F. van Iersel, A. Vrij, and W. B. Russel, *J. Chem. Phys.,* 1985, **83**, 4717.
10. J. M. Krieger, *Adv. Colloid Interface Sci.,* 1972, **3**, 111.

Mechanical Properties of Concentrated Starch Systems during Heating, Cooling, and Storage

By C. J. A. M. Keetels and T. van Vliet

DEPARTMENT OF FOOD SCIENCE, WAGENINGEN AGRICULTURAL UNIVERSITY. PO BOX 8129, 6700 EV WAGENINGEN, THE NETHERLANDS

1 Introduction

Retrogradation of starch plays an important role in the firming of products of high starch content. For example, the staling of bread has been ascribed, at least in part, to starch retrogradation. As bread is a complex system, studies on retrogradation have usually been done on simpler starch–water systems. Several techniques, such as X-ray diffraction, differential scanning calorimetry (DSC), and dynamic mechanical analysis, have been used to study retrogradation of concentrated starch gels.[1,2]

To obtain a better understanding of the mechanism of starch retrogradation and its consequences for product properties, the mechanical properties of 30 wt% potato and wheat starch systems have been determined during heating, cooling, and storage. As consumer perception of changes in product properties is mainly related to changes in mechanical properties at large deformations, the behaviour of these starch gels during storage was also studied in uniaxial compression.

2 Materials and Methods

In order to make homogeneous starch gels, dispersions of 3 wt% potato starch (AVEBE, the Netherlands) or 8 wt% wheat starch (Latenstein BV, the Netherlands) in water (containing 0.02% thiomersal as preservative), were heated to 65 °C, while gently stirring. After cooling to room temperature, sufficient starch was added to obtain suspensions with 30% dry matter.

Mechanical properties of these suspensions at small deformations were measured with a Bohlin VOR Rheometer, equipped with a cup and bob system and a torque bar of 90 g cm. The suspensions were heated from 20 °C to 90 °C, kept at 90 °C for 15 minutes, cooled to 20 °C, and finally kept at 20 °C. Heating and cooling were performed at a rate of 2 °C min^{-1}. Sequential measurements were made every 30 s. Oscillations were performed at a frequency of 0.1 Hz and a strain of 0.01. At this strain all samples showed 'linear' behaviour. To prevent evaporation, samples were covered with paraffin oil.

For the large deformation experiments, test pieces were prepared by filling teflon cylindrical moulds with the suspensions. These were heated at 95 °C for 90 minutes. After cooling to room temperature, the gels were removed from the moulds and kept in paraffin oil at 20 °C to prevent them from drying out. Uniaxial compression tests were performed with a Zwick material testing machine, using a 2000 N load cell. Cylinders of 20 mm height and 15 mm diameter were compressed between perspex plates, and the compressive stress σ and the Hencky strain ε_h were determined.[3] From the stress–strain curve, the Young modulus E $[= (d\sigma/d\varepsilon)_{\varepsilon \to 0}]$ could be calculated. Measurements were made at storage times of 0, 2, 5.5, 24, 144, and 384 hours.

3 Results and Discussion

Changes in the storage modulus (G') and the loss tangent (tan δ) of 30% potato and wheat starch suspensions, measured during heating and cooling, are shown in Figure 1. It is seen that the storage moduli reach maximum values at temperatures of 64 °C for wheat starch and 67 °C for potato starch. At this maximum temperature the granules must be closely packed. Because of the high concentration, the swelling of the individual granules is limited by the available volume.

With a further increase in the temperature, some melting occurs (presumably of remaining crystallites) as determined by DSC. This would be expected to result in softer granules, and indeed the storage moduli were found to decrease.

During and after heating the starch suspensions, higher moduli were found for the wheat starch than for the potato starch. This difference is probably caused by the swollen wheat starch granules being more rigid, maybe due to more internal cross-links in these granules. When fully swollen, potato starch granules also appeared to be less rigid than wheat starch granules.[4]

Figure 1 shows that, for both suspensions, tan δ ($= G''/G'$) decreases in the region where G' increases strongly. Because tan δ of the wheat starch gels decreases to 0.05 during heating and remains that low at a temperature of 90 °C, we suppose that an elastic gel is formed during heating. The somewhat larger value of tan δ for the potato starch gel during heating, cooling, and short storage times is ascribed primarily to the granules themselves being less rigid, thus exhibiting a more viscous behaviour.

Properties of 30% potato and wheat starch gels at large deformations are rather different (Figure 2), with potato starch gels fracturing at a larger

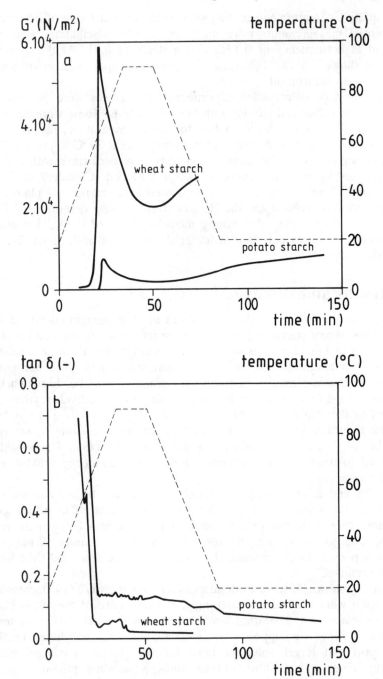

Figure 1 *Changes in the rheology of* 30 wt % *starch suspensions during a heating and cooling cycle: (a) the storage modulus* G′ *is plotted against time; (b) the quantity tan* δ *is plotted against time. The dashed line shows temperature against time*

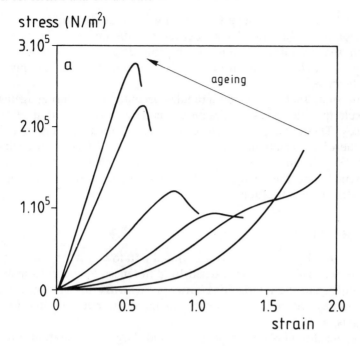

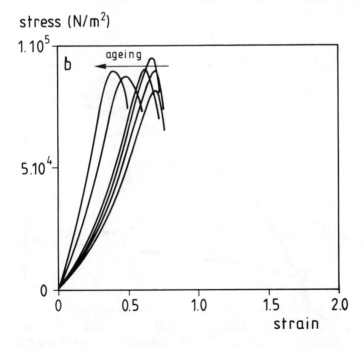

Figure 2 *Stress–strain curves of (a) a 30 wt % potato starch gel and (b) a 30 wt % wheat starch gel after storage times of 0, 2, 5.5, 24, 144, and 384 hours. Fracture occurs at the maximum of the stress–strain curve*

stress and strain. We suppose that this is due to differences in the deformability of the granules as well as to differences in the interactions between the granules. Because freshly made potato starch gels do not fracture, we believe that strong bonds between the granules must form during heating.

We presume that the swollen granules become firmer during ageing, with the result that the stress at fracture increases and the strain at fracture decreases. These changes are more pronounced for the potato starch gels, which therefore results in a 'cross-over' in E as a function of ageing time (Figure 3). In an earlier paper[5] we showed that properties at large deformations and the changes in these during ageing were also dependent on the starch concentration.

4 Conclusion

The mechanical properties of concentrated potato and wheat starch systems are different. This is mainly caused by differences in the deformability of the granules and the interactions between them. During ageing, changes in gel properties are more pronounced for potato starch gels than for wheat starch gels.

These results illustrate that small and large deformation properties reflect different, not easily related, aspects of retrogradation.

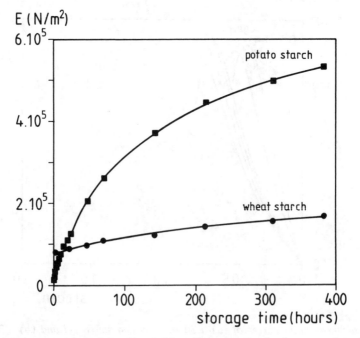

Figure 3 *The Young modulus* E *of* 30 wt % *potato and wheat starch gels as a function of storage time at* 20 °C

Acknowledgements

This work was supported by the Netherlands Technology Foundation (STW). The authors thank Prof. P. Walstra and Dr. A. H. Bloksma for valuable discussions.

References

1. P. D. Orford, S. G. Ring, V. Carroll, M. J. Miles, and V. J. Morris, *J. Sci. Food Agric.*, 1987, **39**, 169.
2. Ph. Roulet, W. M. MacInnes, P. Würsch, R. M. Sanchez, and A. Raemy, *Food Hydrocolloids*, 1988, **2**, 381.
3. H. Luyten, T. van Vliet, and P. Walstra, *Neth. Milk Dairy J.*, 1991, **45**, 33.
4. P. A. M. Steeneken, *Carbohydr. Polym.*, 1989, **11**, 23.
5. C. J. A. M. Keetels and T. van Vliet, in 'Gums and Stabilisers for the Food Industry', ed. G. O. Phillips, D. J. Wedlock, and P. A. Williams, Oxford University Press, 1992, Vol. 6, p. 141.

Relevance of Biaxial Strain Hardening to the Gas Retention of Dough

By Ton van Vliet, Anita J. J. Kokelaar, and Anke M. Janssen[1]

DEPARTMENT OF FOOD SCIENCE, WAGENINGEN AGRICULTURAL
UNIVERSITY, PO BOX 8129, 6700 EV WAGENINGEN, THE NETHERLANDS
[1]DEPARTMENT OF BIOCHEMISTRY, UNIVERSITY OF GRONINGEN,
NIJENBORGH 16, 9747 AG GRONINGEN, THE NETHERLANDS

1 Introduction

Recently a new criterion for the stability of gas cells against coalescence during bread-making has been developed.[1] It is based on the fact that the deformation of dough around a growing gas cell is a biaxial extension; that is, the dough is extended in two directions parallel to the surface of the gas cell, and compressed in the direction perpendicular to the surface of the gas cell. After the whole dough is transformed into a foam with polyhedral gas cells, the relevant deformation is still biaxial extension.

2 The Model

A film between two gas cells will be stable against rupture if the force per unit length for further thinning of a part that is, by accident, thinner than its surroundings is larger than that of the thicker parts. The change in the biaxial strain ε_b on further thinning is

$$d\varepsilon_b = - \tfrac{1}{2} \, dh/h(t) \tag{1}$$

where dh is taken to be negative and small. When a film of a thickness h is locally extended to a thickness $h + dh$, the thinning process will stop if

$$\{h + dh\} \cdot \sigma(h + dh) > h \cdot \sigma(h) \tag{2}$$

where σ is the stress. From equations (1) and (2), we get the condition for stability:

$$d \ln \sigma/d\varepsilon_b > 2 \tag{3}$$

So, the thinner part of a biaxially extended dough film has a larger resistance to further thinning than the thicker part if the ratio of the relative increase of the stress to the accompanying increase in ε_b is higher than 2. If the increase is less, the thinner part of the film will be extended more than its surroundings and eventually the dough film will rupture, impairing gas retention. A complicating factor is that the extension rate of the film also affects the forces involved, and this rate in turn depends on the baking stage. A faster rate causes a higher resistance to extension due to the visco-elastic properties of dough and so provides an additional means of stabilization of a dough film. However, the size of the gas cells at both sides of a dough film also affects the extension rate. As a consequence, the proper criterion for stability of dough films during baking is somewhat stricter than given in equation (3) during the fermentation stage and probably somewhat less strict during the baking stage.

3 Results and Discussion

Biaxial extensional parameters of a wheat flour (Katepwa) having a satisfactory baking behaviour and of a rye flour dough were determined by lubricated uniaxial compression experiments.[1-4] Some results are shown in Figure 1. From a set of such graphs, the compressive stress σ as a function

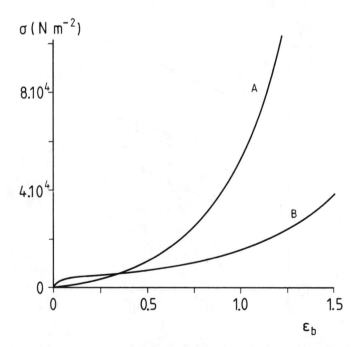

Figure 1 *Compressive stress σ as a function of biaxial Hencky strain ε_b for (A) a wheat flour dough and (B) a rye flour dough. The initial strain rate is $5 \times 10^{-3}\,\mathrm{s}^{-1}$*

of the biaxial strain-rate $\dot{\varepsilon}_b$ for various biaxial strains ε_b has been calculated (Figure 2). As can be seen, σ increases with increasing strain at constant biaxial strain-rate (the elastic behaviour of dough), and also with increasing biaxial strain-rate at constant strain (the viscous behaviour). Starting with Figure 2, a plot of $\ln \sigma$ as a function of ε_b can be drawn (Figure 3). It turns out that the value of $d\ln\sigma/d\varepsilon_b$ is 2.9 for the wheat flour dough (with a satisfactory baking performance) but only 1.25 for the rye flour dough (which shows little gas retention).

4 Conclusion

The results given above show that the difference in the potential for gas retention between a wheat flour dough and a rye flour dough may be at least partly due to a difference in strain hardening on biaxial extension. Results for doughs of different wheat cultivars show that probably this principle can also play a part in determining the potential for gas retention by these doughs.[1] However, besides the criterion described, other parameters, such as the strain at which the dough film fractures in biaxial extension, may also be important.

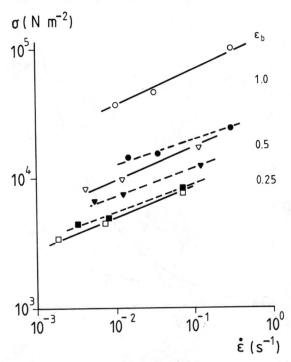

Figure 2 *Compressive stress σ as a function of the biaxial strain-rate $\dot{\varepsilon}_b$:* $\bigcirc$, $\bigtriangledown$, $\square$, ———, *wheat flour dough (Katepwa);* $\bullet$, $\blacktriangledown$, $\blacksquare$, - - - -, *rye flour dough. Values of the biaxial strain ε_b are indicated*

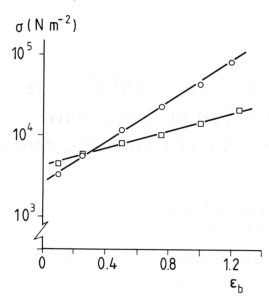

Figure 3 *Compressive stress σ as a function of the biaxial strain ε_b for a constant biaxial strain-rate of 2×10^{-2} s^{-1}: O, wheat flour dough; □, rye flour dough*

References

1. T. van Vliet, A. M. Janssen, A. H. Bloksma, and P. Walstra, *J. Texture Stud.*, accepted for publication.
2. S. H. Chatraei, C. W. Makosko, and H. H. Winter, *J. Rheol.*, 1981, **25**, 433.
3. O. H. Campanella, L. M. Popplewell, J. R. Rosenau, and M. Peleg, *J. Food Sci.*, 1987, **52**, 1249.
4. E. B. Bagley, D. D. Christianson, and J. A. Martindale, *J. Texture Stud.*, 1988, **19**, 289.

Studies of Water-in-Oil-in-Water (W/O/W) Multiple Emulsions: Stabilization and Controlled Nutrient Release

By Eric Dickinson, Jane Evison, Richard K. Owusu, and Qinghong Zhu

PROCTER DEPARTMENT OF FOOD SCIENCE, UNIVERSITY OF LEEDS, LEEDS LS2 9JT, UK

1 Introduction

Water-in-oil-in-water (w/o/w) multiple emulsions are potentially useful in food systems for protecting and controlling the release of nutrients.[1,2] We present initial results relating to the encapsulation and controlled-release of L-tryptophan (L-Trp) and riboflavin-5'-phosphate (vitamin B_2) from coarse multiple emulsions.

Very little work has been done on preparing fine protein-stabilized w/o/w emulsions.[3] One of the major problems often encountered when preparing such fine w/o/w emulsions is that the intensity of homogenization needed to obtain the fine droplets is so severe that it can break the w/o emulsion and cause a reduction in the yield of multiple emulsion droplets.

The entrapment of proteins within the internal aqueous phase has been reported to increase both the yield and stability of multiple emulsions.[4] As part of a study into the possible mechanisms of this stabilization, the effect of bovine serum albumin (BSA) on the primary w/o emulsion stability has been examined.

2 Material and Methods

W/o/w multiple emulsions were prepared by a two-stage procedure using a mineral oil as the oil phase, and Span 80 and Tween 20 as the primary and secondary emulsifiers. All these materials were purchased from Fluka Chemicals, UK. Coarse or fine multiple emulsions were prepared using a high-speed laboratory mixer or a high-pressure jet homogenizer (designed and built in the Procter department).[5] The primary w/o emulsion contained glucose (1.8%), L-Trp (0.25%), or vitamin B_2 (0.25%). Nutrient transport and multiple emulsion yield analysis involved the dialysis approach.[6] L-Trp and vitamin B_2 were analysed by UV absorption at 280 nm and 270 nm, respectively. The neocuproin method[7] was used to detect glucose.

The effect of BSA (0–1 wt%) within the aqueous phase of the primary w/o emulsion on the droplet-size distribution over a 7 day period was investigated. The emulsion size distributions were determined using the Malvern Mastersizer (calculated with a presentation code 0405).

3 Results and Discussion

The coarse w/o/w multiple emulsions employed in the transport studies were produced with a 95–98% yield, and mean diameters of 0.5 μm (w/o emulsion droplets) and 20 μm (w/o/w droplets).

The rate of L-Trp and vitamin B_2 mass transfer from the inner phase of coarse w/o/w emulsions to the external phase was found to follow first-order kinetics at pH 2–10. That is, the kinetics can be expressed in terms of a single time constant. The rate of L-Trp release from the multiple emulsions was reduced compared to the release from a bulk solution in a dialysis bag by a factor of 368, 72, and 28-fold at pH 2, pH 6, and pH 10, respectively. For vitamin B_2 the corresponding relative degrees of controlled-release were 508, 107, and 84-fold, respectively (see Figure 1). The rate of nutrient release was found to be slow at $pH < pK$ ($pK = 5.8$ for L-Trp, and $pK = 4.9$ for vitamin B_2).

BSA (0–1.0 wt%) was entrapped within w/o emulsions. The droplet-size distributions of these emulsions were determined immediately and after 7 days. The $D_{(3,2)}$ values change little in the 7-day period or on changing the BSA concentration. However, the $D_{(4,3)}$ index emphasizes the sizes of the larger droplets which are particularly important with respect to stability. The degree of change that the $D_{(4,3)}$ parameter underwent in the 7 days is plotted against the BSA concentration in Figure 2. Although the $D_{(4,3)}$ values tend to be rather sensitive to random errors, these results do suggest that BSA stabilization of w/o/w emulsions may involve some stabilization of the w/o primary emulsion against coalescence. Previous workers[4] have found that BSA in combination with Span 80 improves the yield and stability of w/o/w emulsions. They concluded that BSA stabilizes w/o/w emulsions as a result of putative 'complex formation' with Span 80. From these results, it may be that the stabilization of multiple emulsions by BSA does involve stabilization of internal w/o droplets against coalescence by the formation of a more visco-elastic mixed adsorbed layer—even though most of the layer thickness is likely to be on the inner aqueous phase side of the interface, and therefore would not be favourable towards stabilization by forming a thick steric barrier around the droplets. More experiments are required to confirm this point.

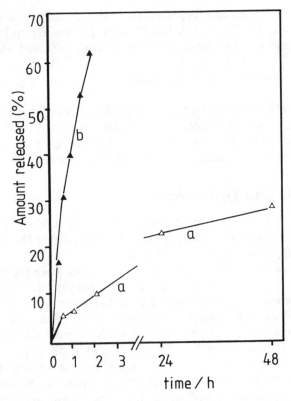

Figure 1 *Time-dependent release of vitamin B_2 at pH 6 from (a) coarse w/o/w emulsion in a dialysis bag and (b) vitamin B_2 solution in a dialysis bag*

4 Conclusions

W/o/w emulsions may be useful to immobilize important nutrients. Diffusion across the oil layer obeys first order kinetics. The rate of release of both L-Trp and vitamin B_2 is influenced by the pH conditions. This suggests that the net charge of the nutrient molecule is an important factor governing the diffusion process.

In the vast majority of food systems, the presence of fine droplets is required to obtain adequate stability. As discussed in the introduction, the second-stage homogenization pressure required to obtain fine w/o/w droplets tends to cause a reduction in the multiple emulsion yield. BSA within the inner aqueous phase could possibly be used to compensate for this loss of multiple emulsion yield.

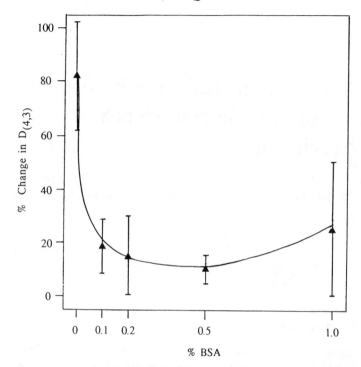

Figure 2 *The concentration of BSA plotted against the percentage increase in* $D_{(4,3)}$ *parameter observed during 7 days storage at* 25 °C

References

1. S. Matsumoto, *J. Texture Stud.*, 1986, **17**, 141.
2. B. de Cindio, G. Grasso, and D. Cacace, *Food Hydrocolloids*, 1991, **4**, 339.
3. E. Dickinson, J. Evison, and R. K. Owusu, *Food Hydrocolloids*, 1991, **5**, 481.
4. J. A. Omotosho, T. K. Law, T. L. Whateley, and A. T. Florence, *Colloids Surf.*, 1986, **20**, 133.
5. I. Burgaud, E. Dickinson, and P. V. Nelson, *Int. J. Food Sci. Technol.*, 1990, **25**, 39.
6. S. Matsumoto, K. Yoshiko, and D. Yonezawa, *J. Colloid Interface Sci.*, 1976, **57**, 353.
7. M. E. Brown and M. S. Boston, *Diabetics*, 1961, **10**, 60.

Modelling Small Intestine Emulsions: Effects of Soluble Non-starch Polysaccharides

By A. J. Fillery-Travis, S. J. Moulson, L. H. Foster, J. M. Gee, G. M. Wortley, S. A. Clark, I. T. Johnson, and M. M. Robins

AFRC INSTITUTE OF FOOD RESEARCH, NORWICH LABORATORY, NORWICH RESEARCH PARK, COLNEY, NORWICH NR4 7UA, UK

1 Introduction

The effects of soluble non-starch polysaccharides (NSP) on lipid metabolism are perceived to be an important property of dietary fibre. At present, however, there is an incomplete understanding of the processes of lipid digestion, particularly in the presence of interacting polymers. Several *in vivo* studies[1,2] indicate that fibre supplements such as guar gum reduce fat absorption. One view is that the high viscosity of the polymers is the important property, but it appears to us that the mechanisms are more complex than those due to viscosity alone.

Digestion of lipid requires its emulsification during passage through the stomach to the duodenum. Using optical fluorescence microscopy, we have identified the lipid droplets in digesta removed from the fundus, corpus, and duodenum of rats fed on a controlled diet containing 8% lipid. The droplets ranged in diameter from 0.2 to 5 μm. The hydrolysis of the triglycerides by pancreatic lipase occurs at the oil–water interface,[3] and so the physicochemical properties of the emulsion system are crucial to lipid digestion.

We present here the first results in a study to examine, *in vitro*, the effect of fibre on the size and stability of small intestine emulsions. The stabilizing properties of two surfactants present in the stomach and small intestine have been investigated in the presence of guar gum. The bile salt, sodium taurocholate (NaT), is an anionic surfactant of fixed molecular weight. The lecithin, L-α-phosphatidylcholine, is a mixture of homologous zwitterionic surfactants. With both surfactant systems guar gum was found to induce coalescence, but the two systems differed in their response to changes in guar concentration.

2 Methods

For each emulsifier a concentrated pre-mix emulsion was prepared of 8 wt% olive oil (highly refined, low acidity oil supplied by Sigma, Poole, UK). The emulsifiers used, 99% L-α-phosphatidylcholine (type XVI) derived from fresh egg yolk and 98% sodium taurocholate (molecular weight 537.7 daltons) were both supplied by Sigma. Emulsification was achieved with a Waring blender using a fixed shear cycle. The lecithin was hydrated for 2 hours at 20 °C before emulsion preparation. The premix was diluted with a continuous phase of guar gum to obtain a series of 2 wt% oil emulsions containing either 6 mM NaT or 0.5 wt% lecithin in an aqueous solution of 0.15 M sodium chloride (AnalaR grade BDH, Poole, UK) and guar gum (Sigma, Poole, UK) at 0.05–0.2 wt%. Prior to use the supernatant of a spun guar gum solution had been freeze dried and dispersed in salt solution by Ultra Turrax for 6 minutes. A constant viscosity was achieved after hydration, with stirring, for 60 hours. The emulsions were kept continually agitated for the next 100 hours on an orbital shaker at 150 r.p.m. At appropriate time intervals, samples were drawn from the centre of the emulsion for size analysis by a Malvern Mastersizer, and oil content determination by density measurement using a Paar DMA602 density meter. The continuous phase viscosity of all emulsions was measured using a Bohlin Rheologi Controlled Stress rheometer at low shear-rate at 20 °C using a double-gap measuring system. Newtonian behaviour was observed in the shear stress range from 4.4×10^{-3} to 1.2×10^{-1} Pa. Although no preservative was added to the emulsions, no significant change in continuous phase viscosity was measured over the timescale of the experiment.

3 Results

All emulsions prepared with each surfactant originated from the same concentrated pre-mix; thus they had the same droplet-size distribution. For the NaT emulsions, size analysis of the original pre-mix gave a mean diameter $d_{43} = 4.3$ μm, whereas with lecithin, under the same shear cycle, a higher value of 10.7 μm was found. In the absence of guar gum the emulsions prepared with either NaT or lecithin were stable to coalescence for over 100 hours.

In the presence of guar gum coalescence was observed in both systems resulting in a dramatic loss of emulsified oil and the formation of a visible oil phase. However, on varying the concentration of guar gum the behaviour of the two surfactant systems differs significantly. Figures 1 and 2 show the percentage by weight of emulsified oil droplets below 10 μm diameter against time for the two systems. For NaT the rate of coalescence is greatest at 0.05 wt%, and then it decreases with increasing guar concentration. Over this range of guar concentrations the continuous phase viscosity displays Newtonian behaviour increasing from 1.1 mPa s without

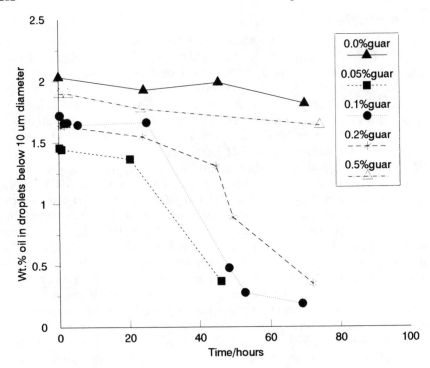

Figure 1 *Weight fraction of oil in droplets below* 10 μm *diameter in sodium taurocholate stabilized emulsions with varying concentrations of guar gum*

guar to 5.7 mPa s at 0.2 wt % guar. Thus viscosity considerations cannot account wholly for the destabilization of the emulsions, although they may contribute to the observed trend. Furthermore, for all NaT emulsions, a lag time of approximately 20 hours was found prior to the onset of measurable coalescence.

With the lecithin stabilized emulsions, coalescence was again observed in the presence of guar gum but no dependence of the observed rate on gum concentration was found. Similarly, there was no lag time before measurable coalescence. Further work with a range of surfactant concentrations will be necessary to judge the generality of this behaviour.

4 Discussion

Guar gum has been shown to destabilize oil-in-water emulsions prepared with physiological concentrations of a bile salt, sodium taurocholate, and a lecithin, L-α-phosphatidylcholine. To determine the possible mechanism of the observed instability will require further work in order to ascertain whether the concentration of emulsifier affects the rate of destabilization of the emulsions. In addition, complementary studies on the measurement of

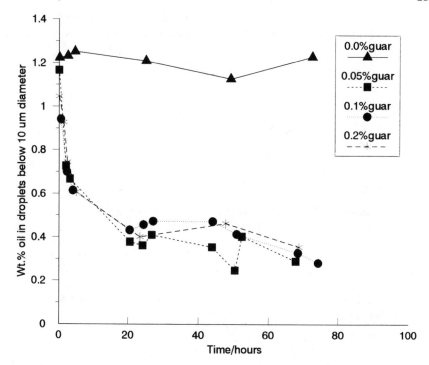

Figure 2 *Weight fraction of oil in droplets below* 10 μm *diameter in lecithin stabilized emulsions with varying concentrations of guar gum*

the surface tension in these systems and the micro-electrophoretic mobility of the droplets is underway.

Direct adsorption of the guar gum onto the surface of the droplets seems unlikely;[4] however, the physiological effects of NSP have been attributed to the binding of surfactants by the polysaccharides.[5] This would effectively reduce their concentration in the continuous phase thereby destabilizing the emulsion. It is perhaps relevant to this mechanism that we have found the degree of hydration of the guar to be critical to the rate of destabilization. After only 2 hours hydration, as compared to the usual 60 hours, 0.1 wt% guar gum completely destabilized a 6 mM NaT emulsion within 60 minutes.

5 Conclusion

This work is part of a study to identify the key steps in the digestion of lipid in the small intestine, to determine the processes that limit the rate of absorption, and to establish how they are modified in the presence of dietary fibre. The multidisciplinary nature of the approach involves *in vivo* studies of lipid metabolism, detailed chemical analysis of dietary sources of soluble NSP, and parallel studies on model emulsion systems. The results reported here are a first step in a project to relate model emulsion studies to the colloidal systems found in the stomach and small intestine during lipid adsorption.

Acknowledgement

This work is funded by the Ministry of Agriculture, Fisheries, and Food.

References

1. S. E. Higham and N. W. Read, *Brit. J. Nutr.*, 1992, **67**, 115.
2. E. Ebihara and B. O. Schneeman, *J. Nutr.*, 1989, **119**, 1100.
3. M. C. Carey, *Annu. Rev. Physiol.*, 1983, **45**, 651.
4. B. Bergenståhl, in 'Gums and Stabilisers for the Food Industry', ed. G. O. Phillips, P. A. Williams, and D. J. Wedlock, IRL Press, Oxford, 1988, Vol. 4, p. 363.
5. J. W. Anderson, D. A. Deakins, T. L. Floore, B. M. Smith, and S. E. Whitis, *Crit. Rev. Food Sci. Nutr.*, 1990, **29**, 95.

Determination of Particle Size Distribution in Tomato Concentrate

By F. W. C. den Ouden and T. van Vliet

DEPARTMENT OF FOOD SCIENCE, WAGENINGEN AGRICULTURAL
UNIVERSITY, PO BOX 8129, 6700 EV WAGENINGEN, THE NETHERLANDS

1 Introduction

The mean particle size and the particle size distribution of tomato concentrate directly affect its rheological properties. Tomato concentrate is made by comminution of the whole fruit except for the skins and the seeds, the obtained juice being subsequently evaporated to a concentrate. Although most of the skin material and seeds are removed during the production process, it is still possible that some residual skin and seed material is present also in the concentrate. The particles in tomato concentrate mainly consist of cells or cell fragments together with some cell aggregates. The cells of tomato concentrate arise from different kinds of tissue:[1] parenchyma (these cells have round anisometric shapes); vascular tissue (xylem and phloem); epidermis (skin tissue), which contains flat, rather small, tightly interlocked cells covered with wax; and seeds.

Several techniques can be used for the determination of the particle-size distribution of tomato concentrate. Kimball[2] and Tanglertpaibul[3] have used the wet-sieving technique for the determination of the size distribution of the particles in tomato juice. In this technique the product is separated into fractions having different particle sizes by allowing it to fall through a series of sieves of decreasing mesh-size, while rinsing with an aqueous phase. Both investigators[2,3] assumed that the particles that passed through one sieve, but not through the next one, had an average particle diameter approximately half way between the sizes of the openings of the two sieves. For the particles which stayed on the sieve with the largest openings, the average effective particle diameter was assumed to be 50% over the size of the openings. In the present study, the wet-sieving technique is compared with two others: (a) microscopic analysis by phase contrast microscopy; (b) a method based on analysis of the forward scattering pattern of laser light by the particles ('laser diffraction'), in which the Fraunhofer diffraction patterns produced are converted into particle-size distributions.

2 Materials and Methods

Wet-sieving

Wet-sieving was carried out with a Fritsch sieve shaker fitted with a set of NEN 2560 sieves with pore sizes of 425, 315, 250, 180, 125, 90, and 45 μm. A 100 g sample of Hot Break tomato concentrate was diluted with water to 1 kg and then poured onto the top sieve. Subsequently the sieves were shaken for 45 min while rinsing with water (20 °C; 150 ml min^{-1}). The retained material on each sieve was determined by dry-weight analysis (16–20 h at 90 °C) and expressed as a percentage of the total amount of water-insoluble solids (WIS).

Light Microscopy

Phase contrast microscopy was applied with a Zeiss–Axiomat microscope having a magnification of *ca.* 160 times. Approximately 200 particles (cells or cell aggregates) per fraction were counted and allocated to particle diameter classes of 100 μm width.

Laser Diffraction

For this purpose a Coulter LS 130 system was used. In this instrument, laser light (wavelength 720 nm) is scattered by the particles, and the generated diffraction pattern which is a composite of the individual diffraction patterns of all the particles, is measured. The instrument converts this composite diffraction pattern to a particle size distribution with 72 classes ranging from 0.4 to 900 μm.

3 Results

A sample of Hot Break tomato concentrate was wet-sieved into 7 fractions, using sieves with pore sizes ranging from 45 to 425 μm. Afterwards, the 0–45 μm fraction was passed over a 32 μm sieve. Figure 1 shows a typical particle size distribution of the industrially prepared Hot Break tomato paste. Subsequently, the five fractions of cells were investigated by phase contrast microscopy and by laser diffraction. The results are shown in Figures 2 and 3. From the microscopic observations, it was clear that particles which had passed through the small sieve pores (45, 90, 125, 180 μm) mostly consisted of separated round-shaped anisometric cells from the parenchyma, whereas particles that had not passed through these sieves consisted mainly of aggregates of cells from skin, seeds, or vascular bundles.

% of WIS

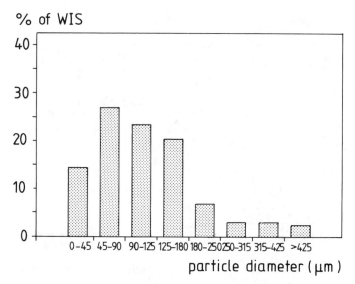

Figure 1 *Particle size distribution as obtained by wet-sieving of a Hot Break tomato paste*

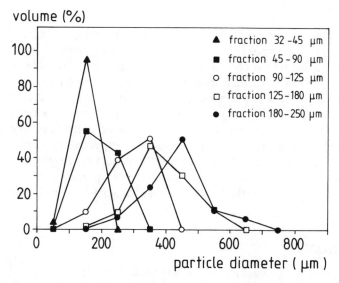

Figure 2 *Approximate volume–frequency distributions of different wet-sieve fractions analysed by microscopy*

4 Discussion and Conclusion

The results obtained by microscopy and laser diffraction, although not in exact agreement, both show that the size of many particles is significantly larger (up to 2–3 times) than the notional average diameter of the pores

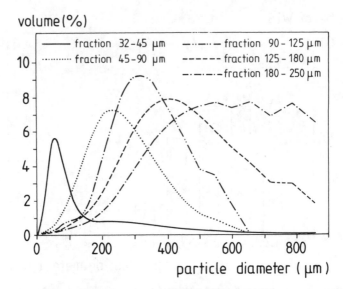

Figure 3 *Approximate volume–frequency distributions of different wet-sieve fractions analysed by laser diffraction*

through which they have passed during wet-sieving. This result is at variance with the suppositions of previous investigators,[2,3] who assumed that the particles which passed through one sieve but not through the next one had an average particle diameter half way between the diameters of the openings of the two sieves. An explanation for the fact that many particles are significantly larger than the openings through which they have passed may be the deformability of the tomato particles. The physical characteristics of tomato particles would be expected to depend on the structure of the original cells and cell walls. A tomato parenchyma cell has a flexible thin wall. During wet-sieving, the particles are apparently deformed in such a way that they can pass through the much smaller sieve pores.

References

1. R. Ilker and A. S. Szczesniak, *J. Texture Stud.*, 1990, **21**, 1.
2. L. B. Kimball and Z. I. Kertesz, *Food Technol.*, 1952, **6**, 68.
3. T. Tanglertpaibul and M. A. Rao, *J. Food Sci.*, 1987, **52**, 141.

Interfacial Phenomena

Dynamic Surface Properties in Relation to Dispersion Stability

By A. Prins and D. J. M. Bergink-Martens

DEPARTMENT OF FOOD SCIENCE, WAGENINGEN AGRICULTURAL UNIVERSITY, PO BOX 8129, 6700 EV WAGENINGEN, THE NETHERLANDS

1 Introduction

Dynamic surface properties come into play in all cases where for some reason a quiescent dispersion is locally not in complete thermodynamic equilibrium, or, where as a result of disturbances applied from outside, the whole system is not in complete equilibrium. This applies to all kinds of dispersions: gas-in-liquid, liquid-in-gas, solid-in-liquid, solid-in-gas, and liquid-in-liquid.

The present paper confines itself to dynamic surface properties present at liquid surfaces in foams, emulsions, or aerosols. This is because the main cause of the dynamic surface properties considered here is connected with the fact that the interface between two fluid phases can move: that is, the interface can locally be expanded or compressed. In the presence of surface-active material, the direct consequence of this movement is that the surface tension can be different at different places. The resulting surface tension gradient is indissolubly connected to velocity gradients exerted in the neighbouring liquid phases because of the demands of mechanical equilibrium. Depending on the conditions, surface tension gradients can be caused by liquid flowing along the interface, or *vice versa* liquid flow can be generated by surface tension gradients.

The stability of foams and emulsions is to a large extent governed by the stability of the thin films between the bubbles or the droplets. Surface-active agents can stabilize such films by adsorption at the interface and by lowering the surface tension. This means that, by stretching the film, the film becomes more unstable because the adsorbed amount goes down and the surface tension goes up. This is the reason why only dilational surface behaviour in relation to film stability will be considered here.

2 Surface Dilational Behaviour

From a physical point of view there are three ways in which a flat liquid surface can be subjected to a dilational disturbance.

(i) *By Means of Moving Barriers in a Langmuir Trough* This area-driven surface disturbance is analogous to what happens when for instance a bubble is blown up.

(ii) *By Means of Liquid Flow along the Surface* This hydrodynamically driven surface disturbance takes place, for instance, in a vertical soap film where, due to the downward liquid flow, the film surface at the top is expanded and that at the bottom is compressed. The generated surface tension gradient can keep up the weight of the film.

(iii) *By Means of a Surface Tension Gradient* The spreading of an oil droplet over the surface of an aqueous solution is an example of a disturbance of a liquid surface driven by the surface tension gradient.

With Newtonian liquids under laminar flow conditions, the hydrodynamic coupling between surface and liquid motion, whether the surface drives the liquid or the liquid drives the surface, is governed by the equation

$$\frac{d\gamma}{dx} = \eta_b \left(\frac{dV_x}{dz}\right)_{z=0} \tag{1}$$

where γ is the surface tension, x is the direction of the liquid motion parallel to the surface, η_b is the bulk viscosity, V_x is the velocity in the x direction, and z is the co-ordinate direction pointing into the liquid perpendicular to the surface.

So far, just flat surfaces have been considered. For curved surfaces, however, dilational disturbances can originate from differences in curvature giving rise to differences in Laplace pressure. These pressure differences induce liquid flow and hence surface tension gradients. An example is a thin liquid film in contact with a Plateau border. Because of the concave surface of the Plateau border, the pressure in the Plateau border is lower than in the film liquid. This results in a difference in surface tension, $\Delta\gamma$, between that for the film surface, γ_f, and that for the Plateau border surface, γ_b, given by

$$\Delta\gamma = \gamma_f - \gamma_b = \Delta\rho \cdot \frac{h}{2} = \rho g H \cdot \frac{h}{2} = \frac{\gamma_b}{R} \cdot \frac{h}{2}, \tag{2}$$

where h is the film thickness, ρ is the density of the liquid, g is the acceleration due to gravity, H is the hydrostatic head of the Plateau border liquid above the zero level, and R is the radius of curvature of the concave Plateau border surface.

For a film thickness of $10\,\mu m$ in contact with a Plateau border at $H = 10$ cm composed of an aqueous surfactant solution of density $\rho = 10^3$ kg m^{-3}, we have $\Delta\gamma = 5$ mN m^{-1}, which is an appreciable increase in surface tension of the film compared to the Plateau border surface tension.

3 Experimental Techniques

There are several experimental techniques available for measuring dilational surface behaviour.

In a Langmuir trough an area-driven surface deformation resulting in a change in surface tension can be realized by moving barriers in the proper way. These measurements can be carried out close to equilibrium in a so-called dynamic experiment by moving the barriers sinusoidally.[1] The resulting infinitesimal change in surface tension dγ is related to the surface dilational modulus E by

$$E = \frac{d\gamma}{d\ln A} \qquad (3)$$

where d$\ln A$ is the relative change in surface area A. In another type of experiment,[2] far from equilibrium, the barriers are moved in a logarithmic way in order to realize a constant rate of deformation d$\ln A/dt$. This, in the steady state, results in a constant change in surface tension, $\Delta\gamma$, which equals the difference between the equilibrium surface tension γ_e and the dynamic surface tension γ_d measured in the steady state, *i.e.*

$$\Delta\gamma = \gamma_d - \gamma_e \qquad (4)$$

This provides a way of measuring the surface dilational viscosity η_s^d defined by

$$\eta_s^d = \frac{\Delta\gamma}{d\ln A/dt} \qquad (5)$$

In another kind of measurement, the deformation of the surface is not applied by moving barriers but is caused by the motion of the liquid in the close neighbourhood of the surface. An example of an experimental technique which makes use of such a hydrodynamically driven surface deformation is the overflowing cylinder technique.[3] In this technique the liquid under investigation is allowed to flow over the top rim of a vertically mounted cylinder, and a continuously radially expanding surface is created at the top of the cylinder. The surface expansion rate follows from the radial velocity distribution of the surface according to

$$\frac{d\ln A}{dt} = \frac{\partial V_r}{\partial r} + \frac{V_r}{r}, \qquad (6)$$

where V_r is the radial surface velocity measured at a distance r from the centre of the circular overflowing surface. From measurements carried out with water and aqueous surfactant solutions, it has been found that the surface velocity increases linearly with the distance to the centre in a considerable part of the surface (see Figure 1). Under these conditions, we have

$$\frac{\mathrm{d}\ln A}{\mathrm{d}t} = 2\cdot\frac{V_r}{r} \qquad (7)$$

Another important observation is that, by introducing a surfactant in the aqueous system, the expansion rate of the surface increases by a factor of 10. The explanation is that the increased stiffness of the surface caused by introducing the surfactant makes it easier to pull on the surface. This pull is exerted on the horizontal surface at the top of the cylinder by means of the vertical film at the outside of the cylinder which falls down in the gravity field. The weight of this film creates a surface tension gradient which pulls the horizontal surface of the liquid over the top rim of the cylinder.

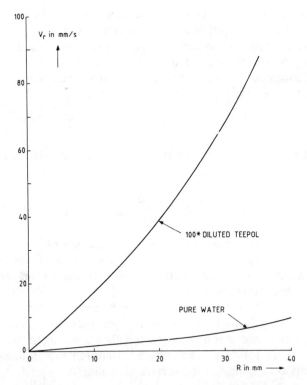

Figure 1 *Radial surface velocity* V_r *as a function of the distance* r *to the centre of the top surface of the overflowing cylinder for water and a hundred-fold diluted aqueous Teepol solution* $(Q = 7.6 \, \mathrm{cm}^3 \, \mathrm{s}^{-1})$

The dynamic surface tension measured at the expanding liquid surface at the top of the cylinder, combined with the measurement of the radial velocity of that surface, gives the value of the surface dilational viscosity η_s^d as defined by equation (5). It is worthwhile to note here that the dynamic surface tension measured in this way has a value which is close to that found using an adapted Wilhelmy plate technique when the surface tension is measured in a *free* falling film made out of the same surfactant solution.[4] One may speculate that this explains the observation that there is a close relationship between the dependency of the dynamic surface tension on the surfactant concentration and the foam stability of these systems.

As expected, the expansion rate of the liquid surface in the overflowing cylinder does not depend only on the surfactant concentration but also on the pumping rate and the height of the film falling downwards at the outside of the cylinder. For both effects, however, the same tendency has been found. For small values of the parameters—pumping rate and film height—the expansion rate of the surface increases with increasing values of the parameters. However, beyond a certain value of the parameters the expansion rate reaches a plateau; for the film height the threshold value is about 3 cm, and for the pumping rate a threshold value of *ca.* 25 cm^3 s^{-1} is found for a cylinder of 4 cm radius. From this it is concluded that the greatest part of the dynamic behaviour of the expanding surface created with the overflowing cylinder technique is governed by the properties of the surfactant solution and not by the operation of the apparatus. In other words the solution itself determines autonomously what is the dynamic surface behaviour, and *not* the applied mechanical agitation.

It is obvious that the conclusions drawn above are restricted to systems in which the surface deformation is driven by hydrodynamic forces. They are not applicable to systems in which the surface is deformed by means of an area-driven disturbance (*e.g.* in a Langmuir trough).

4 Application of Dynamic Surface Properties to Foaming Behaviour

As an example of how dynamic dilational surface properties are related to the practical behaviour of foams and emulsions, some general principles of foaming will be considered here in more detail. (Similar mechanisms may play a role in emulsion behaviour, and we hope to report on that subject in the future as we have found that the overflowing cylinder technique can also be used in oil–water systems simply by replacing the air by oil. This provides the opportunity for studying oil–water interfaces far from equilibrium when they are in a state of continuous expansion.)

In the overview given in Table 1, foaming behaviour is split up into processes which take place during foam making, foam handling, and foam keeping. This is in line with the common practical situation: first the foam is made by applying a severe mechanical treatment, then it is transported,

Table 1

	Process mechanism	Driven By	Expansion and/or Compression	Close/Far from Equilibrium
Making	blowing bubbles	area	expansion	far
	bubble growth	area	expansion	far
	whipping	area viscous curvature	expansion compression	far
Handling	bubble growth	area	expansion	far
	rising bubble	viscous (curvature)	expansion compression	far
	bouncing bubbles	viscous	expansion	far
	pulsating bubble	area	expansion compression	around
	bubble disruption	viscous	expansion compression	far
	expanding bubble	area	expansion	far

Table 1 *Continued*

	Process mechanism	Driven By	Expansion and/or Compression	Close/Far from Equilibrium
	film drainage Marg. Reg.	viscous curvature	expansion compression	close
	border suction	curvature	expansion	far
Foam Keeping	dispropor-tionation	area	compression	far
	rearrange-ment	area	expansion	far
Film breaking by particles				
	hydrophobic	curvature	expansion	far
	spreading	surface tension gradient viscous	expansion	far

and finally it is kept under quiescent conditions. In the table it is indicated what is the main cause of the dilational disturbance: is it area-driven, viscosity-driven, or curvature-driven? In addition, we indicate whether during the process the surface is mainly in the expanded or the compressed state. Finally, we indicate whether the applied disturbance of the surface is far from or close to equilibrium.

It is apparent from Table 1, therefore, that most disturbances of the surface take place far from equilibrium. Expansions of surfaces and films are known to promote the breaking of the films in a foam especially when the expansion is applied far from equilibrium. Because the overflowing cylinder technique is an attempt to imitate the behaviour of a surfactant solution when its surface is subjected to a continuous expansion far from equilibrium under rather well defined physical conditions, its mode of operation will be considered now more in detail.

5 The Overflowing Cylinder Technique

The surface phenomena taking place at the expanding surface at the top of the cylinder are mainly the result of two opposing processes. First, there is a constant discharge of surface-active material by the radial movement of the surface over the top rim of the cylinder. For the sake of simplicity, we can assume that the relative expansion rate $\mathrm{d}\ln A/\mathrm{d}t$ is uniform over the whole surface and that the surfactant adsorbed amount Γ is also uniform over the surface. Under these conditions the amount of surfactant discharged per unit area and per unit time is given by $\Gamma \mathrm{d}\ln A/\mathrm{d}t$.

Second there is a constant supply of surface active material from the solution to the expanding surface by means of convection and diffusion in an attempt by the system to restore the equilibrium situation.

Under the steady state conditions by which the overflowing cylinder technique is performed, both processes keep each other in equilibrium: the amount transported to the surface equals the amount that flows away over the rim of the cylinder. Again for the sake of simplicity, it will be assumed that the amount of surfactant transported to the expanding surface can be calculated by applying simple penetration theory, which means that it is assumed that the transport to the surface is governed by diffusion only.

The penetration theory describes the relation between the distance l over which a substance is transported by means of diffusion in a time t when the diffusion coefficient is given by D:

$$l^2 = \pi D t \tag{8}$$

This theory can be applied to the expanding surface by taking into account that the distance l over which a surfactant, having a concentration c in the solution, has to travel, in order to realize an adsorption Γ at the surface, is given by

$$l = \frac{\Gamma}{c} \tag{9}$$

By combining equations (8) and (9) the result is:

$$\frac{\Gamma^2}{c^2} = \pi D t \tag{10}$$

or

$$\frac{\Gamma}{t} = \frac{c^2 \pi D}{\Gamma} \tag{11}$$

The value of Γ/t is taken as an order of magnitude estimate of the potential transport rate of surfactant to the expanding surface. Equation

(11) shows, as expected, that the transport rate becomes smaller when Γ increases because when Γ is larger the surfactant has to travel a longer distance. Even more important is that the transport rate increases in proportion to the square of the concentration.

Analogous to the dimensionless Fourier number used in describing mass transfer phenomena, which is the ratio of diffusion rate to accumulation rate, we define a dimensionless surface Fourier number Fo_s which is the ratio of the supply rate of surfactant to the surface to the discharge of surfactant at the expanding surface:

$$Fo_s = \frac{supply\ rate}{discharge\ rate} \tag{12}$$

Substituting for the potential supply rate and the discharge rate, the surface Fourier number can be written as

$$Fo_s = c^2 \frac{\pi D}{\Gamma^2 \mathrm{dln}A/\mathrm{d}t} \tag{13}$$

A high value of Fo_s means that the supply rate of surfactant to the expanding surface is larger than the discharge rate, which will result in a relatively low value of the surface tension, and consequently good foaming behaviour of such a system is to be expected. When shaking a cylinder partly filled with an aqueous surfactant solution, the amount of foam produced will benefit from a low surface tension of the expanding surface.

6 Application to the Foamability of a Practical System

For aqueous Teepol solutions the equilibrium and dynamic surface tension as measured by means of the overflowing cylinder technique are given as a function of the surfactant concentration in Figure 1 of ref. 5. Also indicated there is how the foamability of these solutions depends on the Teepol concentration. The foamability is taken as the amount of foam formed in the steady state by shaking a graduated cylinder partly filled with the aqueous solution. The increase in foamability seems to correlate with the decrease in the dynamic surface tension and not with the equilibrium surface tension. This indicates the importance of a low surface tension of an expanding surface for a good foamability.

In Figure 2, the foamability is plotted as a function of the surface Fourier number calculated by means of equation (13) in which the following values for the various parameters have been used: $D = 10^{-9}\ \mathrm{m^2\,s^{-1}}$, $\Gamma = 9.3 \times 10^{-6}\ \mathrm{mol\,m^{-2}}$, $\mathrm{dln}A/\mathrm{d}t = 10\ \mathrm{s^{-1}}$. Further it is assumed that the stock Teepol solution contains 15 wt % surface-active material and that the molecular weight of the surfactant is 300 $\mathrm{g\,mol^{-1}}$.

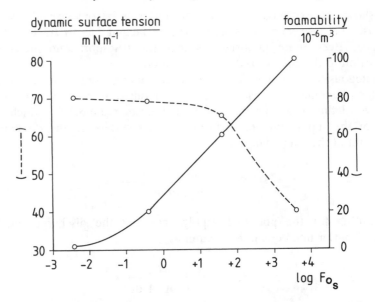

Figure 2 *Dynamic surface tension and foamability* versus *logarithm of surface Fourier number* Fo_s

From Figure 2 it appears that, along the curve that indicates the increase in foamability when the surfactant concentration is increased, the value of Fo_s increases from a value as low as 10^{-3} to a value as high as 10^4. This observation, and the fact that the dynamic surface tension of the expanding surface is related to the value of Fo_s, are taken as arguments in favour of the hypothesis that the foamability of a surfactant solution is governed by the ability of the surfactant solution to adsorb at an expanding surface in such an amount that the surface tension reduces to a low value.

The surface Fourier number, being a measure of the transport efficiency of a surfactant to an expanding surface, seems to be an important parameter in relating the dynamic surface tension of an expanding surface to the foamability of a surfactant solution.

References

1. J. Lucassen and M. van den Tempel, *Chem. Eng. Sci.*, 1972, **27**, 1283; J. Lucassen and M. van den Tempel, *J. Colloid Interface Sci.*, 1972, **41**, 491.
2. F. Voorst Vader, Th.F. van Erkens, and M. van den Tempel, *Trans. Faraday Soc.*, 1964, **60**, 1170.
3. D. J. M. Bergink-Martens, H. J. Bos, A. Prins, and B. C. Schulte, *J. Colloid Interface Sci.*, 1990, **138**, 1.
4. D. J. M. Bergink-Martens, C. G. J. Bisperink, H. J. Bos, A. Prins, and A. F. Zuidberg, *Colloids Surf.*, 1992, **65**, 191.
5. D. J. M. Bergink-Martens, H. J. Bos, H. K. A. I. van Kalsbeek, and A. Prins, this volume, p. 376.

Molecular Dynamics Simulation of Competitive Adsorption at a Planar Fluid Interface

By Edward George Pelan and Eric Dickinson[1]

UNILEVER RESEARCH, COLWORTH HOUSE, SHARNBROOK, BEDFORD MK44 1LQ, UK
[1]PROCTER DEPARTMENT OF FOOD SCIENCE, UNIVERSITY OF LEEDS, LEEDS LS2 9JT, UK

1 Introduction

An understanding of the structure and dynamics of adsorbed layers is essential for the prediction of the properties of food colloids stabilized by surface-active particles and macromolecules. In dairy-type emulsions, for instance, the character of adsorbed layers of protein particles (casein micelles) or globular proteins (*e.g.* β-lactoglobulin) has an important effect on the formation and stability behaviour.[1-4] The interfacial composition in these complex colloidal systems is determined by competitive adsorption between protein adsorbates of different surface activity[5,6] and between protein and small-molecule surfactants.[7,8]

Computer simulation of competitive adsorption provides a precise way of understanding how the properties of an adsorbed layer are determined by the nature and interactions of the surface-active species. Recent research at Leeds has used Monte Carlo simulation to study the equilibrium aspects of competitive adsorption between interacting macromolecules and surfactants[9,10] and between compact deformable particles (globular proteins) and so-called 'fractal polymers' (flexible proteins or polysaccharides).[11] However, in order to study the mobility or lifetimes of adsorbed structures, it is necessary to use a dynamical technique such as molecular dynamics or Brownian dynamics.[12] Recently, we presented[13] a molecular dynamics simulation of an assembly of Lennard–Jones (LJ) particles under the influence of an externally applied local field which was intended to represent the location of a plane fluid interface at which the particles could adsorb. Though developed originally to simulate atomic and molecular liquids, the molecular dynamics (MD) technique has been used successfully in connection with colloidal systems to model the dynamics of collective particle motion in shear flow[14,15] and the effect of polydispersity on the order–disorder transition.[16] The present report extends the previous work[13] to a binary mixture of LJ particles of equal size. Attention here is directed toward how the properties of the adsorbed layer are affected by (i) differences in the LJ unlike particle interaction energy parameter and (ii) differences in the effective adsorption energies for the two kinds of particle.

2 Molecular Dynamics Model and Methodology

The full methodology has been given in our earlier paper,[13] and so only the necessary extensions for modelling the binary mixture are described here.

Pairs of identical particles of each fluid component interact via a standard LJ self-potential of the form

$$\phi(r) = 4\,\varepsilon\,[(\sigma/r)^{12} - (\sigma/r)^6] \tag{1}$$

where ε is the well depth, σ is the collision diameter, and r is the centre-to-centre separation of the particles. In addition to the two self-potentials, we need to introduce a mutual-potential, again of the LJ form, which describes the interaction between pairs of particles from different components, *i.e.*

$$\phi_m(r) = 4\,\varepsilon\xi[(\sigma/r)^{12} - (\sigma/r)^6] \tag{2}$$

where the m-subscript denotes the mutual interaction of well depth $\varepsilon\xi$, and the factor ξ is adjusted to make the mutual attractive interaction stronger ($\xi > 1$) or weaker ($\xi < 1$) relative to the self-potential. As is usual with MD simulations, measured quantities are expressed in terms of LJ reduced units (denoted by a superscript asterisk): σ for distance, ε for energy, ε/k for temperature, $\sigma(m/\varepsilon)^{1/2}$ for time, and $N\sigma^3/V$ for density, where k is Boltzmann's constant, m the particle mass, N the number of particles in the simulation cell, and V is the cell volume.

An adsorbed layer is created in the middle of the cell localized at $z = 0$ by subjecting each component to a symmetrical external potential field perpendicular to the z direction. The external potential is of the Gaussian form,

$$v_{e1}(r_{z1}) = -d_{e1}\exp{(-r_{z1}^2/2w_1^2)} \tag{3}$$

where r_{z1} is the value of the z coordinate of a particle of component 1, d_{e1} is the well depth for type 1 particles, and w_1 is the Gaussian half-width for particles of component 1; a similar expression is used to describe the interaction of particles of component 2. These attractive potentials serve to trap particles in a narrow band in the centre of the simulation cell parallel to the xy-plane. By carefully choosing appropriate values of d_e and w for both components, the density of trapped particles and the physical thickness and composition of the adsorbed layer can be selectively controlled.

Particles were moved using a modified Verlet 'leapfrog' algorithm[17] in which the accelerations were directly accumulated onto the velocities in the force loop in order to save storage at the expense of averaging the previous and current velocities to obtain the kinetic energy. Simulations were performed on a Convex C220 machine. Each run is started from a face-centred cube (FCC) lattice with a lattice spacing chosen to give the required liquid state density after 'melting'. The initial velocities are generated from a random Gaussian distribution weighted to the required temperature. With no external field applied ($d_{e1} = d_{e2} = 0$), each simulation run is first allowed to 'melt' from the initial FCC configuration for 10^3 timesteps. Then, with the selected external field conditions switched on ($d_{e1}, d_{e2} \neq 0$), the simulation is run for a further 10^3 timesteps to bring the notional interface into dynamic equilibrium with the bulk. Following equilibration, each simulation is run for a period of 2×10^4 timesteps during which time the static and dynamic properties are computed. From the preliminary work, it was found[13] that the optimum value for the half-width w to produce a monolayer with appreciable opportunity for particle motion perpendicular to the interface is $w = 0.25$. This value was adopted in all subsequent simulations reported here.

Adsorbed layer structure is expressed in terms of a 'two-dimensional' pair distribution function $g^{xy}(r)$ corresponding to the positional correlations of particle centres projected on the xy-plane at $z = 0$. We here require three such functions to completely describe the mixed monolayer structure, two self distribution functions to account for the separate contributions of each of the components (denoted by $g_{11}(r)$ and $g_{22}(r)$ for components 1 and 2, respectively), and a mutual distribution function $g_{12}(r)$ to describe the distribution of component 1 particles with respect to component 2 particles. For the purposes of calculating these three 'two-dimensional' distribution functions, the adsorbed layer is assumed to have a thickness of σ (i.e. all particles lying between $z = -\sigma/2$ and $z = +\sigma/2$ are included in the statistical averaging). The distribution functions are normalized such that $g_{11}(r) = g_{22}(r) = g_{12}(r) = 1$ as $r \to \infty$ in a fluid-like adsorbed layer. The extent of adsorption is expressed in terms of the particle number density $n(z)$ perpendicular to the adsorbed layer, where $n(z)$ can be expressed as the sum of the numbers of particles of each of the two components, $n_1(z)$ and $n_2(z)$. The symmetry of the model means that, on average, $n(z)$ is equal to $n(-z)$, within the expected statistical fluctuations.

Dynamic information is obtained by monitoring each particle at every timestep using its own private clock. When a particle enters the adsorbed layer ($-\sigma/2 \leqslant z \leqslant \sigma/2$) for the first time, its position co-ordinates are stored and its clock started. As the particle moves within the layer, its projected position in the xy-plane is monitored and the clock updated at each timestep. When the particle eventually leaves the monolayer ($|z| > \sigma/2$), the clock is stopped and the particle lifetime in the adsorbed layer is recorded. A 'two-dimensional' translation diffusion coefficient D_{xy} is calculated from the mean-square displacement of projected positions of particles in the adsorbed layer:

$$4D_{xy}t = \langle |r_i(t) - r_i(0)|^2 \rangle \qquad (i = 1, N) \qquad (4)$$

The quantity $r_i(t)$ is the particle position in the xy-plane at time t, where $t = 0$ is the time when the particle enters the adsorbed layer. Strictly speaking, equation (4) is valid only in the limit $t \rightarrow \infty$ for Einstein-type diffusion, but since each particle has a finite lifetime in the layer, D_{xy} must in practice be calculated for a finite timescale. An upper time limit of $1.5t^*$ was chosen, where t^* is the time for which the probability of a particle remaining in the monolayer is $1/e$.

3 Results and Discussion

Numerical results have been obtained for binary equimolar LJ fluid systems of 500 particles at a density of $\rho^* = 0.5$ and a temperature of $T^* = 2.5$. That is, there are 250 particles of component 1 and 250 particles of component 2 with $\varepsilon_1 = \varepsilon_2$. Various values of the well depth parameter d_e of the external potential for each component have been studied. Results reported here are limited to d_e values ranging between $2\ kT$ and $8\ kT$ as a function of several values of the mutual potential parameter ξ which varies from 0.6 to 1.5. Competition between each of the components for the interface appears when unequal values of the well-depth parameter are used.

Structural information perpendicular to (not presented here) and in the plane of the adsorbed layer for a system with $d_{e1} = d_{e2} = 2\ kT$ was obtained. On examination of the data, we see that there is an associated monolayer located around $z = 0$. The density profile $n(z)$ quickly reduces to the bulk density on either side of the interface. At this low value of well-depth parameter, there is little evidence of any close-packed structure within the monolayer as the radial distribution functions (Figures 1–3) quickly decay to their bulk value of unity after a separation distance greater than two diameters (2σ), irrespective of the value of ξ. Figure 4 shows the density profile $n(z)$ for the system with $d_{e1} = d_{e2} = 6\ kT$ and $\xi = 1.0$. Under these conditions, a further layer of LJ particles can be seen condensing onto the monolayer in the form of a secondary or pre-wetting layer; indeed there is even evidence of a third layer of structure as more of the bulk phase wets the interface. This is shown in the pair distribution plot for this system, Figure 5, where the plot has a higher primary maximum (relative to its $2\ kT$ counterpart), and it also shows a well-defined secondary maximum, indicative of a nearest-neighbour shell (close-packing) around each particle in the interface. A variation in structure is

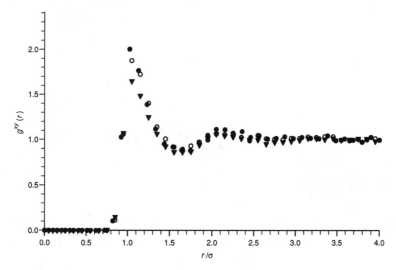

Figure 1 *Radial distribution function* $g^{xy}(r)$ *as a function of particle pair separation* r *for* $\rho^* = 0.5$, $T^* = 2.5$, $\xi = 0.8$, $d_{e1} = d_{e2} = 2\,kT$: $\bullet = g_{11}$; $\bigcirc = g_{22}$; $\blacktriangledown = g_{12}$

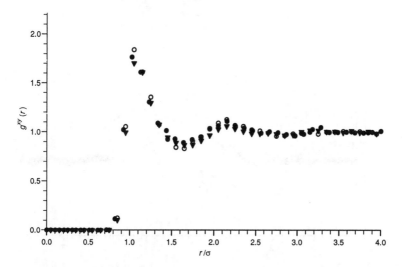

Figure 2 *Radial distribution function* $g^{xy}(r)$ *as a function of particle pair separation* r *for* $\rho^* = 0.5$, $T^* = 2.5$, $\xi = 1.0$, $d_{e1} = d_{e2} = 2\,kT$: $\bullet = g_{11}$; $\bigcirc = g_{22}$; $\blacktriangledown = g_{12}$

also apparent between different values of ξ for this system: the highest primary maximum occurs with the mutual pair distribution function $g_{12}(r)$ for $\xi = 1.2$. In this case there is a stronger net attractive potential between the unlike components compared with their self-interactions. Correspondingly, the lowest value of $g_{12}(r)$ corresponds to $\xi = 0.8$, which is a weaker effective unlike interaction relative to the respective self-interactions.

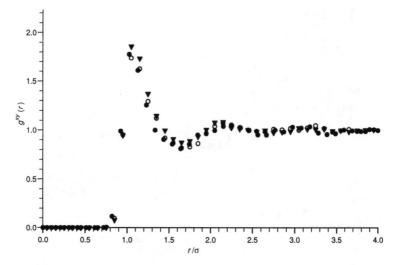

Figure 3 *Radial distribution function* $g^{xy}(r)$ *as a function of particle pair separation* r *for* $\rho^* = 0.5$, $T^* = 2.5$, $\xi = 1.2$, $d_{e1} = d_{e2} = 2$ kT: $\bullet = g_{11}$; $\bigcirc = g_{22}$; $\blacktriangledown = g_{12}$

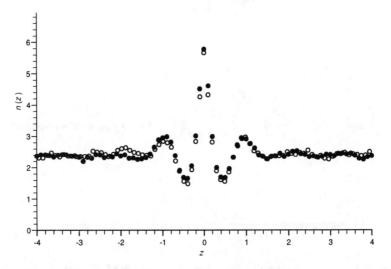

Figure 4 *Particle number density* $n(z)$ *as a function of* z *for* $\rho^* = 0.5$, $T^* = 2.5$, $\xi = 1.0$, $d_{e1} = d_{e2} = 6$ kT: $\bullet = component\ 1$; $\bigcirc = component\ 2$

The number density profile in Figure 6 illustrates the competitive displacement from the interface of one component (type 1) by the other (type 2). This occurs when one component has a different well-depth parameter (or binding affinity for the interface), in this case $d_{e1} = 4\ kT$ and $d_{e2} = 8\ kT$. The more weakly bound component (type 1) is strongly displaced from the interface, residing mainly in the bulk phase, where its

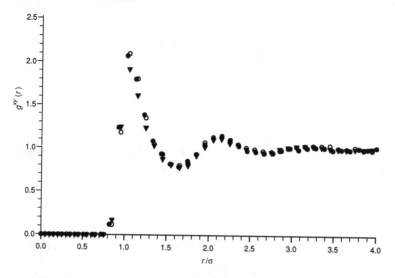

Figure 5 *Radial distribution function* $g^{xy}(r)$ *as a function of particle pair separation* r
for $\rho^* = 0.5$, $T^* = 2.5$, $\xi = 0.8$, $d_{e1} = d_{e2} = 6$ kT: $\bullet = g_{11}$; $\bigcirc = g_{22}$;
$\blacktriangledown = g_{12}$

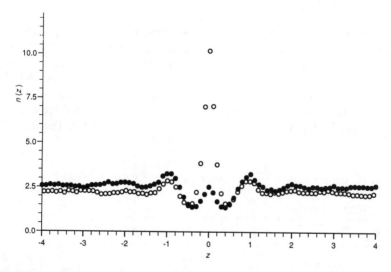

Figure 6 *Particle number density* n(z) *as a function of z for* $\rho^* = 0.5$, $T^* = 2.5$,
$\xi = 1.2$, $d_{e1} = 4$ kT, $d_{e2} = 8$ kT: $\bullet = component\ 1$; $\bigcirc = component\ 2$

density is marginally greater than that of the more strongly adsorbed
component. Again, the density profile is insensitive to the value of the
mutual interaction parameter ξ. The pair-distribution function for this
system at $\xi = 1.2$, Figure 7, shows that the interface under such conditions
is rather closed-packed. The effect of a net attractive mutual potential is
clearly seen since the g_{12} values are the highest on the plot, followed by

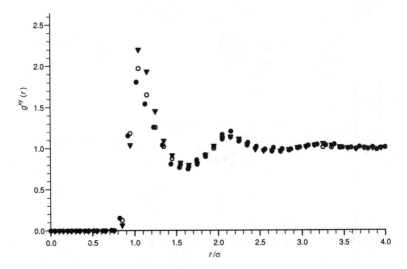

Figure 7 *Radial distribution function* $g^{xy}(r)$ *as a function of particle pair separation* r *for* $\rho^* = 0.5$, $T^* = 2.5$, $\xi = 1.2$, $d_{e1} = 4\,kT$, $d_{e2} = 8\,kT$: $\bullet = g_{11}$; $\bigcirc = g_{22}$; $\blacktriangledown = g_{12}$

those for the more strongly adsorbed species, *i.e.* g_{22} for type 2 at $8\,kT$, with the displaced component 1 showing the least structure at the interface.

Table 1 gives the mean lifetime t^* of the particles in the monolayer (measured in reduced units) as a function of various values of well-depth d_e for each component phase for $T^* = 2.5$ and a bulk density of $\rho^* = 0.5$. The mean lifetimes increase substantially as d_e is increased from $2\,kT$ to $6\,kT$, showing that the particles are more strongly held in the monolayer

Table 1 *Computed monolayer densities* ρ^* *and monolayer lifetimes* t^* *as a function of well-depth* d_e *for* $T^* = 2.5$ *and bulk density* $\rho^* = 0.5$ *for each component (types 1 and 2)*

d_{e1}/kT	d_{e2}/kT	ξ	ρ_1^{*a}	ρ_2^{*a}	t_1^{*b}	t_2^{*b}
2	2	0.8	0.16	0.16	0.96	0.96
2	2	1.0	0.16	0.16	0.98	0.96
2	2	1.2	0.16	0.16	0.95	0.95
6	6	0.8	0.26	0.25	2.09	2.09
6	6	1.0	0.23	0.24	1.54	1.54
6	6	1.2	0.22	0.24	1.49	1.49
4	8	0.8	0.11	0.38	0.97	2.32
4	8	1.0	0.11	0.39	0.86	2.47
4	8	1.2	0.12	0.37	0.98	2.15

[a] Calculated assuming a monolayer thickness of unity in reduced units.
[b] The monolayer lifetime t^* is taken as the time for which the probability distribution $p(t)$ of individual particle lifetimes in the monolayer drops to $1/e$ of the value $p(t = 0)$.

by the higher field strength. The value of the mutual interaction parameter ξ also has an effect on the lifetime of the particles in the monolayer, but in a subtle way, as it reflects the distribution of each species at the interface as evidenced by the phase-separation of the components in Figure 8. This segregation of species for $\xi < 1$, or, conversely, mixing of the species for $\xi > 1$, will thus result in random, non-systematic configurational effects on the lifetimes of the components at the interface. Diffusion coefficients calculated in the monolayer are subject to significant statistical uncertainty. This was also found[13] to be the case with the single component LJ fluid, where diffusion coefficients could only be determined to an accuracy of 15%. In the current work on binary fluids, the problem is even more accentuated, since we now have only half the statistical population in each component compared to the single fluid; consequently, no numerical values of diffusion coefficients are presented here. It is noted, however, that for the competitive adsorption systems (i.e. $d_{e1} \neq d_{e2}$), the values of diffusion coefficients obtained differ from those obtained when the equivalent single component system is simulated at $d_e = d_{e1}$ and $d_e = d_{e2}$ for the same parameters. For example, results obtained for $d_{e1} = 4\ kT$, $d_{e2} = 8\ kT$ differ from the values obtained when the single component fluid is simulated at, respectively, $d_e = 4\ kT$ and $8\ kT$.

Figures 8 and 9 represent 'snapshots' of the static structure in the monolayer during a simulation run. Figure 9 shows a system with $d_{e1} = d_{e2} = 8\ kT$ and $\xi = 1.5$. We see that for this system, the two

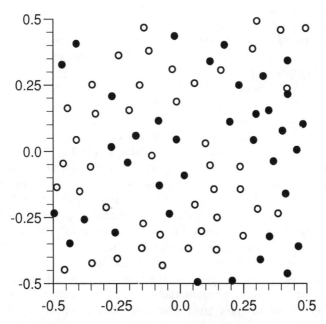

Figure 8 *Snapshot of positions of centres of particles at the interface for* $\rho^* = 0.5$, $T^* = 2.5$, $\xi = 0.6$, $d_{e1} = d_{e2} = 8\ kT$: $\bullet$ = *component 1;* $\bigcirc$ = *component 2*

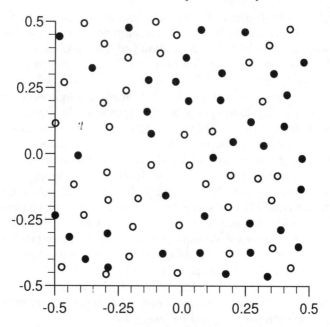

Figure 9 *Snapshot of positions of centres of particles at the interface for* $\rho^* = 0.5$, $T^* = 2.5$, $\xi = 1.5$, $d_{e1} = d_{e2} = 8\,kT$: ● = *component 1*; ○ = *component 2*

components are uniformly dispersed, or mixed, within the monolayer. Figure 8 shows the equivalent system with $\xi = 0.6$; here we observe similar concentrations of the two components in the monolayer, but there is a definite segregation, or phase-separation of the particles, into regions rich in one component or the other.

Finally, it is appropriate to note that, although the mixed adsorbed layers generated in this study are thin (of the order of the particle diameter), they are not purely two-dimensional structures. Whilst the diffusional motion in the plane of the interface strongly resembles that of a molecular dynamics simulation[18] of a genuinely two-dimensional system, an additional mechanism for mass transport in the interfacial region simulated in the present study is the 'hopping' of individual particles in and out of the adsorbed layer. Where significant secondary layers exist (*e.g.* Figure 4), there is a rapid rate of exchange of particles between primary and secondary layers. This makes it likely that a particle jumping out of the monolayer will diffuse more rapidly in the plane of the less dense secondary layer, before inserting itself back into the primary layer. In competitive adsorption of components with different adsorption energies (Figure 6), we note that, whereas the most strongly adsorbing species may predominate in the primary layer, the more weakly adsorbing species may predominate in the secondary layer.

It is proposed to extend the molecular dynamics model described here to model the competitive adsorption from a mixture of adsorbates of different sizes and concentrations. Such simulations should lead to improved understanding of the factors affecting the structure and dynamics of mixed protein and surfactant films.

References

1. E. Dickinson and G. Stainsby, 'Colloids in Food', Applied Science, London, 1982.
2. H. Mulder and P. Walstra, 'The Milk Fat Globule', Pudoc, Wageningen, The Netherlands, 1974.
3. D. F. Darling and R. J. Birkett, in 'Food Emulsions and Foams', ed. E. Dickinson, Special Publication No. 58, Royal Society of Chemistry, London, 1987, p. 1.
4. A. Lips, I. J. Campbell, and E. G. Pelan, in 'Food Polymers, Gels and Colloids', ed. E. Dickinson, Special Publication No. 82, Royal Society of Chemistry, Cambridge, 1991, p. 1.
5. E. Dickinson, *Food Hydrocolloids*, 1986, **1**, 3.
6. E. Dickinson, *ACS Symp. Ser.*, 1991, **448**, 114.
7. J. A. de Feijter, J. Benjamins, and M. Tamboer, *Colloids Surf.*, 1987, **27**, 243.
8. J.-L. Courthaudon, E. Dickinson, and D. G. Dalgleish, *J. Colloid Interface Sci.*, 1991, **145**, 390.
9. E. Dickinson, *Molec. Phys.*, 1988, **65**, 895.
10. E. Dickinson and S. R. Euston, *Molec. Phys.*, 1989, **68**, 407.
11. E. Dickinson and S. R. Euston, *J. Colloid Interface Sci.*, 1992, **152**, 562.
12. E. Dickinson, in 'The Structure, Dynamics and Equilibrium Properties of Colloidal Systems', eds. D. M. Bloor and E. Wyn-Jones, Kluwer, The Netherlands, 1990, p. 707.
13. E. Dickinson and E. G. Pelan, *Molec. Phys.*, 1991, **74**, 1115.
14. L. V. Woodcock, *Chem. Phys. Lett.*, 1984, **111**, 455.
15. D. M. Heyes, G. P. Morriss, and D. J. Evans, *J. Chem. Phys.*, 1985, **83**, 4750.
16. E. Dickinson, R. Parker, and M. Lal, *Chem. Phys. Lett.*, 1981, **79**, 578.
17. D. Fincham and D. M. Heyes, *Information Quarterly for MD and MC Simulations*, 1982, **6**, 4.
18. A. A. Clifford and E. Dickinson, *Molec. Phys.*, 1977, **34**, 875.

Competitive Adsorption in Protein-stabilized Emulsions Containing Oil-soluble and Water-soluble Surfactants

By Eric Dickinson, Graeme Iveson, and Sumio Tanai

PROCTER DEPARTMENT OF FOOD SCIENCE, UNIVERSITY OF LEEDS, LEEDS LS2 9JT, UK

1 Introduction

There are two kinds of molecular species present in food colloids that have a strong tendency to adsorb at the oil–water interface: proteins (from milk, eggs, *etc.*) and small-molecule surfactants (polar lipids and derivatives thereof). In oil-in-water emulsions containing food proteins and low-molecular-weight emulsifiers, it is known[1-6] that the surface concentration of protein in the stabilizing layer around the droplets is dependent on competitive adsorption between the species during or following emulsification. This competitive adsorption has important implications for the interactions between the droplets, which in turn has an influence on emulsion stability and rheology.[7-9]

The vast majority of surfactants (emulsifiers) used as food ingredients are either non-ionic or zwitterionic. Depending on the molecular hydrophile–lipophile balance (HLB), a surfactant may be predominantly oil-soluble (low HLB) or water-soluble (high HLB). The most ubiquitous of surface-active lipids are the monoglycerides, which are present as impurities in all food-grade triglyceride oils. Commercial glycerol monostearate (GMS) is probably the most widely used food emulsifier in product formulation. The most important zwitterionic surfactants are the two naturally occurring phospholipids, phosphatidylcholine (lecithin) and phosphatidylethanolamine. The oil-soluble phosphatidylethanolamine has a tendency to form reversed micelles, in contrast to phosphatidylcholine which can form lamellar structures and vesicles in water. Most of the genuinely non-ionic water-soluble surfactants used in food colloids are produced synthetically, *e.g.* polyoxyethylene sorbitan esters (Tweens) from reaction of sorbitol with first fatty acids and then ethylene oxide. Such micelle-forming water-soluble surfactants are often found in products which also contain oil-soluble surfactants (notably monoglycerides). The nature of the adsorbed layer in these mixed systems is largely unknown.

Our fundamental knowledge of adsorbed protein films is primarily derived from experiments carried out at hydrocarbon oil–water interfaces, which are substantially different in surface free energy from the triglyceride oil–water interfaces encountered in food emulsions.[10] In order to bridge this gap between model systems and real food colloids, it seems appropriate to perform experiments with the same surface-active species adsorbing at the two different oil–water interfaces. Previous work in our laboratory[5,6,11-14] has considered the competitive adsorption of single oil-soluble or water-soluble polyoxyethylene surfactants and milk proteins at the n-alkane–water interface at neutral pH, as well as the competitive adsorption of lecithin and β-casein at n-alkane–water and triglyceride oil–water interfaces.[15] In agreement with theory[2] and computer simulation,[12] it is found that the presence of a high bulk surfactant concentration substantially reduces the protein surface coverage at the oil–water interface, although only with a water-soluble surfactant (either pure C_mE_n surfactant or food-grade Tween) is the protein completely displaced from the interface. In the present paper, we present recent new measurements for emulsions containing pure milk protein and a mixture of both an oil-soluble surfactant ($C_{12}E_2$ or GMS) and a water-soluble surfactant ($C_{12}E_8$ or Tween 20). In addition, we present new results for the competitive adsorption of phosphatidylethanolamine plus milk protein for comparison with measurements made previously[15] involving phosphatidylcholine. In each case, two different oil phases were used to prepare the emulsions: n-tetradecane and purified soya oil.

The influence of temperature on competitive adsorption of milk proteins and emulsifiers is relevant to the processing of various dairy colloids, especially ice-cream.[7] In order to provide experimental data relevant to this issue, the present study includes measurements of the surface coverage of β-casein on emulsion droplets as the temperature is lowered from 20 °C to 0 °C.

2 Systems Containing n-Tetradecane, β-Casein, $C_{12}E_8$, and $C_{12}E_2$

Oil-in-water emulsions (20 wt % oil, pH 7) were prepared by high-pressure homogenization of an aqueous solution of β-casein (0.5 wt %) and n-tetradecane containing oil-soluble surfactant $C_{12}E_2$ (diethylene glycol n-dodecylether, >99% pure). Known amounts of the water-soluble surfactant $C_{12}E_8$ (octaethylene glycol n-dodecylether, >99% pure) were added after emulsion formation. The protein surface coverage was determined after storing the emulsion for 2 hours at 20 °C. The experimental procedures for making the emulsions and determining the protein surface coverage have been described in detail elsewhere.[16]

The average droplet size of the freshly made emulsion depends on the amount of $C_{12}E_2$ present during emulsification. For 0, 0.1, and 0.5 wt % $C_{12}E_2$ (expressed as wt % of *whole* emulsion), the average volume–surface diameter is $d_{32} = 0.98$ μm, 0.66 μm, and 0.41 μm, respectively. There is no discernible change in d_{32} following addition of $C_{12}E_8$.

The protein surface coverage at the n-tetradecane–water interface in the absence of surfactants is $\Gamma = 2.3 \pm 0.1$ mg m^{-2}. Figure 1 shows the change in surface coverage at 20 °C as a function of the $C_{12}E_8/\beta$-casein molar ratio R. The data for emulsions with no oil-soluble surfactant present agree with our previous results[14] in showing that β-casein is completely displaced from the interface by $C_{12}E_8$ at $R \geqslant 20$. In emulsions containing 0.1 or 0.5 wt % $C_{12}E_2$ (*i.e.* a $C_{12}E_2/\beta$-casein molar ratio of approximately 20:1 or 100:1, respectively), the surface coverage Γ is lowered (for $R < 20$), though the relative amount of the water-soluble surfactant required for complete protein displacement remains roughly the same. It can be inferred that, while the presence of $C_{12}E_2$ reduces the total quantity of β-casein that can be accommodated at the interface, the effective adsorption energy of those protein molecules that are adsorbed is approximately the same as in the absence of oil-soluble surfactant, as it still takes the same relative amount of $C_{12}E_8$ to remove β-casein completely from the oil–water interface.

3 Systems Containing Triglyceride Oil, β-Casein, $C_{12}E_8$, and $C_{12}E_2$

The β-casein surface coverage at the triglyceride oil–water interface is lower than at the hydrocarbon oil–water interface, and, whereas the latter appears constant over the temperature range 0–20 °C, the former is temperature dependent.[17] In emulsions made with soya oil, which had

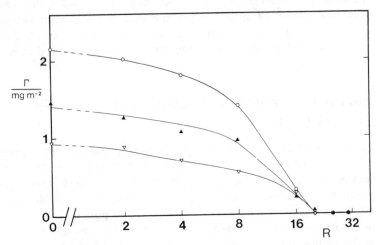

Figure 1 *Influence of $C_{12}E_8$ added after emulsification on the protein surface coverage in n-tetradecane-in-water emulsions (0.4 wt % β-casein, 20 wt % oil, pH 7) containing $C_{12}E_2$ dissolved in the oil phase. The surface concentration Γ at 20 °C is plotted against the $C_{12}E_8/\beta$-casein molar ratio R for various concentrations of $C_{12}E_2$ (expressed as wt % of whole emulsion): O, 0 wt %; ▲, 0.1 wt %; ▽, 0.5 wt %*

previously been made free of surface-active contaminants[18] by twice passing through a Florisil column, it was found[17] that the protein surface concentration is reduced from $\Gamma = 1.9$ mg m^{-2} at 20 °C to $\Gamma = 1.5$ mg m^{-2} at 10 °C, only to increase again as the temperature is further lowered to 0 °C.

Figure 2 shows the β-casein surface concentration as a function of the $C_{12}E_8/\beta$-casein molar ratio R in oil-in-water emulsions (20 wt% purified soya oil, 0.5 wt% protein, pH 7) containing no oil-soluble surfactant at 0 °C, 5 °C, 10 °C, and 20 °C. The data obtained with emulsions cooled to 10 °C, and then left for 2 hours prior to protein surface coverage determination,[17] show the lowest Γ values. The displacement curves obtained at 0 °C and 20 °C are rather similar.

Figure 3 shows the same $\Gamma(R)$ data as for Figure 1 except with purified soya oil replacing n-tetradecane as the dispersed phase. At 20 °C, complete protein displacement is achieved at $R \geqslant 10$ in the absence of oil-soluble surfactant, and at a slightly lower value of R when the emulsion contains 0.5 wt% $C_{12}E_2$ (*i.e.* the oil phase contains 2.5 wt% $C_{12}E_2$). It would appear, then, that the protein is more readily displaced by $C_{12}E_8$ from the triglyceride oil–water interface than from the hydrocarbon oil–water interface. Possibly, the hydrophobic residues of the adsorbed β-casein molecules are less strongly solvated by the more polar triglyceride 'solvent' than by the n-alkane 'solvent'. Hence, in the soya oil-in-water emulsions, it takes a lower water-soluble surfactant concentration to reduce the interfacial free energy to the critical value required for complete protein displacement.[12]

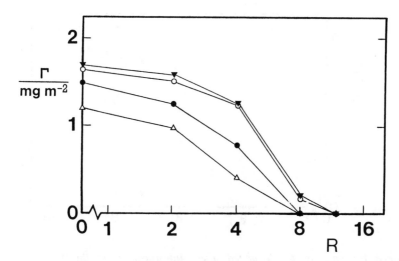

Figure 2 *Temperature dependence of the competitive displacement of protein by $C_{12}E_8$ from the oil–water interface in soya oil-in-water emulsions (0.4 wt% β-casein, 20 wt% oil, pH 7). The protein surface concentration Γ is plotted against the $C_{12}E_8/\beta$-casein molar ratio R: $\bigcirc$, 20 °C; $\triangle$, 10 °C; $\bullet$, 5 °C; $\blacktriangledown$, 0 °C*

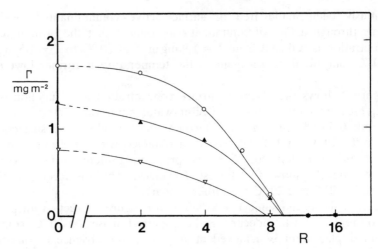

Figure 3 *Influence of $C_{12}E_8$ added after emulsification on the protein surface coverage in soya oil-in-water emulsions (0.4 wt % β-casein, 20 wt % oil, pH 7) containing $C_{12}E_2$ dissolved in the oil phase. The surface concentration at 20 °C is plotted against the $C_{12}E_8/β$-casein molar ratio R for various concentrations of $C_{12}E_2$ (expressed as wt % of whole emulsion):* ○, 0 wt %, ▲, 0.1 wt %; ▽, 0.5 wt %

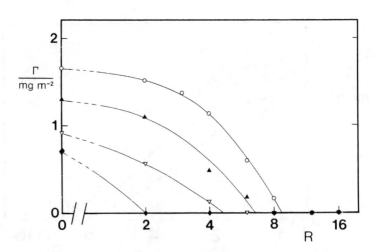

Figure 4 *Influence of $C_{12}E_8$ added after emulsification on the protein surface coverage in soya oil-in-water emulsions (0.4 wt % β-casein, 20 wt % oil, pH 7) containing GMS dissolved in the oil phase. The surface concentration Γ at 20 °C is plotted against the $C_{12}E_8/β$-casein molar ratio R for various concentrations of GMS (expressed as wt % of whole emulsion):* ○, 0 wt %; ▲, 0.05 wt %; ▽, 0.2 wt %; ◆, 0.5 wt %

4 Systems Containing Triglyceride Oil, β-Casein, C₁₂E₈, and GMS

We now consider emulsions containing pure GMS (>99% pure) dissolved in the triglyceride oil instead of $C_{12}E_2$. A comparison of the results in Figure 4 with those in Figure 3 indicates that the monoglyceride is more effective in facilitating protein displacement from the interface by $C_{12}E_8$ than is the oil-soluble polyoxyethylene surfactant. Addition of GMS to soya oil substantially reduces the amount of water-soluble surfactant required for complete protein displacement.[16] With 2.5 wt% GMS in the oil phase (*i.e.* 0.5 wt% in the total emulsion system, corresponding to a 90:1 molar ratio of GMS to β-casein), the protein surface concentration is $\Gamma = 0.7\ \mathrm{mg\,m^{-2}}$ with no water-soluble surfactant added, and a $C_{12}E_8$/β-casein ratio of $R = 2$ is enough to reduce the protein surface coverage to zero.

Figure 5 shows the effect of temperature on the surface coverage Γ for

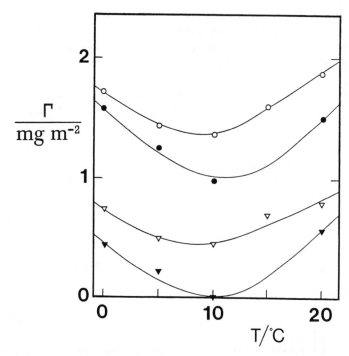

Figure 5 *Temperature dependence of protein surface coverage in soya oil-in-water emulsions (0.4 wt% β-casein, 20 wt% oil, pH 7) containing GMS dissolved in the oil phase with water-soluble surfactant ($C_{12}E_8$ or Tween 20) added after emulsification (surfactant/protein molar ratio R = 2). The surface concentration is plotted against the temperature T at which the water-soluble surfactant was added:* ●, ○, *pure oil;* ▼, ▽, *oil containing 1 wt% GMS (i.e. 0.2 wt% in whole emulsion). Closed symbols (●, ▼) refer to $C_{12}E_8$. Open symbols (○, ▽) refer to Tween 20*

emulsions containing water-soluble surfactant ($R = 2$) added after homogenization and subsequent cooling.[17] It can be seen that there is a minimum surface coverage around 10 °C, and that, with 0.2 wt % GMS present in the emulsion (*i.e.* 1 wt % in the oil), the β-casein is completely displaced from the interface.

5 Systems Containing Triglyceride Oil, β-Casein, Tween 20, and GMS

We now consider the replacement of the pure water-soluble surfactant $C_{12}E_8$ by the water-soluble food-grade surfactant Tween 20 (polyoxyethylene sorbitan monolaurate). The data in Figure 6 for competitive displacement of β-casein by Tween 20 show the same qualitative trends as found with $C_{12}E_8$ (Figure 4). But there are significant quantitative differences between the two sets of data. In the absence of GMS, the molar amount of water-soluble surfactant required for complete displacement is roughly twice that for $C_{12}E_8$; when GMS is present, however, the difference is reduced. In emulsions containing 0.2 wt % GMS, the value $R \approx 6$ for complete protein displacement at 20 °C is similar for Tween 20 and $C_{12}E_8$. This result provides further confirmation of the considerable ability of GMS to displace protein from the triglyceride oil–water interface.

The influence of temperature on the competitive adsorption of β-casein and Tween 20 ($R = 2$) is shown in Figure 5. As with $C_{12}E_8$ there is a

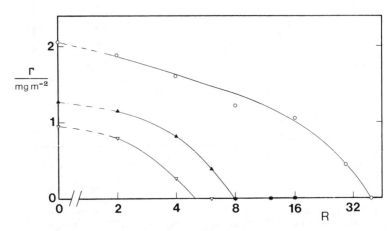

Figure 6 *Influence of Tween 20 added after emulsification on the protein surface coverage in soya oil-in-water emulsions (0.4 wt % β-casein, 20 wt % oil, pH 7) containing GMS dissolved in the oil phase. The surface concentration Γ at 20 °C is plotted against the Tween 20/β-casein molar ratio R for various concentrations of GMS (expressed as wt % of whole emulsion):* $\bigcirc$, 0 wt %; $\blacktriangle$, 0.05 wt %; $\triangledown$, 0.2 wt %

minimum value of Γ around 10 °C, although with the food-grade emulsifier the overall protein surface coverage is higher.

6 Systems Containing n-Tetradecane, β-Lactoglobulin, and Phosphatidylethanolamine

Figure 7 shows experimental results obtained recently[19] for oil-in-water emulsions made with β-lactoglobulin and phosphatidylethanolamine (>99% pure, derived from soya oil). Dissolving phospholipid in the hydrocarbon oil prior to homogenization leads to a reduction in the mean droplet size and in the proportion of the total β-lactoglobulin present that is adsorbed. As noted previously[15] for emulsions containing β-casein and phosphatidylcholine, the effect of a high concentration of phospholipid present during emulsification is to reduce the protein surface coverage to a substantial extent. The adsorbed protein fraction is reduced from $F_p = 0.71$ at $R = 0$ to $F_p = 0.30$ at $R = 100$, and the average droplet diameter from $d_{32} = 0.65\ \mu$m to $d_{32} = 0.48\ \mu$m. This corresponds to a decrease in the β-lactoglobulin surface concentration Γ from $ca.$ 1.0 mg m^{-2} at low phospholipid contents to $ca.$ 0.4 mg m^{-2} at high contents (>1 wt%).

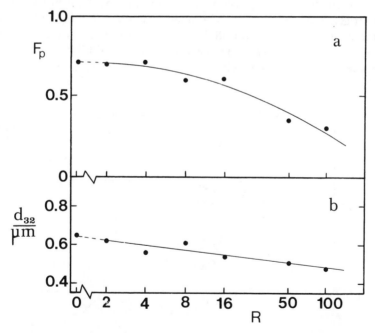

Figure 7 *Influence of phosphatidylethanolamine present during emulsification on the adsorbed protein fraction and droplet size in n-tetradecane-in-water emulsions (0.4 wt% β-lactoglobulin, 20 wt% oil, pH 7). (a) The fraction F_p of the protein adsorbed is plotted against the phosphatidylethanolamine/protein molar ratio R. (b) The average droplet diameter d_{32} is plotted against the molar ratio R*

7 Distribution of Lecithin between Emulsion Droplets and Continuous Phase

Phosphatidylcholine (lecithin) is a zwitterionic amphiphile which has a tendency to form lamellar mesophases and vesicles in aqueous media[9] and to form lipid–protein complexes with excellent emulsifying properties.[20] These factors, together with the competitive adsorption of lecithin and protein at the oil–water interface, make it difficult to predict the distribution of the phospholipid betwen the aqueous and non-aqueous regions of a protein-stabilized emulsion.

Using the analytical procedure described previously,[15] we have determined the quantity of lecithin associated with the droplets in oil-in-water emulsions containing n-tetradecane or soya oil emulsified with a mixture of pure milk protein (β-casein or β-lactoglobulin) and phosphatidylcholine (>99% pure, derived from egg yolk). Figure 8 shows the influence of the lecithin/protein molar ratio R on the lecithin fraction F_L associated with the oil droplets. As results for emulsions made with β-casein and β-lactoglobulin are quantitatively similar, it seems that the distribution of phospholipid is not dependent on the nature of the protein. Nevertheless, the results obtained with the two types of oil are distinctly different, due presumably to the differing solubilities of the phospholipid in the two oils. In the triglyceride systems, only a small fraction ($F_L \leqslant 0.1$) is associated with the emulsion droplets for all R. In the hydrocarbon systems, however, there is a consistent increase in the proportion of lecithin associated with the droplets as R increases, with the value of F_L levelling off in the range

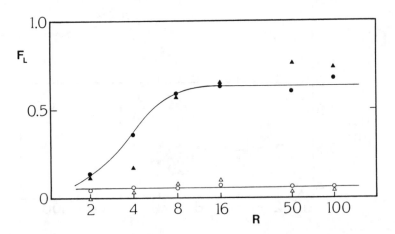

Figure 8 *Distribution of phosphatidylcholine in protein-stabilized emulsions (0.4 wt % protein, 20 wt % oil, pH 7). The lecithin fraction F_L associated with the oil droplets is plotted as a function of the lecithin/protein molar ratio R:* ●, *n-tetradecane, β-casein;* ▲, *n-tetradecane, β-lactoglobulin;* ○, *soya oil, β-casein;* △, *soya oil, β-lactoglobulin. The curves are drawn through the data for the β-casein systems*[15]

0.6–0.7 for $R \geqslant 10$. It should be noted that the quantity F_L does not distinguish between lecithin solubilized in the oil phase and lecithin (strongly) adsorbed at the oil-water interface. Lecithin bilayers weakly associated with the droplet surface might be swept from the interfacial region during the emulsion washing and centrifugal separation processes.

8 Conclusions

(1) The presence of oil-soluble surfactant in the dispersed phase during emulsification reduces the protein surface coverage at the oil–water interface. The protein surface concentration is further reduced on addition of water-soluble non-ionic surfactant after emulsion formation.

(2) Whether the presence of oil-soluble surfactant reduces the amount of water-soluble surfactant required for *complete* protein displacement seems to depend on the chemical nature of both the oil-soluble surfactant and the oil phase.

(3) Glycerol monostearate (GMS) in triglyceride oil-in-water emulsions is particularly effective in reducing the amount of water-soluble surfactant ($C_{12}E_8$ or Tween 20) required for complete protein displacement.

(4) In the presence or absence of GMS, the surface coverage of β-casein in triglyceride oil-in-water emulsions shows a substantial temperature dependence in the range 0–20 °C. It appears that the reduction in milk protein surface load observed[21,22] on cooling of ice-cream emulsions may not be wholly attributable to temperature-dependent phase changes of the emulsifier or the fat since some β-casein displacement is observed in our model systems on cooling from 20 °C to 10 °C even though the droplets remain liquid and no GMS is present.

(5) A moderately high concentration of phosphatidylethanolamine present during emulsification leads to a substantial partial displacement of β-lactoglobulin from the oil–water interface.

(6) The distribution of phosphatidylcholine between the emulsion droplets and the aqueous phase in milk protein-stabilized emulsions depends on the nature of the oil phase (hydrocarbon or triglyceride) but not on the nature of the protein (β-casein or β-lactoglobulin).

References

1. J. A. de Feijter, J. Benjamins, and M. Tamboer, *Colloids Surf.*, 1987, **27**, 243.
2. E. Dickinson and C. M. Woskett, in 'Food Colloids', ed. R. D. Bee, P. Richmond, and J. Mingins, Special Publication No. 75, Royal Society of Chemistry, Cambridge, 1989, p. 74.
3. E. Dickinson, S. K. Narhan, and G. Stainsby, *J. Food Sci.*, 1989, **54**, 77.

4. I. Heertje, J. Nederlof, H. A. C. M. Hendrickx, and E. H. Lucassen-Reynders, *Food Structure*, 1990, **9**, 305.

5. J.-L. Courthaudon, E. Dickinson, Y. Matsumura, and A. Williams, *Food Structure*, 1991, **10**, 109.

6. E. Dickinson and J.-L. Gelin, *Colloids Surf.*, 1992, **63**, 329.

7. D. F. Darling and R. J. Birkett, in 'Food Emulsions and Foams', ed. E. Dickinson, Special Publication No. 58, Royal Society of Chemistry, London, 1987, p. 1.

8. E. Dickinson, *Colloids Surf.*, 1989, **42**, 191.

9. B. A. Bergenståhl and P. M. Claesson, in 'Food Emulsions', 2nd edn, ed. K. Larsson and S. E. Friberg, Marcel Dekker, New York, 1990, p. 41.

10. L. R. Fisher and N. S. Parker, in 'Advances in Food Emulsions and Foams', ed. E. Dickinson and G. Stainsby, Elsevier Applied Science, London, 1988, p. 45.

11. E. Dickinson, A. Mauffret, S. E. Rolfe, and C. M. Woskett, *J. Soc. Dairy Technol.*, 1989, **42**, 18.

12. E. Dickinson, S. R. Euston, and C. M. Woskett, *Prog. Colloid Polym. Sci.*, 1990, **82**, 65.

13. J.-L. Courthaudon, E. Dickinson, Y. Matsumura, and D. C. Clark, *Colloids Surf.*, 1991, **56**, 293.

14. J.-L. Courthaudon, E. Dickinson, and D. G. Dalgleish, *J. Colloid Interface Sci.*, 1991, **145**, 390.

15. J.-L. Courthaudon, E. Dickinson, and W. W. Christie, *J. Agric. Food Chem.*, 1991, **39**, 1365.

16. E. Dickinson and S. Tanai, *J. Agric. Food Chem.*, 1992, **40**, 179.

17. E. Dickinson and S. Tanai, *Food Hydrocolloids*, 1992, **6**, 163.

18. A. G. Gaonkar, *J. Amer. Oil Chem. Soc.*, 1989, **66**, 1090.

19. E. Dickinson, G. Iveson, and W. W. Christie, unpublished.

20. R. Nakamura, R. Mizutani, M. Yano, and S. Hayakawa, *J. Agric. Food Chem.*, 1988, **36**, 729.

21. N. Krog, *ACS Symp. Ser.*, 1991, **448**, 138.

22. N. M. Barfod, N. Krog, G. Larsen, and W. Buchheim, *Fat Sci. Technol.*, 1991, **93**, 24.

Surface Interactions in Oil Continuous Food Dispersions

By Dorota Johansson and Björn Bergenståhl

INSTITUTE FOR SURFACE CHEMISTRY, PO BOX 5607, S-114 86 STOCKHOLM, SWEDEN

1 Introduction

Surface interactions are important in determining the properties of oil continuous food products containing solid particles, for instance, chocolate (a dispersion of fat and sugar crystals in oil),[1] and margarine or butter (emulsions with a network of fat particles in the continuous oil phase).[2] Consistency and mouthfeel depend on rheology and crystal interactions. Emulsion stability during storage is also dependent on the fat network strength, since the network prevents the droplets from coalescing. Emulsion stability during production is determined by the ability of the emulsifier to adsorb to the oil–water interface and by the crystal position at the oil–water interface.

The effect of fat crystals, called the Pickering effect, has been studied in oil continuous food systems by Lucassen-Reynders[3] and Campbell,[4] and in water continuous systems by van Boekel,[5] Darling,[6] and Boode.[7]

The purpose of the present paper is to survey different surface interactions in oil continuous foods. We have started by studying the adsorption of food emulsifiers to crystals. Further, we have examined the influence of these emulsifiers on the crystal interactions in sedimentation and rheological experiments. We have also studied the influence of traces of water on emulsifier adsorption and on crystal interactions. The detailed results are published separately in a series of articles.[8-10] We are continuing our work with the investigation of crystal position at the oil–water interface (contact angle measurements) in the presence of different emulsifiers in the oil. Finally, we are considering the influence of temperature on the fat network strength.

2 Emulsifier–Crystal Interactions

Adsorption to Crystals in the Oil Phase

The emulsifier–crystal interactions are studied by adsorption measurements, where the adsorbed amount (in $mol\,m^{-2}$) is determined as a function of the emulsifier concentration in the oil. We have found that most food emulsifiers (lecithins, monoglycerides, their esters, and esters of fatty acids) adsorb both to polar surfaces (like sugar crystals) and to nonpolar surfaces (like fat crystals) in the oil.[6] The adsorption strength increases with increasing polarity and increasing saturation of the emulsifier, as well as with increasing polarity of the surface. Thus, the adsorption is stronger to sugar than to fat crystals. Lecithins adsorb most strongly and form multilayers on both types of crystals. Unsaturated monoglycerides adsorb weakest, and form loosely packed layers, which become denser at high concentrations. Most esters of monoglycerides and fatty acids tend to form tightly packed monolayers, with the hydrocarbon chains directed to the oil in the case of sugar and with the hydrocarbon chains directed to the crystal surface in the case of fat.

Crystal Wetting at the Oil–Water Interface

Wetting of fat crystals has been determined by contact angle (Θ) measurements using a method developed at Unilever[4,6] as illustrated in Figure 1. The contact angles were measured through the oil phase. The oil phase has been adjusted by adding different food emulsifiers. We have measured both the advancing contact angle Θ_a (corresponding to the fat crystals approaching the oil–water interface from the water side) and also the

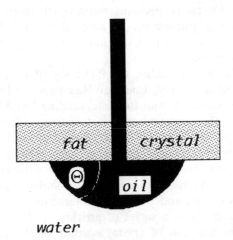

Figure 1 *An arrangement for contact angle measurements at the fat crystal/oil/water interface. The contact angle Θ is measured through the oil phase. Both advancing (Θ_a) and receding angles (Θ_r) are determined*

receding contact angles Θ_r (corresponding to the fat crystals approaching the oil–water interface from the oil side).

Wetting of fat crystals is very important for the stability of emulsions.[3-7,11] When crystals are totally wetted by oil ($\Theta \approx 0°$) or water ($\Theta \approx 180°$), they do not influence the emulsion stability. Crystals stabilize oil-in-water emulsions if they are better wetted by water than by oil and water-in-oil emulsions if they are better wetted by oil than by water.

Wetting behaviour of fat crystals strongly depends on the emulsifier present in the oil. Contact angle as a function of concentration can be determined for each emulsifier. The resulting contact angle isotherm is directly related to the adsorption isotherm and to the sedimentation behaviour as indicated in Figure 2. The adsorption isotherm shows that

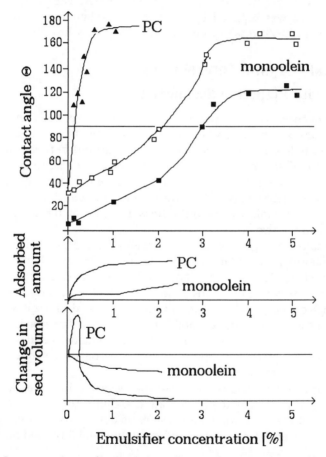

Figure 2 *Contact angles to β′-type fat crystals as a function of emulsifier concentration in the oil. The oil phase contains soybean phosphatidylcholine, PC (▲-both Θ_a and Θ_r), and mono-olein (■-Θ_r and □-Θ_a), respectively. The 'contact angle isotherm' (top) is compared with the adsorption isotherm (middle, in a simplified form) and the sedimentation data (bottom, in a simplified form)*

soybean phosphatidylcholine (PC) adsorbs relatively strongly to fat crystals and changes the crystal surface to a polar one (Θ_a, $\Theta_r > 90°$) even at low concentrations. The crystals are equally polar, either when they are located in the oil phase ($\Theta_r > 90°$), or when they are forced into the water phase ($\Theta_a > 90°$). Mono-olein adsorbs weakly and changes the crystal polarity only slightly in the oil up to concentrations of about 3%. Crystals with adsorbed mono-olein are more polar when they are forced into the water phase ($\Theta_a > \Theta_r$). This contact angle hysteresis occurs with most of the emulsifiers.

The most polar emulsifiers adsorb the strongest, resulting in the most polar fat crystals. We have found a correlation between both advancing and receding contact angles and the HLB value for the emulsifiers. The emulsifiers used here were specially purified samples. The phosphatidylcholine (>98 wt %) was supplied by Lucas Meyer (Hamburg, Germany) and the mono-olein (>99 wt %) by Grindsted Products (Århus, Denmark).

3 Crystal–Crystal Interactions

Stability with Respect to Sedimentation

The consequences of adsorption onto crystals have been evaluated in sedimentation experiments, where an increased sediment volume indicates an increased adhesion (stronger attractive interaction) between the crystals and a decreased sediment volume indicates decreased adhesion.[12] Generally, we have found that the adsorption of emulsifiers influences the crystal–crystal interactions. The effects are different for fat and sugar crystals, which is indicative of the different nature of the adsorption and interactions in the two cases.[8]

Some results for sugar crystals in soybean oil are presented in Figure 3. Soybean phosphatidylcholine gives a decrease in crystal adhesion—the sediment volume decreases by about 30%. The same effect is given by polyglycerol esters of fatty acids (and most of other esters of fatty acids and monoglycerides) at high concentrations.[8] The decreased adhesion is probably due to a weak steric stabilization by the hydrocarbon tails of the adsorbed emulsifiers. Mono-olein (and other monoglycerides and their lactic acid esters) gives the opposite effect—a stronger network and better stability with respect to sedimentation.

The effects of fat crystals are somewhat weaker. Lecithins and monoglycerides give a decreased sedimentation stability (due to a weak steric stabilization of crystals), while most esters of fatty acids or monoglycerides at high concentrations give better sedimentation stability (additional adhesion).

Influence of Water

Traces of water influence the surface interactions in oil continuous systems. Sedimentation experiments with different amounts of water present have

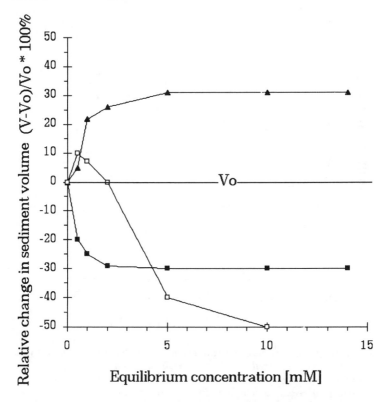

Figure 3 *Sedimentation of sucrose crystals dispersed in soybean oil at room tem-*
perature. A relative change in sediment volume (in %) is plotted as a
function of the equilibrium concentration of emulsifiers. The line labelled
Vo represents the sample without any emulsifier. The following emulsifiers
are examined: ■, *soybean phosphatidylcholine;* ▲, *mono-olein;* □, *poly-*
glycerol esters of fatty acids

shown that water increases the adhesion between both fat and sugar
crystals in oils, as manifested by higher sediment volumes. Water also
increases the adsorption of oil-soluble emulsifiers onto the crystals.[10]

An example of results for sugar dispersions is presented in Figure 4. The
sediment volume of the crystals increases as the water content increases.
Further, the effect of emulsifiers on the sediment volume (an increase in
the case of mono-olein, but a decrease in the case of phosphatidylcholine)
is more pronounced when the water content is increased.

Rheology

In a series of strain sweep experiments, we have determined yield stresses
and elastic moduli for fat and sugar dispersions without and with food
emulsifiers.[9] We have found that emulsifiers influence the rheological
properties of oil dispersions. The magnitude of the changes is larger for

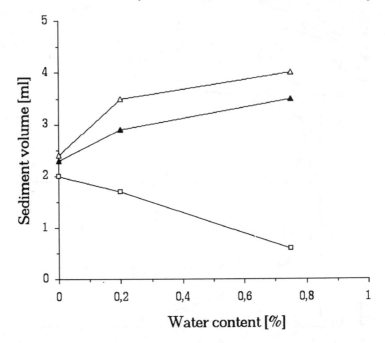

Figure 4 *The sediment volumes of sugar crystals as a function of added water. The crystals are dispersed in (i) pure refined soybean oil (▲), (ii) a 1 mM solution of mono-olein in soybean oil (△), and (iii) a 1 mM solution of soybean phosphatidylcholine (PC) in soybean oil (□)*

sugar crystals than for fat crystals. Lecithins give the greatest decrease in the yield stress. The largest increase is caused by saturated monoglycerides, which partly precipitate, and probably form a network together with crystals in the oil. Since an increased adhesion between the crystals results in an increased elasticity and yield stress of the dispersion on the one hand, and also in an increased sedimentation stability on the other hand, a relationship between the sedimentation and the rheological results should exist. Such a correlation, presented in Figure 5, confirms our interpretation.

We have evaluated the rheological parameters[9] by means of the modified network model of van den Tempel[13] and the yield stress correlation with the interparticle potential by Gillespie[14] and Tadros.[15] The results indicate that van der Waals forces alone cannot be responsible for the relatively high network strength. Other stronger attractive interactions (*e.g.* water bridges) have also to be present.

Influence of Temperature

Fat crystals with a narrow melting point dispersed in oil show an unexpected maximum in adhesion at high temperature. This maximum can be

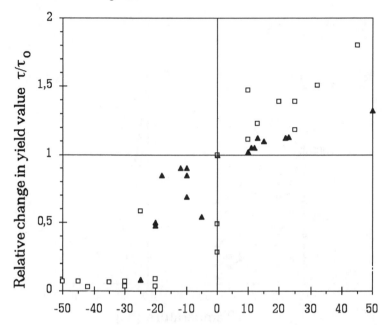

Relative change in sediment volume (V-Vo)/Vo * 100%

Figure 5 *A comparison of the sedimentation results (relative change in sediment volume in %) with the rheological results (relative change in yield stress) for fat (▲) and sugar (□) crystal dispersions in soybean oil. An increase in sediment volume corresponds to an increase in yield stress, interpreted as an increase in crystal adhesion; and a decrease in sediment volume corresponds to a decrease in yield stress, interpreted as a decrease in crystal adhesion*

detected by rheological measurements, as shown in Figure 6 for β'-type fat crystals with a melting point of 57–59 °C. The maximum, which is more pronounced for the modulus than for the Bingham yield stress, occurs between 30 and 40 °C. The increased interaction can be attributed to greater stickiness of fat crystal surfaces close to the melting temperature. The maximum disappears when fats with intermediate melting points are present in the system.

4 Conclusions

Food emulsifiers are surface active in vegetable oils. They adsorb to both polar and nonpolar solid surfaces, and cause a change in the crystal–crystal interactions. As a result, various properties of oil continuous products are strongly influenced by the emulsifiers present. Thus, the proper choice of the emulsifier and its concentration is a critical issue in product development work. Another important parameter is the water activity in the oil.

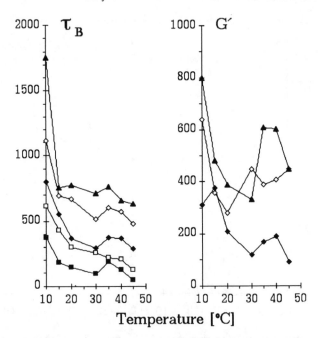

Figure 6 *The temperature dependence of the rheological parameters (Bingham yield stress τ_B and elastic shear modulus G'). The dispersions examined consist of β'-type fat crystals with a narrow melting point (57–59 °C) in soybean oil without any emulsifier added. The following volume fractions were examined:* ■, ~10%; □, ~15%; ◆, ~20%; ◇, ~25%; ▲, ~30%. *An unexpected maximum appears around 30–40 °C*

References

1. B. W. Minifie, 'Chocolate, Cocoa and Confectionary: Science and Technology', AVI Publishing Company, Westport, Connecticut, 1982.
2. H. Mulder and P. Walstra, 'The Milk Fat Globule', Centre for Agricultural Publishing and Documentation, Wageningen, 1974.
3. E. H. Lucassen-Reynders, 'Stabilization of water-in-oil emulsions by solid particles', Ph.D. thesis, Wageningen Agricultural University, 1962.
4. I. J. Campbell, in 'Food Colloids', ed. R. D. Bee, P. Richmond, and J. Mingins, Special Publication No. 75, Royal Society of Chemistry, Cambridge, 1989, p. 272.
5. M. A. J. S. van Boekel, 'Influence of fat crystals in the oil phase on stability of oil-in-water emulsions', Ph.D. thesis, Wageningen Agricultural University, 1980.
6. D. F. Darling, *J. Dairy Res.*, 1982, **49**, 695.
7. K. Boode, 'Partial coalescence in oil-in-water emulsions', Ph.D. thesis, Wageningen Agricultural University, 1992.
8. D. Johansson, and B. Bergenståhl, *J. Am. Oil Chem. Soc.*, 1992, **69**, 705.
9. D. Johansson, and B. Bergenståhl, *J. Am. Oil Chem. Soc.*, 1992, **69**, 718.

10. D. Johansson, and B. Bergenståhl, *J. Am. Oil Chem. Soc.*, 1992, **69**, 728.
11. 'Food Emulsions', ed. S. Friberg, Marcel Dekker, New York, 1976.
12. F. M. Tiller and Z. Khatib, *J. Colloid Interface Sci.*, 1984, **100**, 55.
13. M. van den Tempel, *J. Colloid Sci.*, 1964, **16**, 284.
14. T. J. Gillespie, *J. Colloid Sci.*, 1960, **15**, 219.
15. Th. F. Tadros, *Langmuir*, 1990, **6**, 28.

Influence of Phosphorylation on the Interfacial Behaviour of β-Casein

By J. Leaver and D. S. Horne

HANNAH RESEARCH INSTITUTE, AYR KA6 5HL, SCOTLAND

1 Introduction

Phosphorylation of the casein proteins at specific serine amino acid residues is important in maintaining the colloidal structure of milk. Partial dephosphorylation of micelles or the incorporation of dephosphorylated casein into artificial micelles has been found to affect adversely both the rennet coagulation time and syneresis.[1] Dephosphorylation of individual caseins has also been found to influence the calcium sensitivity of both α_s- and β-caseins.[2,3]

As a result of their relatively open, unfolded structure and their possession of discrete hydrophobic and hydrophilic regions, the caseins (in particular the α_s- and β-caseins) are efficient emulsifiers. By using the proteolytic enzyme trypsin to probe the structure of β-casein at such interfaces, it has been inferred[4] that the protein adopts a relatively ordered configuration with the most hydrophilic phosphorylated N-terminal region projecting into the aqueous phase as a loop or tail and effectively shielding the remainder of the molecule from proteolytic attack. Proteolysis has also shown[5] that the nature of the oil phase affects the secondary structure of the protein presumably as a result of interactions between the oil and certain of the more hydrophobic amino acid side chains. β-Casein also binds readily to solid–water interfaces such as polystyrene latices,[6] and studies using small-angle X-ray scattering together with photon correlation spectroscopy (PCS) have shown that most of the protein molecule lies near the surface with only part of the chain extending into the aqueous phase.[7] Trypsinolysis of the protein on the latex results in a biphasic decrease in the thickness of the adsorbed layer, the initial rapid stage being followed by a slower second stage.[8] The rapid stage has been interpreted as resulting from the removal of the tail/loop region, but this suggestion is still speculative. Adsorption of dephosphorylated β-casein to the latex gives a layer which is significantly thinner than that observed with the fully phosphorylated molecule.[6] This may be as a result of the dephosphorylated loop/tail region not projecting so far into the aqueous phase.

This paper reports a more detailed investigation of the effects of phosphorylation of β-casein on protein behaviour at both oil–water and solid–water interfaces.

2 Experimental

Polystyrene latex (nominal diameter 91 nm), trypsin (TPCK treated), potato acid phosphatase, soya oil, and n-tetradecane were purchased from Sigma Chemical Co. Ltd. (Poole, Dorset, UK).

Bovine β-casein was purified from bulk milk from the Institute herd by chromatography on Fast Flow Sepharose S (Pharmacia Ltd., Milton Keynes, Bucks., UK) in 6M-urea at pH 5.0 using a sodium chloride gradient. The protein was completely dephosphorylated using potato acid phosphatase.[6,9] The purity of the proteins and the completeness of the dephosphorylation were determined by FPLC on a Mono-Q column[10] and by urea–polyacrylamide gel electrophoresis.[11] Preparation of the β-casein–tetradecane emulsions, trypsinolysis, extraction and separation of peptides, and their identification were as detailed elsewhere.[4,5]

Changes in the hydrodynamic radius of the polystyrene latex in the presence of phosphorylated and dephosphorylated β-casein in imidazole buffer (20 mM; pH 7.0) were measured as detailed by Dalgleish[6] with the exception of a Malvern 7032 correlator being used for the signal analysis. The amount of protein adsorbed on the latex was determined by centrifuging to pellet the latex and assaying the concentration of protein remaining in the supernatant.[12]

3 Results and Discussion

The amino acid sequence of bovine β-casein is shown in Figure 1 together with the sites of phosphorylation and the trypsin-sensitive lysyl and arginyl peptide bonds. The phosphorylation sites occur in a cluster in the hydrophilic region near the N-terminal end of the molecule. It appears likely, therefore, that the removal of these phosphate groups would significantly reduce the hydrophilicity of this part of the protein molecule.

The increase in the average hydrodynamic radius of the polystyrene latex particles as a function of the concentration of added phosphorylated and dephosphorylated β-casein is shown in Figure 2. The maximum thickness of the phosphorylated β-casein layer was 16 nm whereas that of the phosphorylated protein was 12 nm. The thickness of the dephosphorylated β-casein layer decreased by approximately 2 nm at still higher protein concentrations. The concentration of added protein at which the maximum layer thickness is attained was slightly lower with the dephosphorylated protein and this is also seen in the plots of layer thickness as a function of protein loading (Figure 3). The differences are, however, relatively minor: the maximum layer thickness of the phosphorylated protein is reached at 2 mg m^{-2} compared to around 1.8 mg m^{-2} with the dephosphorylated.

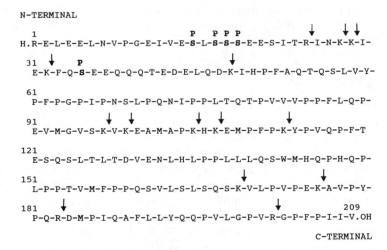

```
N-TERMINAL

 1                                        P   P P P               ↓       ↓ ↓
H.R-E-L-E-E-L-N-V-P-G-E-I-V-E-S-L-S-S-S-E-E-S-I-T-R-I-N-K-K-I-

31 ↓      P                              ↓
E-K-F-Q-S-E-E-Q-Q-Q-T-E-D-E-L-Q-D-K-I-H-P-F-A-Q-T-Q-S-L-V-Y-

61
P-F-P-G-P-I-P-N-S-L-P-Q-N-I-P-P-L-T-Q-T-P-V-V-V-P-P-F-L-Q-P-

91          ↓    ↓         ↓    ↓          ↓
E-V-M-G-V-S-K-V-K-E-A-M-A-P-K-H-K-E-M-P-F-P-K-Y-P-V-Q-P-F-T

121
E-S-Q-S-L-T-L-T-D-V-E-N-L-H-L-P-P-L-L-L-Q-S-W-M-H-Q-P-H-Q-P-

151                                      ↓              ↓
L-P-P-T-V-M-F-P-P-Q-S-V-L-S-L-S-Q-S-K-V-L-P-V-P-E-K-A-V-P-Y-

181 ↓                                               ↓        209
P-Q-R-D-M-P-I-Q-A-F-L-L-Y-Q-Q-P-V-L-G-P-V-R-G-P-F-P-I-I-V.OH

                                                    C-TERMINAL
```

Figure 1 *Amino acid sequence of bovine β-casein A² showing phosphorylated residues (P) and trypsin-sensitive peptide bonds (↓)*

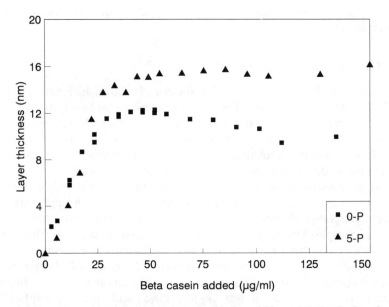

Figure 2 *Hydrodynamic thickness of layers of native (▲) and fully dephosphorylated (■) β-casein adsorbed onto polystyrene latex particles (diameter 91 nm; volume fraction 0.001%) as a function of added protein concentration in imidazole buffer (20 mM; pH 7.0) at 20 °C. (Full details of measuring procedures are given in reference 6)*

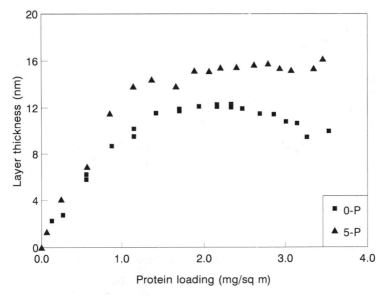

Figure 3 *Hydrodynamic layer thickness on polystyrene latex as a function of protein loading:* ▲, *native β-casein;* ■, *dephosphorylated β-casein*

With both forms of the protein, the loading continues to increase with no accompanying increase in the layer thickness. This is presumably due to the additional protein molecules fitting into gaps in the existing layer. The decrease in the layer thickness which was detected with the dephosphorylated molecule occurs despite the protein loading continuing to increase. This may result from a change in the conformation of the protein molecules at these high molecular densities. The absence of this decrease in thickness with the phosphorylated molecule suggests that the phosphate groups are important in preventing the change. This may be due to charge repulsion by the phosphate clusters preventing the protein molecules, particularly in the N-terminal region, coming so close to each other.

Addition of trypsin to the protein-coated latex causes the thickness of the protein layers to decrease. Figure 4 shows the biphasic behaviour which has been noted previously with phosphorylated β-casein on latex[8]; it consists of a rapid primary stage followed by a slower second stage. The rate of the decrease in thickness in the initial stage was almost three times as fast with the dephosphorylated protein (Table 1), although the rate of the second phase was similar with both proteins. The final thickness of the proteolysed layer was the same in both cases (about 5 nm).

Elution profiles of the tryptic peptides extracted from digests of β-casein-stabilized n-tetradecane-in-water emulsions at early and late stages of the proteolysis are shown in Figure 5. Changes in the retention times of some of the peptides are as a result of removing the phosphate groups. The decrease in the hydrophilicity of these peptides causes an increase in the

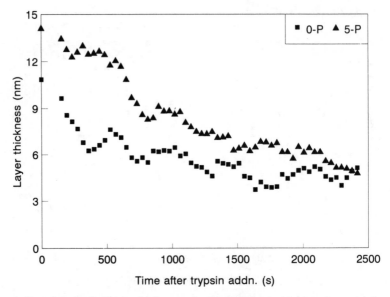

Figure 4 *Decrease in the layer thickness of adsorbed β-casein on polystyrene latex as the protein is hydrolysed by trypsin. Symbols and conditions as in Figure 2. Applied β-casein concentration = 50 μg ml^{-1}; casein/trypsin ratio = 375. Plotted data have been smoothed by calculating a moving average with a 'window' of 5 points*

Table 1 *Rates of trypsinolysis as measured by N-terminal peptide formation in n-tetradecane-in-water emulsions and the decrease in particle layer thickness of loaded latex*

Protein	Initial Rates	
	Emulsion (% min^{-1})	Latex (s^{-1})
Phosphorylated	4.2	9×10^{-3}
Dephosphorylated	7.7	2.5×10^{-2}

retention time on the reverse phase column. Apart from these shifts, the nature and identity of the tryptic peptides are unchanged. This indicates that dephosphorylation does not cause significant changes to the secondary structure of the bound protein.

Initially, proteolytic attack is concentrated on the N-terminal region of the protein, since, in both cases after 10 minutes hydrolysis, the predominant peptides detected are 1-25 and 1-28. A measure of the accessibility of the two ends of the β-casein molecule when it is adsorbed at an oil–water interface can be gained by monitoring the kinetics of formation of the N- and C-terminal peptides. The N-terminal peptide is taken as being the

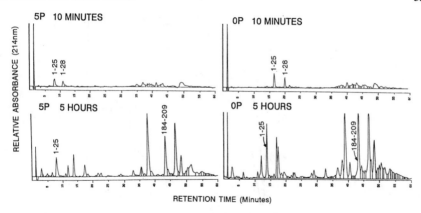

Figure 5 *Reverse phase HPLC elution profiles of tryptic peptides formed from native (5P) and dephosphorylated (0P) β-casein hydrolysed at the n-tetradecane–water interface at 24 °C*

sum of peptides 1-25 and 1-28 since both are formed rapidly in the trypsin-catalysed proteolysis and 1-28 is then further hydrolysed to 1-25. The C-terminal peptide consists of residues 184–209, since at the n-tetra-decane–water interface, the arginyl residue at position 202 is not cleaved under these experimental conditions.[5] The results are shown in Figure 6 together with the same results obtained with the phosphorylated β-casein in a soya oil–water emulsion. All three systems show the rapid formation of the N-terminal peptide but there are clear differences in the rates of attack at the C-terminal region. Whereas with the phosphorylated protein at the soya oil–water interface the plot is hyperbolic, at the n-tetradecane–water interface the plot is sigmoidal with relatively little attack at the C-terminal region until most of the N-terminal peptides have been formed. However, the rate of formation of the C-terminal peptide from the dephosphorylated protein in the n-tetradecane–water system appears to be intermediate between that for the two other systems, having an initial rapid hyperbolic phase at a rate similar to that seen in the soya oil–phosphoryl-ated protein system, followed by a plateau region which in turn is followed by a further increase at a rate similar to that observed in the n-tetradecane–phosphorylated protein system. This suggests that there are two popula-tions of C-terminal regions, one of which is more exposed to proteolytic attack than the other. Measurements of the protein surface loading on n-tetradecane droplets showed that the dephosphorylated protein covers a slightly greater surface area than the phosphorylated one, but only by about 10%. This is in accordance with the view that the loop/tail lies somewhat flatter at the interface as a result of dephosphorylation (*cf.* the aforementioned results with the latex). There is none of the loss of secondary structure which is seen at the soya oil–water interface[5] and which results in a twofold increase in the area which the protein molecule

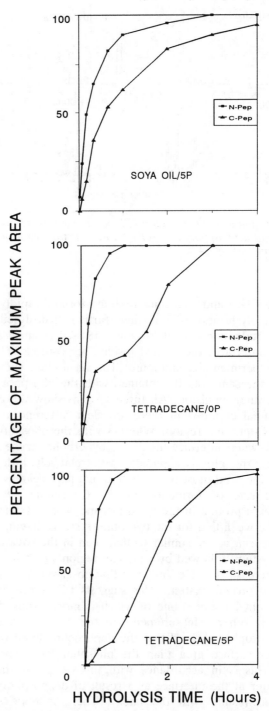

Figure 6 *Time course of the formation of the N-and C-terminal tryptic peptides from various β-casein-stabilized emulsions*

covers and also in formation of novel peptides. A possible explanation for this apparent heterogeneity is that, as a result of the N-terminal region loop/tail not projecting as far from the interface, part of the C-terminal population becomes accessible to the protease after a smaller proportion of the N-terminal region has been cleaved. However, some of the relatively hydrophobic C-terminal regions are still shielded due to the surface density of the casein molecules which is not relieved to any great extent by dephosphorylation.

Comparison of the rates of formation of the N-terminal peptide from the two forms of the protein at the n-tetradecane–water interface (Table 1) shows that this is almost twice as fast with the dephosphorylated as it is with the fully phosphorylated protein. This difference, together with that in the rates of the initial decrease in protein layer thickness on latex, supports the idea that the initial decrease in the thickness of the adsorbed protein layer on latex represents cleavage at the tail/loop region. Why dephosphorylation should enhance the rate of formation of the N-terminal peptide is not yet clear, since it would be expected that the further the relevant peptide bonds were from the surface, the more accessible they would be to the protease. One explanation is that removal of these bulky, negatively charged groups on either side of the trypsin sensitive bonds, allows easier access of the protease.

Attempts to combine the two aspects of this paper by studying both the decrease in the thickness of the protein layers and the peptide formation with the same latex samples has proved not to be possible due to interference in the chromatographic separations by material present in the latex suspension which is co-extracted with the peptides. We are, however, currently studying the binding of protein to unilamellar liposomes; this system should be 'cleaner', and should also permit more protein binding than does the latex.

Acknowledgements

This work was funded by The Scottish Office Agriculture and Fisheries Department. The skilled assistance of Mrs C. M. Davidson in performing the experiments with the polystyrene latex is much appreciated.

References

1. M. J. Pearse, P. M. Linklater, R. J. Hall, and A. G. Mackinlay, *J. Dairy Res.*, 1986, **53**, 381.
2. E. W. Bingham, H. M. Farrell, and K. J. Dahl, *Biochim. Biophys. Acta*, 1976, **429**, 448.
3. M. Yoshikawa, M. Tamaki, E. Sugimoto, and H. Chiba, *Agric. Biol. Chem.*, 1974, **38**, 2051.
4. J. Leaver and D. G. Dalgleish, *Biochim. Biophys. Acta*, 1991, **1041**, 217.
5. J. Leaver and D. G. Dalgleish, *J. Colloid Interface Sci.*, 1992, **149**, 49.

6. D. G. Dalgleish, *Colloids Surf.*, 1990, **46**, 141.
7. A. R. Mackie, J. Mingins, and A. N. North, *J. Chem. Soc., Faraday Trans.*, 1991, **87**, 3043.
8. D. G. Dalgleish and J. Leaver, *J. Colloid Interface Sci.*, 1991, **141**, 288.
9. E. W. Bingham, H. M. Farrell, and R. J. Carroll, *Biochemistry*, 1972, **11**, 2450.
10. D. T. Davies and A. J. R. Law, *J. Dairy Res.*, 1987, **54**, 369.
11. D. T. Davies and A. J. R. Law, *J. Dairy Res.*, 1977, **44**, 213.
12. M. M. Bradford, *Anal. Biochem.*, 1976, **72**, 249.

Food Surfactants at Triglyceride–Water and Decane–Water Interfaces

By R. D. Bee, J. Hoogland,[1] and R. H. Ottewill[1]

UNILEVER RESEARCH, COLWORTH HOUSE, SHARNBROOK, BEDFORD MK44 1LQ, UK
[1]SCHOOL OF CHEMISTRY, UNIVERSITY OF BRISTOL, BRISTOL BS8 1TS, UK

1 Introduction

Monoglycerides were introduced as commercial food emulsifiers more than fifty years ago.[1] Since then the food ingredients industry has broadened the properties, and hence the applications, of these materials by chemical modifications to the glyceride head-group. Food emulsifiers of increasing hydrophilic character have been created by esterification with hydrophilic organic acids or by polymerization of the glycerol unit.[2] As with simple monoglycerides, these modified monoglycerides do not solely function by adsorbing at oil–water interfaces. They also, for example, modify crystallization of fats, bind to proteins or polysaccharides, and compete with polymeric surfactants at the air–water interface.[3,4]

Despite these complex contributions, their interfacial properties remain of central importance in those food applications where oil is dispersed in an aqueous phase, or *vice versa*. In this paper we compare the interfacial behaviour of six chemically modified monoglycerides covering HLB values in the range 5–9. Particular attention is drawn to the effect of sodium chloride addition ($0.5 \, \text{mol} \, \text{dm}^{-3}$) to the aqueous phase and to the marked dissimilarity in behaviour caused by the decrease in polarity when triglyceride oil is substituted by a paraffin oil. Electrophoretic mobility measurements obtained as a function of sodium chloride concentration on oil-in-water emulsion droplets are used to aid the interpretation of the interfacial tension behaviour.

2 Experimental

The organic acid esters of monoglycerides and the polyglycerol monostearate were of commercial grade, purchased from Grindsted, and used without further purification. Diglycerol monopalmitate (>85% pure) was

donated by Dr. P. C. Harries, Unilever Research, Colworth. Samples were
sealed from the light and stored at 277 K.

De-ionized water was distilled from alkaline permanganate in a Pyrex
still; its conductivity was $<10^{-4} \Omega^{-1} m^{-1}$ and it had a surface tension
>72 mN m^{-1} at 298 K. n-Decane and miglyol® 812 (Aldrich Limited and
Dynamit Nobel Chemical Limited) were used as purchased. The n-decane
was $>99\%$ by GLC. Miglyol® was confirmed by GLC to be a mixture of
octanoic and decanoic acid triglycerides. Sodium chloride was BDH
AnalaR-grade material.

Interfacial Tension Measurements

A spinning drop interfacial tensiometer (designed at the University of
Bristol, constructed by V. Bailey, Prototype Engineering, Windsor) was
used throughout the work. Interfacial tensions γ were calculated from the
relationship derived by Vonnegut[5,6] for elongated drops,

$$\gamma = (\rho_1 - \rho_2)\, \omega^2 R^3 / 4 \qquad (1)$$

where R is the radius of the drop, ρ_1 is the density of the continuous
phase, ρ_2 is the density of the oil phase, and $\omega = 2\pi n$ is the angular
frequency with n the number of revolutions per minute.

The oil phase was prepared by mixing into it the appropriate amount of
each surfactant and warming the oil to 333 K, at which temperature all the
surfactants were fully soluble. The oil and the aqueous phases were
equilibrated separately in a water bath at 333 K for several hours. The
tensiometer cell was then loaded with the aqueous phase at 333 K and a
drop of the oil, *ca.* 10 μl, was added from a microsyringe. The drop
diameter ($2R$) was recorded at 15 minute intervals to ensure that a steady
state had been achieved.

Emulsions for Electrophoretic Mobility Measurements

The surfactant solutions were prepared by dispersing known weights of the
surfactant in approximately 1 ml of oil and heating the oil to 333 K. A
coarse emulsion was established by adding water (99 ml) at 333 K and
agitating with a Silverson LR4 mixer using an emulsifying screen at full
speed for 15 minutes. The resulting pre-emulsion was finally homogenized
by two passes through a homogenizer (Multispec M500B, nominal homogenization pressure 180 bar). Oil droplets with diameters of the order of
1 μm were obtained with all the surfactants.

Electrophoretic Mobility Measurements

Samples at an oil phase concentration of *ca.* 0.5 vol% were prepared for
measurement by mixing equal volumes of sodium chloride solution of

known concentration with the emulsions prepared as described above. Mixing was achieved by shaking, and then the samples were allowed to stand for 10 minutes. Electrophoretic mobility measurements were performed on a Pen Kem System 3000 Automated Electrokinetics Analyser. Measurements were made at a temperature of 298 K.

3 Results

The surfactants used in the present work were all derivatives of monoglycerides. Their basic chemical formulae and HLB numbers are given in Figure 1. The compounds were not homogeneous. The dominant alkyl chain was reported to be heptadecyl ($C_{17}H_{35}-$), but some other chain lengths were certainly present and it is possible that some diesters were also present.

Interfacial Tension Measurements

First, the behaviour of the surfactants is compared at water–decane and water–miglyol interfaces.

Figure 2 shows the results obtained with two acidic materials. The diacetyl tartaric acid ester at the decane–water interface shows two linear portions which intersect at a concentration of 0.24 g $(100 \text{ ml})^{-1}$. The interfacial tension at the decane–water interface at concentrations above the intersection point is $0.35 \pm 0.1 \text{ mN m}^{-1}$. However, with miglyol, the lowest interfacial tension achieved is only *ca.* 3.0 mN m^{-1}. The shape of

DERIVATIVES OF MONOGLYCERIDES

$$CH_2.O.CO.(CH_2)_{16}.CH_3$$
$$|$$
$$CH.OH$$
$$|$$
$$CH_2.O.R_2$$

R_2	SPECIES	HLB
Acetic acid ester	.CO.CH$_3$	5.4
Lactic acid ester	.O.CO.CH$_2$OH.CH$_3$	6.1
Polyglycerol	.[CH$_2$.CHOH.CH$_2$-O]$_n$H	7.1
Diglycerol	.[CH$_2$.CHOH.CH$_2$-O]$_2$H	8.3
Citric acid ester	.COCH$_2$.COH.COOH.CH$_2$.COOH	8.0
Diacetyl tartaric acid ester	O.CO.CH$_3$ | .CO.CH.C.COOH | O.CO.CH$_2$	9.1

Figure 1 *Chemical formula and HLB number of each of the surfactant materials used*

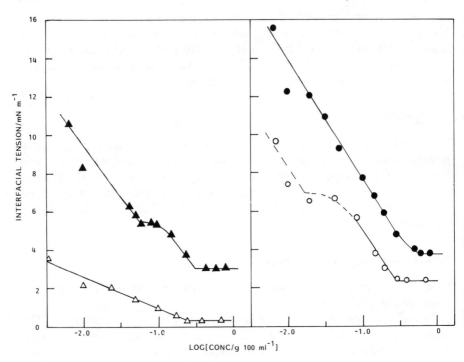

Figure 2 *Plot of interfacial tension against log [concentration (g 100 ml⁻¹)⁻¹] with open symbols denoting decane as the oil phase and closed symbols miglyol. The surfactants are:* ○, *citric acid ester;* △, *diacetyl tartaric ester*

this curve is more complex, being linear at the lowest concentrations, then exhibiting a small plateau followed by a linear portion until a sharp intersection occurs at 0.33 g $(100\,\text{ml})^{-1}$. At higher concentrations the interfacial tension remains constant at *ca.* $3.0\,\text{mN m}^{-1}$. The citric acid ester (Figure 2) is less effective in lowering the interfacial tension at the decane–water interface, the lowest value reached being *ca.* $2.4 \pm 0.1\,\text{mN m}^{-1}$ with an intersection point at 0.29 g $(100\,\text{ml})^{-1}$. With miglyol the lowering of the interfacial tension is even less effective, the lowest reached values being $3.70 \pm 0.1\,\text{mN m}^{-1}$, with an intersection at 0.57 g $(100\,\text{ml})^{-1}$.

The results obtained for the acetic acid and lactic acid esters are shown in Figure 3. Neither the acetic acid ester nor the lactic acid ester was found to be very effective in lowering the interfacial tension of the miglyol–water interface. In the case of the former the lowest value obtained is *ca.* $17.3\,\text{mN m}^{-1}$, and for the latter $15.0\,\text{mN m}^{-1}$. Both materials are more effective as surfactants at the decane–water interface: the lowest value reached with the lactic acid ester was about $3.6\,\text{mN m}^{-1}$, and with the acetic acid ester $7.40\,\text{mN m}^{-1}$, a constant value being achieved for a short concentration range above 0.32 g $(100\,\text{ml})^{-1}$; an intersection occurred at the latter value.

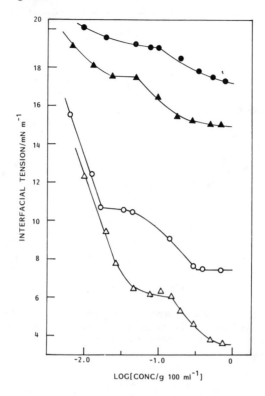

Figure 3 *Plot of interfacial tension against log [concentration (g 100 ml⁻¹)⁻¹] with open symbols denoting decane as the oil phase and closed symbols miglyol. The surfactants are*: ○, *acetic acid ester*; △, *lactic acid ester*

The results for diglycerol and polyglycerol derivatives are illustrated in Figure 4. The interfacial tension against log (concentration) curve in both cases has a rather complex form. The material with the larger polymer head-group gives a plateau in surface tension of *ca.* 4.0 mN m⁻¹ with decane. With miglyol the tension continues to decrease with concentration, the lowest value recorded being 2.4 mN m⁻¹. With the diglycerol material the shapes of the curves are rather similar. The lowest interfacial tension reached at the water–decane interface is 3.3 mN m⁻¹ and with miglyol is *ca.* 2.0 mN m⁻¹. In the case of these materials, the comparison is made between a comparatively pure compound, the diglycerol, and a commercial polyglycerol. However, it should be noted that the interfacial tension behaviour is remarkably similar.

Effects of Salt

The effect on the interfacial tension at the oil–water interface of the various surfactants was also investigated using a 0.5 mol dm⁻³ solution of sodium chloride as the aqueous phase. The effects observed, which were

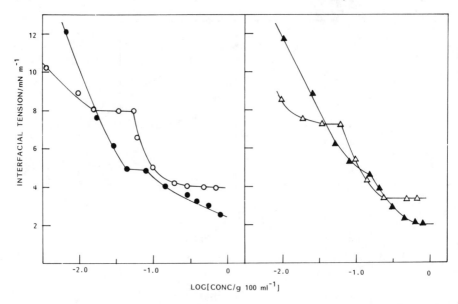

Figure 4 *Plot of interfacial tension against log [concentration (g 100 ml⁻¹)⁻¹] with open symbols denoting decane as the oil phase and closed symbols miglyol. The surfactants are polyglycerol derivatives with (○) n > 2 groups and (△) n = 2 groups*

quite marked, are shown in Figures 5–7 with decane as the oil phase and in Figure 8 with miglyol as the oil phase.

The results essentially fall into two different groups. With the diacetyl tartaric acid, citric acid, diglycerol, and polyglycerol derivatives (Figures 5 and 7), typical surfactant behaviour[7] is observed, a good linear plot of decreasing interfacial tension being obtained at lower concentrations of surfactant, followed by a sharp intersection, after which the interfacial tension remains constant. With all four of these surfactants, the intersection point was found to decrease with the salt concentration, and the interfacial tension was lowered. With the diacetyl tartaric acid ester the lowering of interfacial tension is less marked (see Figure 5).

With the acetic acid and lactic acid esters, the behaviour is quite different, in that the curve obtained with the aqueous salt solution is in both cases moved to a region of higher concentration (Figure 7). Good linear plots are obtained as a function of the logarithm of the concentration and, in the case of the lactic acid ester, a well marked intersection point is obtained at 0.5 g (100 ml)⁻¹ and the interfacial tension, *ca.* 2.33 mN m⁻¹, is lower than against water (3.6 mN m⁻¹). With the acetic acid ester, a plateau region is not reached when the aqueous phase contains salt, at least over the range examined.

Only two sets of results were obtained with miglyol as the oil phase and these are shown in Figure 8. With the citric acid ester, a marked reduction

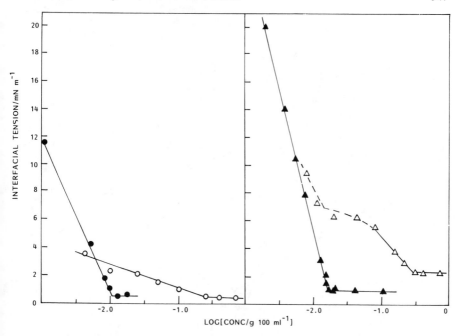

Figure 5 *Plot of interfacial tension against log [concentration* $(g\,100\,ml^{-1})^{-1}]$; ○, *diacetyl tartaric ester;* △, *citric acid ester. Open symbols, decane–water; filled symbols, decane–0.5* $mol\,dm^{-3}$ *sodium chloride solution*

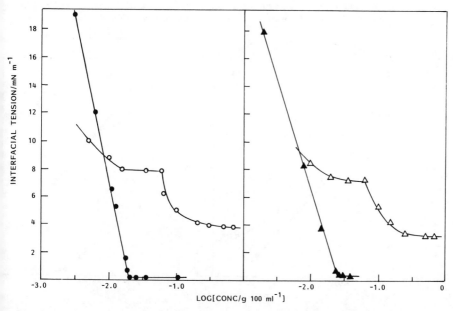

Figure 6 *Plot of interfacial tension against log [concentration* $(g\,100\,ml^{-1})^{-1}]$; ○, *polyglycerol* $(n > 2)$; △, *diglycerol. Open symbols, decane–water; filled symbols, decane–0.5* $mol\,dm^{-3}$ *sodium chloride solution*

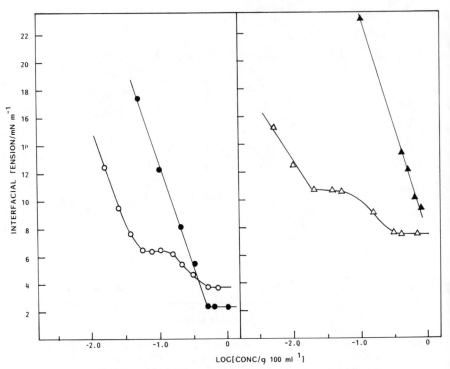

Figure 7 *Plot of interfacial tension against log [concentration* $(g\,100\,ml^{-1})^{-1}]$; $\bigcirc$, *lactic acid ester*; $\triangle$, *acetic acid ester. Open symbols, decane–water; filled symbols, decane–0.5 mol dm*$^{-3}$ *sodium chloride solution*

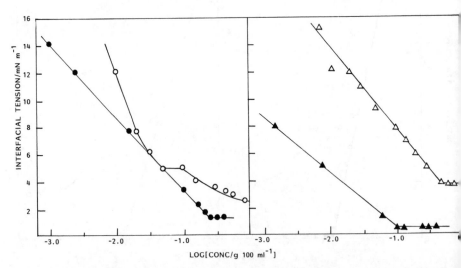

Figure 8 *Plot of interfacial tension against log [concentration* $(g\,100\,ml^{-1})^{-1}]$; $\bigcirc$, *polyglycerol*; $\triangle$, *citric acid ester. Open symbols, miglyol–water; filled symbols, miglyol–0.5 mol dm*$^{-3}$ *sodium chloride solutions*

in interfacial tension occurs against the salt solution and the interfacial tension is reduced to *ca.* $0.4 \, \text{mN m}^{-1}$; also the reduction commences at a lower surfactant concentration. With the polyglycerol derivative a much more linear relationship of interfacial tension against log (concentration) is obtained and a sharp intersection point is found at 0.24 g $(100 \, \text{ml})^{-1}$. The interfacial tension is reduced to $1.2 \pm 0.1 \, \text{mN m}^{-1}$. Again the interfacial tension is lower in the presence of salt.

Electrophoretic Mobility of Emulsion Droplets

Oil-in-water emulsions with both decane and miglyol as the oil phase were prepared with each of the surfactants, and the electrophoretic mobilities were determined as a function of sodium chloride concentration. The results are presented in Figure 9. The two surfactants with free carboxyl groups, the diacetyl tartaric acid ester and the citric acid ester, show the highest mobility values. In addition, the electrophoretic mobilities of the droplets stabilized with these two compounds exhibit maxima in mobility in the sodium chloride range from 5×10^{-2} to $5 \times 10^{-3} \, \text{mol dm}^{-3}$. These

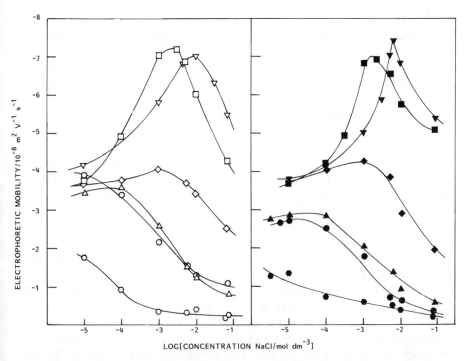

Figure 9 *Plot of electrophoretic mobility against* log [*sodium chloride concentration* $(\text{mol dm}^{-3})^{-1}$] *for various surfactants:* ∇, *diacetyl tartaric acid ester;* $\square$, *citric acid ester;* $\diamond$, *lactic acid ester;* $\bigcirc$, *acetic acid ester;* $\triangle$, *polyglycerol* $(n > 2)$; $\bigcirc$, *diglycerol; open symbol denotes decane as the oil phase and closed symbol denotes miglyol as the oil phase*

maxima in electrophoretic mobility are well known and frequently occur with highly charged particles. The cause of the maximum is often attributed to surface conductance effects occurring as a consequence of the higher concentration of counter-ions in the diffuse double-layer surrounding the particles.[8] The difference observed between decane as the oil phase and miglyol as the oil phase is relatively small.

The results obtained with droplets stabilized by the acetic acid ester and the lactic acid ester show a small increase in mobility with increase in salt concentration followed by a gradual fall-off with further increase in salt. Qualitatively, however, the two curves are very similar using both decane and miglyol as the oil phase.

The lowest electrophoretic mobility values were obtained with the polyglycerol derivatives. Interestingly, the material with the larger number of glycerol units gives somewhat higher mobility values. This is contrary to expectations since, if, as seems reasonable to assume, the polyglycerol chains project into the aqueous phase, this would move the electrokinetic slipping plane further away from the oil–water interface and hence lead to a decrease in mobility compared with the diglycerol ester.

4 Discussion

The modified monoglyceride surfactants investigated here can be subdivided into 3 groups: those with head-groups containing essentially only an ester functionality, the most hydrophobic; those with a polyglycerol functionality; and those with free carboxylic acid groups in the head group, the most hydrophilic. The latter two groups show a number of similar characteristics, whereas the first group behaves quite differently.

In the case of the diacetyl tartaric acid and citric acid esterified monoglycerides, the electrophoretic mobility results show them to be negatively charged. At the lower salt concentrations the mobility increases to a maximum and then decreases. The reason for the increase in mobility could be the increasing degree of ionization of the carboxyl groups at the interface with increasing ionic strength;[9,10] surface conductance also affects the results for highly charged particles in this region. The high mobility values indicate a highly charged interface, and the decrease of mobility at the higher salt concentrations is as anticipated on the basis of electrical double-layer theory.

The changes in the interfacial tension values on moving from water to $0.5 \, mol \, dm^{-3}$ sodium chloride as the continuous phase, that is to a linear plot with a lower intersection point and a lower interfacial tension, are consistent with a better packing of the charged anionic head-groups at the interface in the presence of salt. This constitutes an example of 'salting-out', a phenomenon observed with many surfactants having ionic head-groups. Taking the well-defined linear slope for the data at the decane– $0.5 \, mol \, dm^{-3}$ salt interfaces (Figures 5, 6, and 7), and applying the Gibbs equation in its simple form, the areas occupied per molecule for the diacetyl tartaric ester, diglycerol ester, lactic acid ester, and the acetic acid ester are calculated to be 83 Å^2, 70 Å^2, 73 Å^2, and 69 Å^2, respectively.

The area occupied per molecule for the citric acid ester and the polyglycerol ester at the 0.5 mol dm^{-3} sodium chloride interfaces with both decane and miglyol have been deduced from Figures 5, 6, and 8. With decane, the citric acid ester and the polyglycerol ester occupy 52 Å^2 and 45 Å^2, respectively, whilst at the miglyol interface the areas are 248 Å^2 and 172 Å^2, respectively. These data lead to the conclusion that the expected influence of the head-group on packing at the interface is observed, but, more importantly, that the overriding influence on the interfacial packing is the nature of the oil phase.

The sharp transition observed in the γ *versus* log (concentration) curve indicates that the activity of the surfactant becomes constant in the bulk phase as a consequence of either association occurring or a solubility limit being reached. Since no precipitation was detected in the presence of salt, it seems most probable that there is some association, with the sharpness of the transition indicating a substantial association number, *i.e.* of the order of 100.[7] Scattering measurements are needed on both the oil and aqueous phases in order to confirm whether association occurs in only one phase or possibly even in both.

At the lower concentrations of several surfactants in the absence of salt, a small region of constant interfacial tension is obtained before the value decreases to the intersection point. The reason for this is currently not clear. It could be a consequence of species of lower surface activity being replaced by those of higher surface activity as found in work on binary surfactant mixtures.[11] However, it should be noted that a well-defined plateau was observed with both the polyglycerol derivatives. In this case, the material with $n = 2$ had been synthesized and purified by distillation before esterification and yet both this and the commercial material showed essentially similar behaviour. An alternative explanation with these complex materials is that some re-orientation of the molecule at the interface occurs as the concentration increases.

The electrophoretic mobilities of emulsion droplets made with the diglycerol derivative are low and decrease rapidly in the presence of salt. Both the decane and miglyol droplets give essentially identical results. These are typical of those found with an organic phase which has no 'in-built' surface charge. It occurs as a consequence of weak association of the more polarizable inorganic anions with the interfacial region accompanied by concomitant negative adsorption of the more hydrated cations.[8,10,12] The electrophoretic behaviour of the polyglycerol ($n > 2$) is, however, quite different. A much higher electrophoretic mobility is exhibited with both decane and miglyol droplets, which suggests the presence of some free charges at the interface, possibly attributable to trace impurities of fatty acid in the original material.

The interfacial tension behaviour of the glycerol derivatives shows a strong 'salting-out' effect in the presence of sodium chloride suggesting that the activity of the saline solution is less favourable towards accommodating the hydrophilic part of the molecule than is water. The presence of high concentrations of sodium and chloride ions in the solution means that many water molecules are involved in hydrating the ions; this means, in principle, that fewer water molecules are available to hydrate the bulky polyglycerol head group of the surfactant.

The behaviour of the acetic acid and lactic acid esters contrasts with the behaviour of the four other monoglyceride derivatives, particularly in that the interfacial tension appears to show 'salting-in', the curves in the presence of salt moving to higher surfactant concentration. Moreover, neither of these compounds are effective in lowering the interfacial tension of the miglyol–water interface. The electrophoretic behaviour of the two compounds is rather different in the case of both decane and miglyol as the dispersed phase. The electrophoretic mobilities of the lactic acid ester emulsion droplets are substantially higher than those of the acetic acid ester, possibly indicating trace ionic impurities in the surfactant. Salting-in suggests that less work needs to be done on the aqueous phase in the presence of salt in order to insert the molecule into a cavity in the water.[13] Since the salt dissolves in the water only this effect must be associated with the aqueous phase. Unlike the other more hydrophilic head groups, the ester functionality makes less demands on water molecules for hydration, and a possible explanation is that it becomes easier to insert the hydrophobic molecule into the aqueous phase once the structure of the water had been modified by the dissolution of salt. Moreover, the chloride ion is considered to be a water-structure-breaking ion,[14,15] which supports this hypothesis.

Alternatively, the apparent 'salting-in' may result from the two most hydrophobic surfactants being located predominantly in the oil phase. The decreased interfacial activity is then caused by the diminished solvating property for the head group of the salt solution compared with pure water.

The basic objective of the present work was to examine the role of monoglyceride surfactants in modifying the interfaces between water and a hydrocarbon and between water and a triglyceride. The results obtained can to some extent be rationalized on the basis of surfactant type. However, a detailed interpretation of the results requires more extensive information on the influence of salts on the solution behaviour of the surfactants in both the oil and aqueous phases. Evidence for micellar association has been obtained but needs to be confirmed by scattering experiments. Moreover, solubility data and partition coefficient data are still lacking for these materials.

Acknowledgement

Our thanks are due to the Agricultural and Food Research Council for studentship support to J. Hoogland during the course of this work.

References

1. E. A. Flack and N. J. Krog, *Food Trade Review*, 1970, August, p. 29.
2. N. J. Krog, in 'Food Emulsions', 2nd edn, ed. K. Larsson and S. E. Friberg, Dekker, New York, 1990, p.127.
3. N. J. Krog, *J. Am. Oil Chem. Soc.*, 1977, **54**, 124.
4. E. Dickinson and G. Stainsby, 'Colloids in Food', Applied Science, London, 1982.
5. B. Vonnegut, *Rev. Sci. Inst.*, 1942, **13**, 6.
6. A. Couper, R. J. Newton, and C. Nunn, *Colloid Polym. Sci.*, 1983, **261**, 371.
7. R. H. Ottewill, in 'Surfactants', ed. Th. F. Tadros, Academic Press, London, 1984, p. 7.
8. R. J. Hunter, 'Zeta Potential in Colloid Science', Academic Press, London, 1981.
9. R. H. Ottewill and J. N. Shaw, *Kolloid Z. Z. Polym.*, 1967, **218**, 34.
10. R. H. Ottewill and D. G. Rance, *Colloid Polym. Sci.*, 1986, **264**, 982.
11. J. H. Clint, *J. Chem. Soc., Faraday Trans. 1*, 1975, **71**, 1327.
12. R. Buscall and R. H. Ottewill, *ACS Adv. Chem. Ser.*, 1975, **144**, 83.
13. M. J. Rosen, 'Surfactants and Interfacial Phenomena', Wiley-Interscience, New York, 1978.
14. J. L. Kavanau, 'Water and Solute–Water Interactions', Holden-Day Inc., London, 1964, p. 66.
15. G. R. Choppin and K. Buijs, *J. Chem. Phys.*, 1963, **39**, 2042.

Surface Dilational Behaviour of a Mixture of Aqueous β-Lactoglobulin and Tween 20 Solutions

By David C. Clark, Peter J. Wilde, Diane J. M. Bergink-Martens[1], Anita J. J. Kokelaar[1], and Albert Prins[1]

AFRC INSTITUTE OF FOOD RESEARCH, NORWICH LABORATORY, NORWICH RESEARCH PARK, COLNEY, NORWICH NR4 7UA, UK
[1]WAGENINGEN AGRICULTURAL UNIVERSITY, DEPARTMENT OF FOOD SCIENCE, LABORATORY OF DAIRYING AND FOOD PHYSICS, PO BOX 8129, 6700 EV WAGENINGEN, THE NETHERLANDS

1 Introduction

Many processed foods contain complex mixtures of surface-active molecules. Principal amongst these are proteins, which are macromolecules and can be highly effective in the stabilization of food foams and emulsions.[1] Low-molecular-weight molecules such as natural polar lipids (lecithins) or synthetic emulsifiers (*e.g.* monoglycerides and polysorbates) constitute another family of surface-active molecules found in food.[2] These two different classes of molecules not only compete for interfacial area[3-6] but can also interact with each other by non-covalent binding.[7] The complexes resulting from this interaction can have different surface behaviour than that expected from the sum of the individual species.[5]

We have studied a two component interacting system of this type in considerable detail. The mixture we have investigated is comprised of the bovine whey protein, β-lactoglobulin, and the high HLB polysorbate emulsifier, Tween 20 (polyoxyethylene 20 sorbitan monolaurate). The properties of the high affinity Tween 20 binding site on the β-lactoglobulin molecule have been re-investigated recently and a dissociation constant has been determined.[8]

In earlier studies we identified a correlation between foam instability in β-lactoglobulin–Tween 20 mixtures and the onset of Tween 20 induced surface diffusion of adsorbed protein using a fluorescence recovery after photobleaching (FRAP) technique.[6] More recently, it has become evident that the β-lactoglobulin–Tween 20 complex is more surface active at the oil–water interface than at the air–water interface.[5,8] In this paper, we describe a further investigation of this mixed system to include measurements of dynamic surface properties including surface dilational elasticity and viscosity[9] and the dynamic surface tension of a continuously expanded liquid surface.[10] Dynamic surface properties in general are particularly important in foams since the thin films between the gas bubbles in these dispersions are subjected to mechanical and thermal disturbances during their production and storage.

2 Materials and Methods

The whey protein β-lactoglobulin was obtained from Sigma Chemical Co. High purity Tween 20 (Surfact-Amp grade) was obtained from Pierce Chemical Co. In this study the value of R refers to the molar ratio of Tween 20 to β-lactoglobulin.

The surface dilational modulus E of a liquid is the ratio between the small change in surface tension ($\Delta\gamma$) and the small relative change in surface area ($\Delta\ln A$):

$$E = \frac{\Delta\gamma}{\Delta\ln A} \qquad (1)$$

Preliminary measurement of the dynamic surface dilational modulus typically involves an oscillatory experiment. Our measurements were made in a 5 cm diameter glass vessel. The experiment involved the positioning of two surface-roughened glass plates at the liquid–air interface; one was present to change the interfacial area, the other to measure the change in surface tension $\Delta\gamma$. A periodical expansion and compression of the interface was achieved by dipping a surface-roughened glass plate into the vessel using a sinusoidal drive. $\Delta\gamma$ was measured by the Wilhelmy plate method as a function of the amplitude of the movement of the dipping plate (change in interfacial area). Data were plotted on an X–Y recorder and appeared in the form of an ellipse. The phase angle difference θ between the area and tension changes can be expressed as

$$\theta = 2\arctan (x/y) \qquad (2)$$

where x and y are the width and length of the ellipse. θ is also referred to as the loss angle. The surface dilational modulus is a complex quantity:

$$E = |E|\cos\theta + i|E|\sin\theta \qquad (3)$$

where $|E|$ represents the amplitude ratio between the stress and the strain. The real part of the surface dilational modulus is the 'storage' modulus ε' (often called the surface dilational elasticity E_d). The second or imaginary part of equation (3) is the loss modulus ε'', which is related to the product of the radial frequency ω ($\mathrm{rad\,s^{-1}}$) and the surface dilational viscosity η_d ($\mathrm{N\,s\,m^{-1}}$). That is, we have

$$\varepsilon' = E_d = |E|\cos\theta \tag{4}$$

$$\varepsilon'' = \eta_d\omega = |E|\sin\theta \tag{5}$$

In our experiments, the linearity of the response was confirmed by plotting the amplitude of the plate movement against $\Delta\gamma$. Only data from the linear region of this plot were used. These preliminary measurements were carried out at a single frequency of $0.628\ \mathrm{rad\,s^{-1}}$.

The dynamic surface tension γ_{dyn} of mixtures of Tween 20 and β-lactoglobulin solutions was examined using an overflowing cylinder (OFC) apparatus, the principles of which have been described previously.[10] The main feature of this apparatus is that the measurements are made at a continuously expanding surface far from equilibrium conditions. A small capacity glass version of the OFC apparatus was used to reduce the volume of liquid required to approximately 600 ml. The OFC apparatus consisted of a vertical glass cylinder with a diameter of 6 cm through which liquid was pumped upwards. The cylinder was approximately 10 cm long and surfactant solution entered the cylinder through a sintered glass plate at its base. The sinter ensured laminar flow of the solution up the cylinder. In these preliminary experiments the rate of expansion of the interface ($d\ln A/dt$) was not determined. The dynamic and equilibrium surface tensions were measured using the Wilhelmy plate technique under flow and non-flow conditions, respectively. A glass plate with a roughened surface was used. The plate was positioned approximately in the centre of the expanding surface at the top of the cylinder. Care was taken to ensure that the liquid surface was horizontal throughout the experiments.

In a typical experiment, the apparatus was filled with 10 mM sodium phosphate buffer pH 7.0 containing 0 or $0.4\ \mathrm{mg\,ml^{-1}}$ β-lactoglobulin (21.7 μM). Measurements of dynamic and equilibrium surface tension were made under conditions of flow and no flow, respectively. Small volumes of $0.1\ \mathrm{kg\,dm^{-3}}$ Tween 20 (81.5 mM) were added sequentially under conditions of flow to achieve rapid mixing and the dynamic and equilibrium surface tension measurements were made after each addition. This experimental protocol reduced the requirement for large quantities of purified β-lacto-globulin.

3 Results

The surface dilational properties of the mixture of Tween 20 and β-lacto-globulin were investigated and the results compared with existing data from FRAP experiments.[6] The β-lactoglobulin concentration was kept constant at $0.2\ \mathrm{mg\,ml^{-1}}$ (10.9 μM) and separate solutions containing different concentrations of Tween 20 were investigated. It was evident from visual evaluation of the raw data [elliptical plots of amplitude of displacement of the plate (*i.e.* area change) against $\Delta\gamma$] that a transition in dilational properties occurred at an R value (molar ratio of Tween 20 to β-lactoglobulin) of approximately 1.0. This was signified by a very sharp transition in the shape of the ellipse from long and thin to short and wide at approximately $R = 1$. The results of a more detailed analysis of the dilational data are given in Figure 1. The surface dilational modulus, the storage modulus, the loss modulus, and the tangent of the loss angle are plotted against R, the molar ratio of the solution components. At $R < 1$, the surface dilational modulus and the storage modulus (dilational elasticity) were large and relatively constant. In contrast, as R was increased from 0 to 1 the loss modulus (dilational viscosity) increased. This signifies that the surface was dominated by the elastic component below $R = 1$, but as R exceeded 1 the viscous component became more important. At approximately $R = 1$ a transition occurred. As R was increased above $R = 1$, the magnitude of the surface dilational modulus and the storage modulus decreased sharply, whereas the loss modulus soon reached a plateau where its value did not change significantly over the range of $R = 1.5$ to 10, and consequently the value of the tangent of the loss angle showed a sharp inflection in the region of $R = 1$.

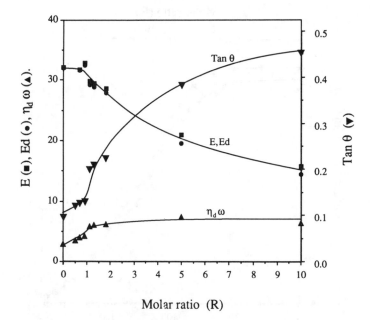

Molar ratio (R)

Figure 1 *A plot of the surface dilational modulus (■), the storage (surface dilational elasticity) modulus (●), the loss (surface dilational viscosity) modulus (▲), and the tangent of the loss angle (▼) as a function of R, the molar ratio of Tween 20 to β-lactoglobulin. The experiments were carried out at a single frequency of 0.628 rad s⁻¹ with a constant protein concentration of 0.2 mg ml⁻¹ (10.9 μM)*

Separate experiments were performed to examine the dynamic surface tension (γ_{dyn}) properties of this two component system. The measurements were performed using a continuously expanding surface and solutions containing Tween 20 alone and Tween 20 in the presence of 0.4 mg ml^{-1} β-lactoglobulin (21.7 μM) and are presented in Figure 2. The γ_{dyn} data curves shown in Figure 2 have similar shapes to data curves obtained under equilibrium conditions (not shown), but are shifted to increased concentration by approximately two orders of magnitude. This effect was reported previously for other systems.[11] The curve for Tween 20 alone has a classical sigmoidal shape. Limited sample availability prohibited determination of the minimum γ_{dyn} for Tween 20. However, the gradient of the curve had become reduced at the highest concentrations examined, indicating that we were approaching the γ_{dyn} minimum. A solution containing 0.4 mg ml^{-1} β-lactoglobulin (21.7 μM) alone caused a small reduction (1.5 mN m^{-1}) in the γ_{dyn} of the water. This remained unaltered upon addition of Tween 20 up to a concentration of approximately 15 μM. Above this concentration a small but significant further reduction in γ_{dyn} was observed. The difference between the γ_{dyn} curves for Tween 20 in the presence and absence of β-lactoglobulin continued to increase until approx-

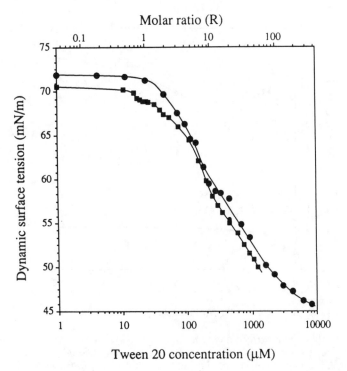

Figure 2 *A plot of dynamic surface tension determined using the overflowing cylinder apparatus for Tween 20 alone (●) and Tween 20 in the presence of 0.4 mg ml^{-1} (21.7 μM) β-lactoglobulin (■)*

imately 40 μM Tween 20 was present. This effect resulted in the appearance of a small but reproducible inflection in the γ_{dyn} curve in the region corresponding to 15–40 μM Tween 20. As the concentration of Tween 20 was further increased, the curve for the mixed sample followed the shape of the curve for Tween 20 alone.

Interpretation of the data from both sets of experiments described here requires knowledge of the precise composition of the solutions under investigation. Since the two components studied here interact, the composition of the solutions was dependent on the Tween 20 binding properties of β-lactoglobulin. A dissociation constant (K_d) of 4.6 μM for the β-lactoglobulin/Tween 20 complex has been reported recently.[8] The K_d may be expressed in the form:

$$K_d = \frac{[P][L]}{[PL]} \tag{6}$$

where [P], [L], and [PL] are the molar concentrations of the free protein (β-lactoglobulin), free ligand (Tween 20) and protein–ligand complex (β-lactoglobulin Tween 20 complex). Since K_d and the initial protein and ligand concentrations ([P]$_{init}$ and [L]$_{init}$) are known, [PL] can be calculated by solving the quadratic equation obtained upon substituting

$$[P] = [P]_{init} - [PL] \tag{7}$$

and

$$[L] = [L]_{init} - [PL] \tag{8}$$

into equation (6) above. Using such an approach, the solution composition present under the sample conditions prevailing in the surface dilational experiments [0.2 mg ml^{-1} (10.9 μM) β-lactoglobulin] is shown in Figure 3(a). Similarly, the solution composition in the OFC experiments (0.4 mg ml^{-1} (21.7 μM) β-lactoglobulin) is shown in Figure 3(b). The concentrations of the three components, free protein, free Tween 20, and protein–Tween 20 complex are plotted as a function of the total concentration of added Tween 20. Comparison of Figure 3(a) and 3(b) demonstrates that doubling the total protein concentration did not result in maintenance of the same solution composition shifted to higher total Tween 20 concentrations. Rather, doubling the protein concentration completely changed the relative concentrations of the three solution species across the complete concentrations range of Tween 20.

4 Discussion

The surface dilational properties of this system were studied in order to allow comparison with existing FRAP data. The area of the interface was

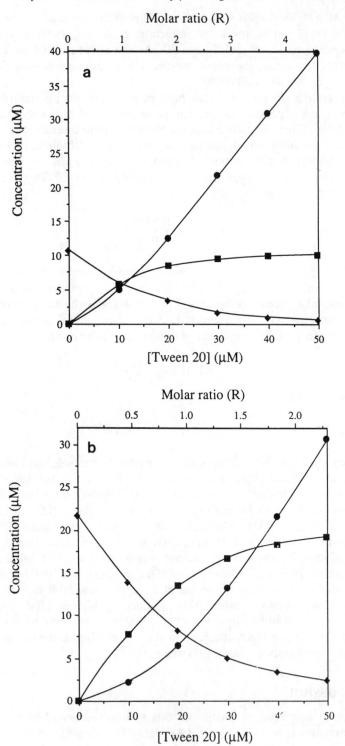

expanded and compressed sinusoidally rather than expanded at a constant rate as was the case in the overflowing cylinder. The raw data were in the form of an ellipse that contained information about the phase angle difference between the area and surface tension changes for each particular sample. Qualitative examination of the raw data revealed a transition in the dilational behaviour that co-incided with the appearance of the complex as the dominant solution component in concentration terms [Figure 3(a)]. This transition correlates precisely with the observed change in surface diffusion properties of the protein as determined by fluorescence recovery after photobleaching (FRAP) described previously.[6] Our interpretation of this transition remains unchanged and is shown schematically in Figure 4. Our hypothesis is that the complex formed between β-lactoglobulin and Tween 20 is incapable of interacting with other molecules of complex or with uncomplexed β-lactoglobulin at the interface. This is due to steric shielding of the protein–protein interaction sites by the polyoxyethylene chains of Tween 20.[12] This moiety is highly hydrated and projects from the surface of the protein into the surrounding solution. Quantitative analysis of the dilational data confirmed this hypothesis, since the magnitude of the surface dilational and storage moduli remained relatively constant in the range of R from 0 to 1. Thus, in this range of R values, the elastic component dominated the dilational properties. This signifies that protein–protein interactions were present. At $R > 1$ there was a decrease in the surface dilational and storage moduli consistent with reduction in the extent of protein–protein interactions at the surface. This behaviour is built into our model (Figure 4). As significant amounts of the complex start to appear at the air–water interface, interactions between this and other molecules of complex are obstructed by the polyoxyethylene chains. In contrast, the loss modulus increases over the range $R = 0$ to 1. This correlates with an observed decrease in the thickness of the interfacial layer in this range of R values.[6] The loss modulus reached a plateau value at an R value of 1.3. This corresponded to the end of the initial increase in the surface diffusion coefficient of the surface adsorbed protein determined by FRAP.[6] We interpret this to indicate the point where the interface is completely populated by complex. We have demonstrated by FRAP that further increases in the surface diffusion coefficient occur at $R > 1.3$ as the surface concentration of complex is reduced by adsorption of increasing amounts of free Tween 20 into the interface. This is the penultimate stage in the schematic representation of our model (Figure 4). There was found

---

Figure 3 *Plots of the composition of solutions of mixtures of β-lactoglobulin and Tween 20 as a function of the total added Tween 20 concentration. The concentrations of free β-lactoglobulin (◆), free Tween 20 (●), and β-lactoglobulin/Tween 20 complex (■) are shown. The solutions contained a total concentration of β-lactoglobulin of (a) 0.2 mg ml⁻¹ (10.9 μM) and (b) 0.4 mg ml⁻¹ (21.7 μM). A dissociation constant of 4.6 μM was used in the calculation*

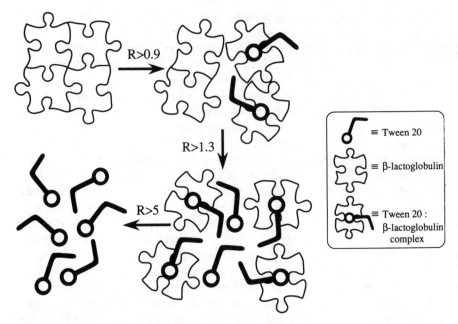

Figure 4 *A schematic representation of the change in the composition of the adsorbed layer at the air–water interface in solutions containing mixtures of Tween 20 and β-lactoglobulin as the total concentration of Tween 20 is increased*

to be insignificant change in the loss modulus over this range of R. Finally, the complex is displaced from the interface by excess Tween 20.

The reduction in dynamic surface tension due to adsorption is diffusion limited. Diffusion of proteins to a freshly created interface is slower than diffusion of low molecular weight surfactants due to their larger size. The importance of diffusion as a controlling factor can be reduced by increasing the concentration of the solution. We believe that this accounts for the observed small reduction in γ_{dyn} seen with the protein in the absence or presence of low concentrations of Tween 20 (0–10 μM). Indeed, the reduction in γ_{dyn} by the protein shown in Figure 2 is as great as that obtained with an equivalent molar concentration of Tween 20. Given the approximately 15-fold difference in molecular weight of these species, it is evident that β-lactoglobulin is highly effective at reducing the γ_{dyn} (Figure 2). After reducing the surface tension by adsorption, proteins can cause further reductions in surface tension by unfolding to reveal previously buried hydrophobic amino-acid side chains. This process, often referred to as surface denaturation, is very slow and can take tens of hours depending on the compactness of the globular structure of the protein.[13] We believe that this process is of little importance with respect to the γ_{dyn} data presented here.

The decrease in the γ_{dyn} curve for the mixture at concentrations of

Tween 20 greater than 10 μM cannot be due to the adsorption of Tween 20 since the concentration of free Tween is reduced in the presence of β-lactoglobulin due to specific binding of this molecule by the protein [Figure 3(b)]. The inflection in the γ_{dyn} curve (Figure 2) for protein with Tween 20 occurred when approximately 15 μM Tween 20 was present. At this point, the protein/Tween 20 complex was the dominant component in the mixture, in terms of relative concentration [Figure 3(b)]. This suggested that the complex was responsible for the additional reduction in γ_{dyn}.

In Figure 2, the greatest difference between the two curves is observed at a Tween 20 concentration of approximately 40 μM. This corresponds to the point [Figure 3(b)] where the protein binding site is almost saturated, where the complex ceases to be the dominant species in terms of relative concentration and where the concentration curves for free Tween 20 and complex cross (*i.e.* the values are similar). Therefore, we attribute the reduction in γ_{dyn} in the region of the inflection (15–40 μM Tween 20) to the adsorption of β-lactoglobulin Tween 20 complex at the expanding interface.

Finally, the γ_{dyn} experiments (Figure 2) indicate that there is very little difference between the surface activity of free Tween 20, β-lactoglobulin, and the Tween 20 β-lactoglobulin complex. This is the case because the γ_{dyn} behaviour of the mixed system is completely dominated by the most highly concentrated component present. In Figure 2, the γ_{dyn} behaviour can be divided into three regions corresponding to 0–15 μM, 15–40 μM, and >40 μM Tween 20. The junctions between these regions correlate exactly with transitions in the dominant species present in the solution [Figure 3(b)]. In the range 0–15 μM Tween 20, β-lactoglobulin is the dominant species; in the range 15–40 μM Tween 20, the Tween 20 β-lactoglobulin complex is the dominant species; above 40 μM Tween 20, free Tween 20 is the dominant species. Therefore, these components must have similar surface activity since the most concentrated component appears to determine γ_{dyn}.

These observations suggest that the β-lactoglobulin/Tween 20 complex has an important effect in reducing γ_{dyn}. This property probably increases the foamability of solutions containing the complex. However, the presence of this species at the interface inhibits formation of interactions between it and adsorbed protein molecules. This is reflected in the major changes observed in the surface dilational modulus. As a consequence, the structural integrity of the adsorbed protein film is reduced, which causes a decrease in the stability of the foam.[6]

5 Conclusions

The surface dilational and dynamic surface tension properties of a solution of two interacting components have been studied. The dilational properties show good correlation with previous FRAP measurements of surface diffusion of the adsorbed species in the same system. The dilational properties and the small but distinct differences observed between the Tween 20 and the β-lactoglobulin/Tween 20 system in the γ_{dyn} data can be explained completely by a simple model that includes all three species present in solution.

Acknowledgement

This collaboration was in part supported by an AFRC International Section Travel Grant which D. C. Clark and P. J. Wilde gratefully acknowledge.

References

1. E. Dickinson, *Colloids Surf.*, 1989, **42**, 191.
2. J. D. Dziezak, *Food Technol.*, 1988, **42**, 172.
3. E. Dickinson, S. R. Euston, and C. M. Woskett, *Prog. Colloid Polym. Sci.*, 1990, **82**, 65.
4. E. Dickinson and C. M. Woskett, in 'Food Colloids', ed. R. D. Bee, P. Richmond, and J. Mingins, Special Publication No. 75, Royal Society of Chemistry, Cambridge, 1989, p. 74.
5. D. C. Clark and P. J. Wilde, in 'Gums and Stabilizers for the Food Industry', ed. G. O. Phillips, D. J. Wedlock, and P. A. Williams, Oxford University Press, 1992, Vol. 6, p. 343.
6. M. Coke, P. J. Wilde, E. J. Russell, and D. C. Clark, *J. Colloid Interface Sci.*, 1990, **138**, 489.
7. N. Krog, *J. Am. Oil Chemists Soc.*, 1977, **54**, 124.
8. P. J. Wilde and D. C. Clark, *J. Colloid Interface Sci.*, 1992, in press.
9. A. J. J. Kokelaar, A. Prins, and M. De Gee, *J. Colloid Interface Sci.*, 1991, **146**, 507.
10. D. J. M. Bergink-Martens, H. J. Bos, A. Prins, and B. C. Schulte, *J. Colloid Interface Sci.*, 1990, **138**, 1.
11. D. J. M. Bergink-Martens, C. G. J. Bisperink, H. J. Bos, A. Prins, and A. F. Zuidberg, *Colloids Surf.*, 1992, **65**, 191.
12. D. C. Clark, P. J. Wilde, and D. R. Wilson, in 'Food Polymers, Gels and Colloids', ed. E. Dickinson, Special Publication No. 82, Royal Society of Chemistry, Cambridge, 1991, p. 272.
13. D. E. Graham and M. C. Phillips, *J. Colloid Interface Sci.*, 1979, **70**, 403.

Nucleation and Growth of Carbon Dioxide Gas Bubbles

By Anne M. Hilton, Michael J. Hey, and Rodney D. Bee[1]

DEPARTMENT OF CHEMISTRY, UNIVERSITY OF NOTTINGHAM, NOTTINGHAM
NG7 2RD, UK
[1]UNILEVER RESEARCH, COLWORTH HOUSE, SHARNBROOK, BEDFORD MK44
1LQ, UK

1 Introduction

The mechanisms whereby carbon dioxide gas bubbles form and grow in supersaturated aqueous solutions are of considerable interest in the food industry, particularly with regard to carbonated beverages. Classical nucleation theory[1-5] provides a framework within which experimental data on both homogeneous and heterogeneous nucleation rates may be rationalized but it cannot yet be used to predict rates in specific systems. There is a need, therefore, for a rapid and simple method of monitoring bubble formation so that the effects of factors such as supersaturation ratio, contact angle, and viscosity, which determine nucleation and growth rates, can be easily studied.

Bubble formation is conventionally studied by saturating a solution under a high pressure of gas and then rapidly releasing the pressure to produce supersaturation. In this paper we describe an alternative method for studying carbon dioxide gas bubbles in which aqueous solutions of sodium bicarbonate and hydrogen chloride are rapidly mixed in a stopped flow apparatus. The short half-life of the reaction means that it is essentially complete before any gas bubbles are formed. Supersaturation ratios are thus readily controlled by choosing appropriate reagent concentrations. The number of bubbles formed is recorded photographically and their growth is followed by monitoring the turbidity in the reaction cell over an interval of *ca.* 200 ms. We report some preliminary results on the effects of (i) changing the supersaturation and viscosity of the solution and (ii) changing the wettability of the cell surface.

2 Experimental

Solutions of hydrochloric acid (AnalaR, BDH) and sodium bicarbonate (AnalaR, 99.5%, BDH) in triply-distilled, de-ionized water were sonicated to ensure complete mixing and then filtered under pressure through 0.2 μm Millipore membrane filters to remove dust particles. All glassware, including the stopped flow apparatus, was cleaned by soaking for 24 h in concentrated nitric acid before being copiously rinsed with triply-distilled, filtered water.

The stopped flow apparatus is shown diagrammatically in Figure 1. Two identical syringes (volume 2 cm^3) were used to rapidly introduce equal

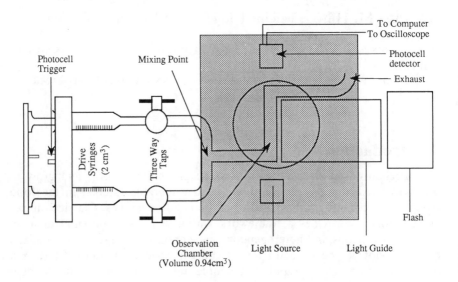

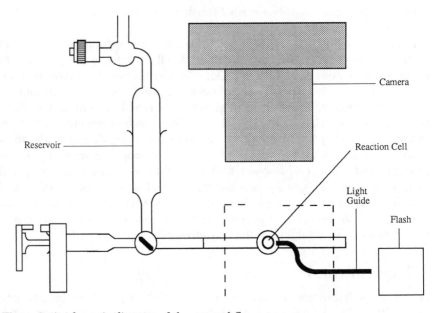

Figure 1 *A schematic diagram of the stopped flow apparatus*

quantities of acid and bicarbonate solutions into a cylindrical glass reaction cell of length 8.53 cm and volume 0.94 cm^3. When the pistons reached the end of their travel, the flow of liquid through the cell stopped, and a small vane interrupted a light beam to trigger the shutter and flash gun of an Olympus OM-2 camera. The camera was mounted vertically over the cell while the light from the flash gun was channelled by a perspex light guide to illuminate the cell horizontally. The delay between the trigger signal and the opening of the camera shutter was measured and found to be 130 ± 5 ms. From the snapshot, the number N of gas bubbles was derived. In order to monitor the turbidity of the solution, a tungsten filament lamp was positioned so that a light beam passed through the cell before falling onto a sensitive, high-speed, large area photocell (RS 303-674). The amplified signal from the photocell went to the analogue input of a microcomputer (Acorn, BBC model B) where analogue to digital conversion was started by the trigger signal, the data being finally presented as values of turbidity as a function of time after mixing.

Variation of the supersaturation ratio (σ) was achieved by altering the reagent concentrations. To obtain the equilibrium solubilities of carbon dioxide, gas at known initial pressures was introduced into a vessel containing a measured volume of solution at 25 °C. The vessel was then shaken vigorously to equilibrate the gas/liquid mixture before the final gas pressure was measured with a pressure transducer. The mass of gas dissolved in the solution at the final pressure was calculated on the assumption that the gas behaved ideally. Quoted supersaturation ratios are based on equilibrium concentrations corresponding to a carbon dioxide pressure of one atmosphere.

Solution viscosities were measured at 25 °C with an Ostwald viscometer using triply-distilled, filtered water as a standard. An allowance was made for the effect of differing solution densities on the efflux times. Surface tensions were determined using Sugden's modification of the maximum bubble pressure method.[6] Two glass capillaries with tip radii of 0.00972 cm and 0.11145 cm were immersed to the same depth in a test solution and the difference in pressure required to form bubbles from the two tips was observed. The surface tension was then calculated by applying the Laplace equation.

The surface of the cell was modified by soaking the clean, dry cell in a perfluorinated hydrocarbon solvent (Freon 113, 3M) containing a perfluorinated acrylate polymer (FC723, 3M) for 1 h. Scanning electron microscopy (JEOL JSM 35C) and contact angle measurements carried out on a glass microscope slide which had been treated identically showed that this treatment produced a smooth, extremely thin, hydrophobic film covering the glass surface. Advancing and receding contact angles were measured with a telescope and goniometer for liquid drops placed on slides tilted at various angles. Equilibrium contact angles were then obtained by extrapolating to zero tilt.

Protein-covered surfaces were prepared by soaking the reaction cell for

3 h with sodium phosphate solutions of β-lactoglobulin (Sigma) or β-casein (90%, Sigma) buffered at either pH 3.0 (hydrophilic surface) or pH 8.9 (hydrophobized surface). Before protein was adsorbed, the surface was preconditioned by leaving the cell filled overnight with the appropriate buffer solution. Concentrations of 0.5 or 1.0 mg ml^{-1} were used, in the expectation that monolayer coverages would be achieved, since Luey *et al.* found[7] that, for β-lactoglobulin, adsorption from aqueous solutions onto either hydrophobic or hydrophilic silicon reaches a plateau value for protein concentrations in excess of 0.25 mg ml^{-1}.

3 Results and Discussion

Gas bubbles in a liquid scatter light imparting a turbidity to the suspension defined by

$$\tau = \frac{\ln (I_0/I)}{l} \tag{1}$$

where I_0 and I are the incident and transmitted light intensities, respectively, of a light beam of path length l. Since the bubble-free aqueous solutions used to generate the carbon dioxide have negligible turbidity, the response of the photocell can be used to calculate the turbidity caused by the bubbles according to

$$\tau = \frac{\ln (S_0/S)}{l} \tag{2}$$

where S_0 is the initial signal from the photocell following mixing of the reagents and S is the signal recorded when bubbles are present. A monodisperse suspension containing N bubbles per unit volume has a turbidity given by

$$\tau = NC \tag{3}$$

where C is the bubble scattering cross-section. For spherical bubbles of radius r, C is related to the geometric cross section by an efficiency factor such that

$$C = Q\pi r^2 \tag{4}$$

Generally, Q is a complex function of bubble size, refractive index of the solution, and the wavelength of the light,[8] but for bubbles large compared to the wavelength of the light and small compared to the cross-sectional area of the beam (as is the case in these experiments), Q approaches a value of 2 (although the existence of a finite angle of acceptance will cause the effective Q to be smaller).

The growth of gas bubbles in supersaturated solutions has been studied theoretically by Epstein and Plesset.[9] If motion of the bubble boundary is neglected, these authors have shown that diffusion-limited growth leads to the relationship:

$$r^2 = r_0^2 + \frac{2D(c_i - c_s)t}{\rho} \tag{5}$$

where t is time, r_0 is the bubble radius at $t = 0$, c_i is the concentration of dissolved gas, c_s is the concentration in a saturated solution, ρ is the gas density in the bubble, and D is the diffusion coefficient of the gas in the liquid. An experimental study[10] of the growth rate of microscopic air bubbles resting underneath a Perspex plate in supersaturated water has confirmed the linear dependence of r^2 on t. Equation (5) can therefore be combined with equations (3) and (4) to give the time dependence of the turbidity of a suspension of growing bubbles, i.e.

$$\frac{\tau}{N} = 2\pi r_0^2 + \frac{4\pi D(c_i - c_s)t}{\rho} \tag{6}$$

Figure 2 shows a plot of (τ/N) against time for solutions with three different initial carbon dioxide concentrations. The linearity of the graphs suggests that the number of bubbles remains contant over the measurement period, implying that nucleation occurs as a short burst in the first few milliseconds following mixing of the reagents. An increase in viscosity is expected to lower the diffusion-controlled growth rate, and Figure 3 shows that this does occur when glycerol is present in a solution with an initial carbon dioxide concentration of 0.3 mol dm^{-3}.

The gradients of the (τ/N) vs. t plots are recorded as a function of initial carbon dioxide concentration in Figure 4. The linear dependence predicted by equation (6) is found. In order to compare the theoretical and experimental values for the growth rates, the carbon dioxide pressure inside the bubbles was assumed to be 1 atmosphere. A diffusion coefficient for CO_2 in water at 293 K of 1.77×10^{-9} m^2 s^{-1} was taken from ref. (11) and corrected to the apropriate temperature and viscosity using the Stokes–Einstein equation. When these values were substituted into equation (6), theoretical growth rates were obtained which were lower than the experimental rates by factors of 8 and 3 for the non-glycerol-containing and glycerol-containing solutions, respectively. The most likely source of this discrepancy lies in the neglect in the theory[9] of the movement of the bubble surface as the bubble grows, since the higher viscosity of the glycerol-containing solution should make this effect less important.

The effect of supersaturation ratio and viscosity on the number of bubbles formed per cm^3 of solution is shown in Figure 5. Increasing supersaturation leads to an increase in bubble formation whilst a viscosity increase reduces the number of bubbles. Extrapolation of the data allows a

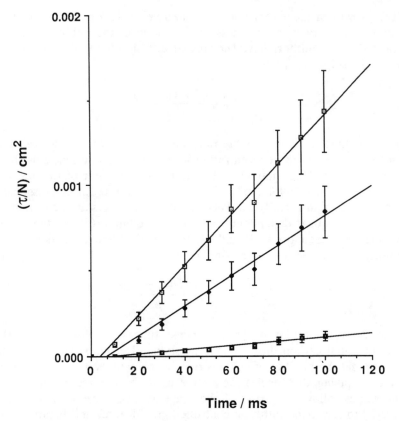

Figure 2 *Turbidity (τ/N) normalized with respect to number of CO_2 gas bubbles per cm^3 of solution following formation of supersaturated solutions. Initial CO_2 concentrations are:* ($\square$) 0.15, ($\blacklozenge$) 0.30, *and* ($\boxdot$) 0.50 mol dm^{-3}

critical supersaturation ratio of 3.3 to be identified. These observations will only be fully understood when more information is available on the nature of the nucleation process. At the levels of supersaturation employed here, nucleation is believed to be heterogeneous.[3] It has been suggested[1,12] that for hydrophilic surfaces nucleation occurs at sites which contain pre-existing gas cells ('Harvey nuclei'). One can envisage that small pockets of air may be trapped in cavities or cracks when the cell is filled with liquid. For liquid surfaces concave on the gas side, an excess pressure will exist on the gas given by the Laplace equation, *e.g.* the excess pressure is approximately 14 bar for a surface tension of 72 mN m^{-1} and a radius of curvature 0.1 μm. The trapped air will not therefore be in equilibrium with a solution saturated with air under atmospheric pressure and over a period of time it will tend to diffuse out of the cavity and into the liquid. Shrinkage of the gas cell will be opposed, however, by an inward flux of carbon dioxide molecules from the supersaturated solution: if this is sufficiently rapid a gas bubble will form and ultimately detach from the site. This model may

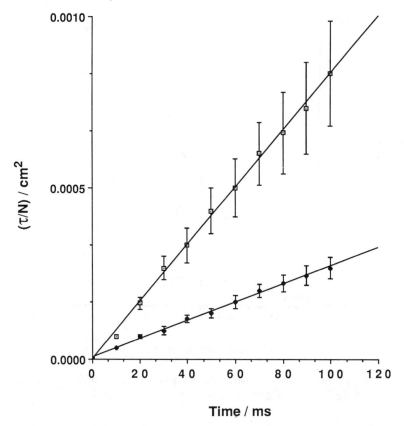

Figure 3 *Effect of added glycerol on the increase in normalized turbidity (τ/N) following formation of a supersaturated solution with an initial CO_2 concentration of 0.30 mol dm^{-3}: (□) no added glycerol, (◆) 25 vol % glycerol added*

account for the effects of supersaturation and viscosity on the number of bubbles formed, since the rate of diffusion of carbon dioxide into the air cells will be critical in determining which cells are active as nucleation sites.

Figure 6 shows the effect on bubble formation of adsorbing a layer of perfluorinated polymer onto the cell surface. This treatment converts the initial hydrophilic surface into a hydrophobic surface since the contact angle for aqueous salt solutions on glass slides treated identically was found to increase from 22° to 103°. It can be seen that the hydrophobic surface nucleates a much greater number of bubbles at a given supersaturation ratio. This effect is expected on the basis of classical nucleation theory which predicts a lower energy barrier to the formation of a critical nucleus on a given surface as the surface becomes less wettable. Alternatively, the observed effect may be related to an increase in the number of cavities in the poorly wetted surface containing entrapped air.

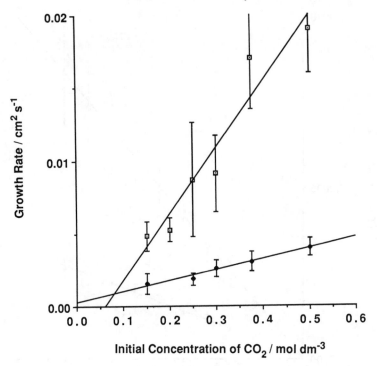

Figure 4 *Growth rates of normalized turbidities plotted against initial CO_2 concentration:* ($\Box$) *no added glycerol,* ($\blacklozenge$) 25 vol% *glycerol added*

To explore further the influence of adsorbed layers on bubble nucleation at both hydrophilic and hydrophobic surfaces, sodium dodecyl sulfate was added to reagent solutions. Table 1 shows the results observed. For both types of surface, increasing the concentration of surfactant lowered the contact angle. This produced the expected decrease in the number of bubbles nucleated for concentrations of surfactant below the critical micelle concentration, but at higher concentrations the number of bubbles began to increase again despite a small continuing decrease in contact angle. This suggests that the micelles provide additional nucleation sites. Otherwise the surfactant action is confined to its effect on contact angle.

Finally, a few preliminary measurements were undertaken on the effects of adsorbed proteins. Two proteins were chosen for study: β-casein, which is a disordered, flexible protein with $pI = 5.2$, and β-lactoglobulin which is a globular protein with $pI = 5.1$. Table 2 shows that surfaces covered with adsorbed protein are less efficient at nucleating bubbles than the corresponding uncovered surfaces. In the case of the hydrophobic surface, a comparison of the two proteins shows that the disordered β-casein is a more effective inhibitor of nucleation than the globular β-lactoglobulin, possibly due to a difference in their modes of adsorption.

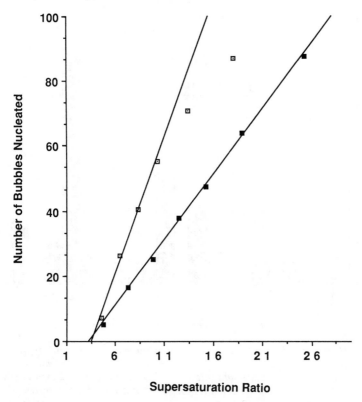

Figure 5 *Number of CO_2 gas bubbles per cm^3 of solution present* 130 ms *after supersaturation is established plotted against supersaturation ratio:* (⊡) *no added glycerol,* (■) 25 vol% *glycerol added. Supersaturations based on CO_2 solubilities at* 298 K *of* 28 mM *and* 20 mM *for solutions containing no glycerol and* 25 vol% *glycerol, respectively*

4 Conclusions

We have devised a simple stopped flow apparatus which allows the number and growth of carbon dioxide gas bubbles in supersaturated solutions to be monitored. After mixing hydrochloric acid and sodium bicarbonate solutions, bubble growth is followed turbidimetrically and the number of bubbles present after *ca.* 0.1 s is recorded photographically. Results obtained for a range of supersaturations typical of carbonated beverages show that nucleation in a glass reaction cell is almost certainly heterogeneous and that the nucleation mechanism probably involves pre-existing gas cells. The number of bubbles nucleated is greatly increased when the cell surface is hydrophobized by the adsorption of a perfluorinated polymer. The effect of adsorbing sodium dodecyl sulfate, β-casein or β-lactoglobulin onto the cell surface is generally to inhibit bubble formation.

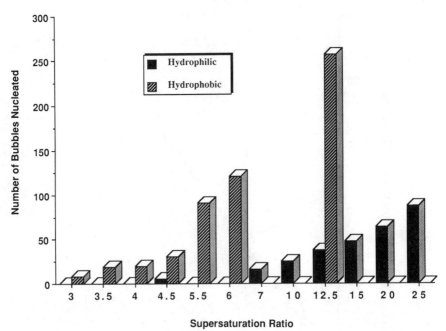

Figure 6 *Histogram illustrating the effect on gas bubble formation of hydrophobizing the cell surface. The number of bubbles per cm^3 present 130 ms after establishment of supersaturation in solutions containing 25 vol% glycerol is shown for various supersaturation ratios and for both the untreated (hydrophilic) and hydrophobized cells*

Table 1 *Effect of sodium dodecyl sulfate on CO_2 gas bubble formation. Numbers of bubbles per cm^3 of solution (N) formed 130 ms after a supersaturation of 4.62 was established in a 0.15 mol dm^{-3} NaCl solution are shown for various surface tensions (γ) and contact angles (θ). Micelles are present for concentrations of SDS below the dashed line*

[SDS] mM	γ mN m^{-1}	Hydrophilic surface θ deg.	N cm^{-3}	Hydrophobic surface θ deg.	N cm^{-3}
0	71.9	33.1	8	99.6	373
0.208	66.8	28.0	2	92.1	172
0.415	45.4	26.0	3	87.8	119
0.830	37.8	17.8	14	79.7	61
1.660	35.3	13.5	49	74.0	86

Table 2 *Effect of adsorbed protein on the number N of CO_2 gas bubbles formed per cm^3 of solution 130 ms after a supersaturation of 6.45 was established. The cell surface was modified by adsorbing protein from buffered solutions containing 0.5 or 1.0 mg ml^{-1} protein at pH 3.0 for the hydrophilic surface and pH 8.9 for the hydrophobic surface*

	N	
	Hydrophilic surface	*Hydrophobic surface*
no protein	60	586
β-lactoglobulin (0.5 mg ml^{-1})	40	–
β-lactoglobulin (1.0 mg ml^{-1})	13	418
β-casein (1.0 mg ml^{-1})	–	137

Acknowledgements

A. M. Hilton thanks the Agricultural and Food Research Council for the award of a Co-operative Studentship sponsored by Unilever Research. The authors gratefully acknowledge the help given by Mr J. Eyett in setting up the photographic equipment.

References

1. R. Cole, *Adv. Heat Transfer*, 1974, **10**, 85.
2. M. Blander, *Adv. Colloid Interface Sci.*, 1979, **10**, 1.
3. P. M. Wilt, *J. Colloid Interface Sci.*, 1986, **112**, 530.
4. M. Blander and J. L. Katz, *AIChE J.*, 1975, **21**, 836.
5. M. Volmer and A. Weber, *Z. Physiol. Chem.*, 1926, **119**, 277.
6. S. Sugden, *J. Chem. Soc.*, 1922, **121**, 862.
7. J.-K. Luey, J. McGuire, and R. D. Sproull, *J. Colloid Interface Sci.*, 1991, **143**, 489.
8. M. Kerker, 'The Scattering of Light and Other Electromagnetic Radiation', Academic Press, New York, 1969.
9. P. S. Epstein and M. S. Plesset, *J. Chem. Phys.*, 1950, **18**, 1505.
10. D. M. J. P. Manley, *Brit. J. Appl. Phys.*, 1960, **11**, 38.
11. A. Prins, in 'Food Emulsions and Foams', ed. E. Dickinson, Special Publication No. 58, Royal Society of Chemistry, London, 1987, p. 36.
12. S. D. Lubetkin, *J. Chem. Soc., Faraday Trans. 1*, 1989, **85**, 1753.

Dilation of Fluid Interfaces in a Vertical Overflowing Cylinder

By D. J. M. Bergink-Martens, H. J. Bos[1], H. K. A. I. van Kalsbeek, and A. Prins

WAGENINGEN AGRICULTURAL UNIVERSITY, DEPARTMENT OF DAIRY SCIENCE AND FOOD PHYSICS, BIOTECHNION, BOMENWEG 2, 6703 HD WAGENINGEN, THE NETHERLANDS
[1]DELFT UNIVERSITY OF TECHNOLOGY, FACULTY OF AEROSPACE ENGINEERING, DELFT, THE NETHERLANDS

1 Introduction

It is important to study the behaviour of expanding liquid interfaces far from equilibrium because such behaviour plays a crucial role in numerous technological processes both inside and outside the food industry. Expanding liquid interfaces can be found during the preparation of foams and emulsions, the nucleation of bubbles, the spreading of droplets, and the application of thin liquid layers like paint, coating, and glue. A better understanding of the behaviour of expanding liquid interfaces in these various cases may lead to improvement of products or production processes.

Experimentally, a pure dilation of a liquid interface can be accomplished far from equilibrium by using the overflowing cylinder technique. The surfactant depletion which occurs in the expanding area generally results in a higher surface tension than the equilibrium value, as demonstrated by measuring the dynamic surface tension by the Wilhelmy plate technique.

The overflowing cylinder technique has been used since the fifties to study expanding liquid–air interfaces. Researchers like Padday[1] and Piccardi and Ferroni[2,3] studied the dilational properties of various fluids at the expanding surface of the overflowing cylinder. More recently, Ronteltap,[4] de Ruiter et al.,[5] and Clark et al.,[6] have successfully used the overflowing cylinder technique in their investigations. In general, the technique is a handy and useful tool in studying surface dilational properties of surfactant solutions. The following example serves to illustrate this statement.

Surface dilational properties are very important during the process of foaming. During this process freshly created thin liquid films are exposed to all kinds of mechanical disturbances. Because these thin liquid films can be made unstable, especially when subjected to expansion, the study of the expanding surface can result in a better understanding of the foaming behaviour of a surfactant solution. This is illustrated by Figures 1 and 2 for the surfactants Teepol and whey protein, respectively. The foam volume, resulting from shaking the aqueous solution in a glass cylinder for ten seconds, is compared to the equilibrium and the dynamic surface tension measured by the overflowing cylinder technique. All three quantities are given as functions of surfactant concentration. Figure 1 shows that, in the concentration region where the equilibrium surface tension has reached its low plateau value, the foamability still increases considerably. The greatest increase in foamability correlates with the decrease in the measured dynamic surface tension. In Figure 2 the foamability is low and constant when the decrease in equilibrium surface tension is the highest, and the foamability increases again when the dynamic surface tension starts to decrease. These data qualitatively demonstrate that the process of foaming benefits from a low surface tension during expansion, which is in agreement with results of Prins[7] who showed that a high surface tension is probably unfavourable for foam stability.

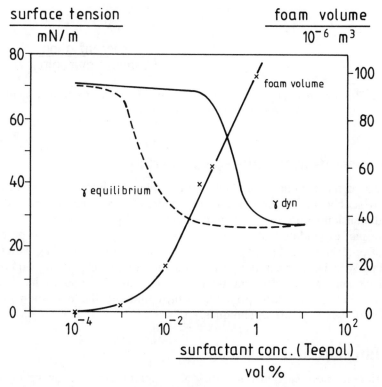

Figure 1 *The dynamic and equilibrium surface tensions and the foam volume as a function of the concentration of Teepol*

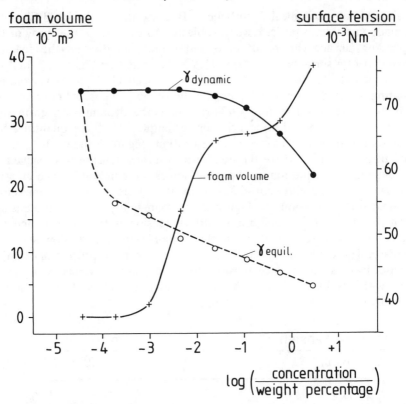

Figure 2 *The dynamic and equilibrium surface tensions and the foam volume as a function of the concentration of whey protein*

Measurements at the Oil–Water Interface

Aspects of the technique applied to the liquid–air interface have already been described in one of our previous papers.[8] Just recently, however, a new dimension to the overflowing cylinder technique has been explored by creating an expanding water–oil interface. This is relevant to the preparation of emulsions. In this paper, the behaviour of the expanding water–oil interface will be compared to the behaviour of the expanding liquid–air interface and the results will be discussed. Finally, conclusions will be drawn about the physical principles which underlie the operation of the two overflowing cylinder techniques.

2 Materials and Methods

Experiments were carried out both on the liquid–air overflowing cylinder (see Figure 3) and on the new water–oil overflowing cylinder (see Figure 4).

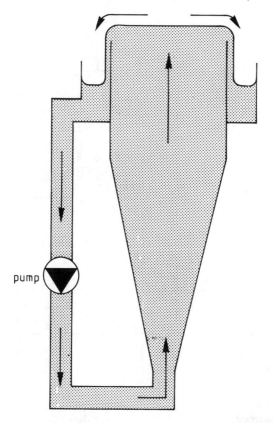

Figure 3 *Cross-section of the liquid–air overflowing cylinder apparatus. Arrows show direction of liquid flow*

The liquid–air overflowing cylinder consists of a temperature-controlled metal cylinder (diameter 8 cm, height 70 cm) through which a liquid is pumped upwards. To ensure laminar flow in the cylinder, the fluid first passes through a conical tube. The liquid is allowed to flow over the perfectly horizontal top rim of the vertical cylinder, causing continuous radial expansion of the circular liquid surface. Figure 5 shows the steady-state surface velocity profile of the expanding surface. Typical relative expansion rates range from about 0.3 s^{-1} for pure water to values at least ten times larger for surfactant solutions. To estimate the magnitude of the surface radial velocity v_r, the motion of small polypropylene particles sprinkled on the surface was followed. The surface tension under these dynamic conditions, γ_{dyn}, was measured in the centre of the circular surface using the Wilhelmy plate technique. The overflowing cylinder forms a closed system containing about 4 litres of fluid. The flow of liquid through the cylinder is kept constant. The length L of the wetting film on the outside of the overflowing cylinder was varied by changing the total

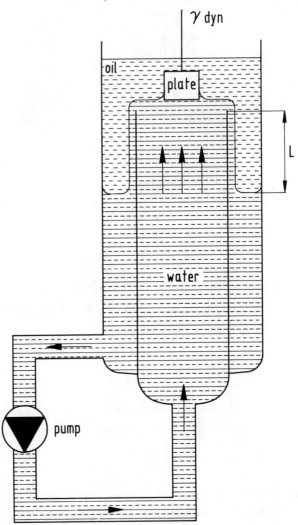

Figure 4 *Cross-section of the water–oil overflowing cylinder. The length* L *denotes the height of the wetting film. The tension* γ_{dyn} *is measured by the Wilhelmy plate technique*

amount of fluid. Both L and the thickness of the wetting film near the rim of the cylinder, δ_0, were measured by means of a screw micrometer.

The water–oil overflowing cylinder consists of an inner glass cylinder (diameter 6 cm, height 50 cm) through which water is pumped upwards. The outer glass cylinder is higher than the inner cylinder, in order to create the possibility of pumping the aqueous phase 'against' an oil phase. The resulting expanding water–oil interface has the same steady state velocity profile as in Figure 5. The mean radial velocity, v_r, of the interface was determined using high density polyethylene particles (density 950 $kg\,m^{-3}$).

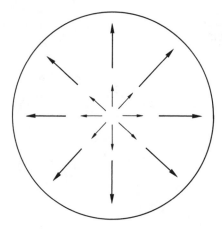

Figure 5 *Topview of the expanding surface of the overflowing cylinder*

The interfacial tension under these dynamic conditions, γ_{dyn}, was again measured by using the Wilhelmy plate technique. The closed system contains about 4 litres of water and 400 ml of oil. The length L of the wetting water film on the outside of the inner cylinder was increased by reducing the total amount of water and adding some extra oil. In the same way, L was decreased by increasing the amount of water, and, if necessary, reducing the amount of oil. Both L and δ_0 were measured as before.

In the liquid–air overflowing cylinder, measurements were carried out with diluted Teepol solution (Shell) in water. In the water–oil overflowing cylinder, a commercially available soya oil (Reddy) was used.

3 Results and Discussion

The Liquid–Air Overflowing Cylinder

The behaviour of the liquid–air surface of the overflowing cylinder is governed by the falling film on its outside. This is demonstrated in Figure 6 for a 0.3 vol% dilution of the Teepol solution. Here the length L of the wetting film was varied, and the corresponding γ_{dyn} was measured. At a length of *ca.* 2.5 cm, a sudden change in behaviour of the overflowing liquid was observed. When the length of the wetting film becomes smaller than 2.5 cm, the decrease in γ_{dyn} amounts to more than 10 mN m^{-1}. This change in behaviour can be explained by considering the boundary layer along the outside wall of the overflowing cylinder. Figure 7 shows the velocity profiles which may be present in the wetting film· at various heights. Following the wetting film on its way down, the boundary layer gradually influences a bigger part of the velocity profile, until the boundary layer reaches the film surface at a length L_{crit}. Thereon, the presence of

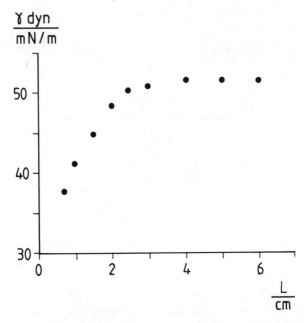

Figure 6 *Dynamic surface tension* γ_{dyn} *of the radially overflowing liquid as a function of the length* L *of the wetting film for a Teepol concentration of 0.3 vol%*

the boundary layer completely determines the velocity profile of the wetting film. In a previous paper the following expression for L_{crit} is derived:[9]

$$L_{crit} \approx \frac{(v_s^0)^2}{2g} \left[\left(\frac{\delta_0^2 \rho g}{\mu v_s^0} \right)^{2/3} - 1 \right] \tag{1}$$

where v_s^0 is the surface velocity of the wetting film near the rim of the cylinder, g is the acceleration due to gravity, and ρ and μ are respectively the density and dynamic viscosity of the liquid phase. For this particular surfactant solution the thickness of the wetting film near the rim of the cylinder, δ_0, was measured to be 0.7 mm. The mean radial surface velocity, v_r, was found to be 8.6 cm s^{-1} at a distance of $r = 2.5$ cm from the centre of the circular surface. From our experiments we know that v_r increases more than linearly with r.[8] We estimate v_r ($r = 4$ cm) to be 18 cm s^{-1} and v_s^0 to be 25 cm s^{-1}. The calculated value of L_{crit} from equation (1) is 2.0 cm. Physically this implies that the boundary layer has become of the same thickness as the wetting film at the point $L = 2.0$ cm. So the wetting film may be considered a free falling film only for $L < 2.0$ cm. This calculated value of L_{crit} corresponds very well with the value of $L = 2.5$ cm where in Figure 6 a sudden change in the behaviour is observed. As L is

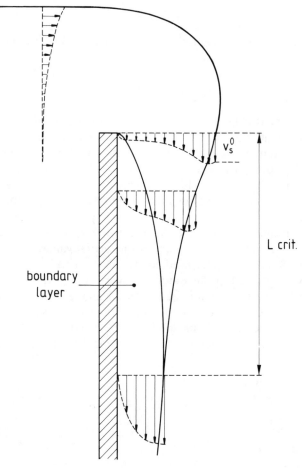

Figure 7 *Cross-section of the wetting film in which the boundary layer and various velocity profiles at different heights have been indicated. (The quantities* v_s^0 *and* L_{crit} *are defined in the text.)*

increased to values higher than 2.5 cm, the length of the free falling part of the film stays the same. As also the measured value of the dynamic surface tension does not change any more from that length on, it is concluded that the value of γ_{dyn} in the centre of the circular surface is dependent on the length of the free falling part of the wetting film.

The Water–Oil Overflowing Cylinder

The question now is whether the behaviour of the water–oil interface is governed by the wetting water film on the outside of the overflowing cylinder in the same way as above? When a liquid is pumped against air, the driving force of the system is the body force

$$F = \rho_w g \qquad (2)$$

the value of γ_{dyn} in the centre of the circular surface is dependent on the length of the free falling part of the wetting film.

$$F = (\rho_w - \rho_o)g \equiv \rho_w g^* \qquad (3)$$

where ρ_o $(0.9 \times 10^3 \, kg\,m^{-3})$ is the density of the oil phase. It should be noted here that the soya oil used was not a purified oil, and so water-soluble impurities from the oil phase could dissolve in the aqueous phase during pumping. So, in the water–oil overflowing cylinder, the aqueous phase was also a dilute surfactant solution. From equation (3) it follows that the two experimental systems are comparable, where the newly defined constant g^* for the water–oil system amounts to $g^* = 0.1 \, g$. Therefore, in the water–oil overflowing cylinder, the gravitational constant may be considered to have one tenth of its actual value.

In Figure 8 data are presented for the case in which the experiment of Figure 7 was repeated in the water–oil overflowing cylinder. The length L of the wetting water film was again varied and γ_{dyn} was measured in the centre of the circular interface. Here, at almost the same length L as in Figure 6, a change in the behaviour of the expanding interface is observed. And again, when the length of the wetting film becomes smaller than

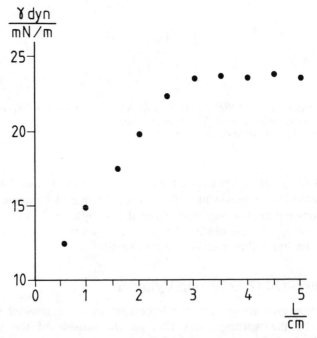

Figure 8 *Dynamic interfacial tension γ_{dyn} of the water–oil interface as a function of the length L of the wetting water film*

2.7 cm, the decrease in γ_{dyn} amounts to more than $10\,\text{mN}\,\text{m}^{-1}$. These results are surprisingly close to the results of the liquid–air overflowing cylinder. However, the thickness of the wetting film near the rim of the cylinder, δ_0, was found to be 2.5 mm, and the mean radial velocity of the interface at a distance $r = 2\,\text{cm}$ from the centre was found to be $0.2\,\text{cm}\,\text{s}^{-1}$, which is almost a factor 50 smaller than the value for the liquid–air surface.

We consider that a good estimate of v_r ($r = 3\,\text{cm}$) is $0.5\,\text{cm}\,\text{s}^{-1}$ and that a good estimate of v_s^0 is $2\,\text{cm}\,\text{s}^{-1}$. This gives $L_{crit} = 1\,\text{cm}$, which is of the same order of magnitude as the value of $L = 2.7\,\text{cm}$ where in Figure 8 the change in behaviour of the expanding interface is observed. This indicates that the boundary layer explanation given for the liquid–air system is also plausible for the water–oil system. This means that, in the case of a water–oil overflowing cylinder, the dynamic interfacial tension is also dependent on the length of the free falling part of the wetting water film. The measured and calculated values of L_{crit} were found to be of the same order of magnitude for both overflowing cylinders, but the values of δ_0 differed considerably, because the values of v_s^0 differed so much. Hence, it may be better to say that, not the length, but actually the weight of the free falling part of the wetting water film determines the behaviour of the expanding interface.

4 Conclusions

The behaviour of the expanding surfaces of the liquid–air and water–oil overflowing cylinder appears to be governed by the wetting film on the outside of each overflowing cylinder. This has been shown experimentally by varying the length of the wetting film and measuring the resulting dynamic interfacial tension in the centre of the circular interface. A physical explanation has been proposed by considering the boundary layer which is built up along the outside wall of the overflowing cylinder. This shows that it is in fact the weight of the free falling part of the wetting film which determines the behaviour of the expanding circular interface.

The value of L_{crit}, the length of the wetting film at which point a change in the behaviour of the overflowing liquid is observed, was found to be of the same order of magnitude for both types of experiment. The same is true also for the decrease in dynamic interfacial tension which occurs when the wetting film becomes smaller than L_{crit}. These results support the statement that the liquid–air and water–oil overflowing cylinders operate according to the same physical principles.

As the free falling part of the wetting film on the outside of the overflowing cylinder determines the behaviour of its expanding surface, it can readily be understood why the data obtained with the overflowing cylinder technique correlate so well with the stability of thin liquid films under dynamic conditions.

Acknowledgements

Part of this research is supported by the Netherlands Technology Foundation (STW).

References

1. J. F. Padday, 'Proceedings of the International Congress on Surface Activity I', London, 1957, p. 1.
2. G. Piccardi and E. Ferroni, *Ann. Chim. (Rome)*, 1951, **41**, 3.
3. G. Piccardi and E. Ferroni, *Ann. Chim. (Rome)*, 1953, **43**, 328.
4. A. D. Ronteltap, 'Beer Foam Physics', Ph.D. Thesis, Wageningen Agricultural University, 1989.
5. H. de Ruiter, A. J. M. Uffing, E. Meinen, and A. Prins, *Weed Science,* 1990, **38**, 567.
6. D. C. Clark, P. J. Wilde, D. J. M. Bergink-Martens, A. J. J. Kokelaar, and A. Prins, this volume, p. 354.
7. A. Prins, in 'Foams', ed. R. J. Akers, Academic Press, London, 1976, p. 51.
8. D. J. M. Bergink-Martens, H. J. Bos, A. Prins, and B. C. Schulte, *J. Colloid Interface Sci.*, 1990, **138**, 1.
9. D. J. M. Bergink-Martens, C. G. J. Bisperink, H. J. Bos, A. Prins, and A. F. Zuidberg, *Colloid Surf.*, 1992, **65**, 191.

X-Ray Photoelectron Spectroscopy of a Protein–Glass Interface

By Muriel Subirade and Albert Lebugle[1]

LABORATOIRE DE BIOCHIMIE ET TECHNOLOGIE DES PROTEINES, INRA, RUE DE LA GERAUDIERE, BP 527, 44026 NANTES CEDEX 03, FRANCE
[1]LABORATOIRE DE PHYSICO-CHIMIE DES SOLIDES, CNRS, 38 RUE DES 36 PONTS, 31400 TOULOUSE, FRANCE

1 Introduction

The interfacial behaviour of proteins at the air–water interface plays an important role in many food and pharmaceutical products in which proteins are used as surface-active agents. The structural properties that are important in controlling the surface activity of proteins are not well elucidated. Most of techniques commonly used to study the protein interfacial structure involve transfer of the floating monolayer to a solid substrate by the Langmuir–Blodgett (L–B) technique.

X-Ray Photoelectron Spectroscopy (XPS), known also as Electron Spectroscopy for Chemical Analysis (ESCA), has been successfully applied for several years to the surfaces analysis of solids. More recently, this technique has been used for studying protein films adsorbed on surfaces in order to provide information on their structural organization.[1-4]

11S Globulin, also known as legumin, is one of the major reserve proteins of pea seeds. A great deal of work has been done on interfacial properties of this protein showing its efficiency as surface active agent.[5,6] Although the interfacial properties of a protein are closely related to its molecular characteristics, there is little information available concerning the detailed physical structure of this protein at an interface.

In the present report the structure of a legumin film deposited onto a glass support using the Langmuir–Blodgett technique has been studied by X-Ray Photoelectron Spectroscopy.

2 Experimental

Materials

Legumin was prepared from pea seeds (*Pisum sativum*) and purified by successive chromatographic separations as described previously.[7]

Formation of Langmuir–Blodgett Films

Monolayer experiments were done using a Langmuir film balance system KSV 3000. An IBM-AT computer-controlled Langmuir trough with a Wilhelmy balance was used to record the π–A isotherm. The surface area A could vary from 720 to 88 cm^2. The L–B films were prepared by spreading the protein solution with a microsyringe onto a phosphate buffer subphase. After one hour of equilibrium, the spread monolayer was compressed at a speed of 10 mm min^{-1} and transferred onto microscope glass slides at various controlled surface pressures using the conventional Langmuir–Blodgett technique.

X-Ray Photoelectron Spectropscopy

XPS measurements were made on an ESCALAB MK II ESCA spectrometer with an AlK$_\alpha$ *X*-ray source (1486.6 eV of photons). Typical operating pressures were 10^{-9} mbar. The glass substrates were cut into pieces (1 cm × 1 cm) for use in the XPS measurements. The binding energy scale was fixed by assigning Eb = 285.0 eV to the –CH$_2$–carbon (1s) peak.

3 Results and Discussion

Compression Isotherm

The compression isotherm, *i.e.* the plot of surface pressure π *versus* mean molecular area A, of the legumin film is given in Figure 1. At the beginning of the compression, below 5 mN m^{-1}, the legumin film shows a little increase in surface pressure with a decrease in surface area. Then the surface pressure increases rapidly from 5 mN m^{-1} to 43 mN m^{-1} as the surface area goes down. Even at this last high pressure, the film does not

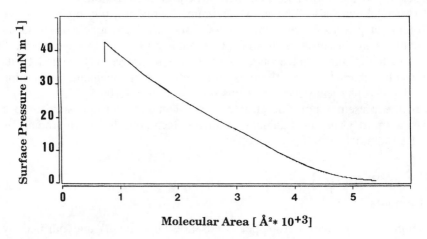

Figure 1 *Surface pressure–area curve of the legumin monolayer*

present sign of collapse. This result expresses the high stability of a legumin film. The mean molecular area is around 5500 $\mathring{A}^2$ at the beginning of the compression and it reaches 800 $\mathring{A}^2$ at 43 mN m^{-1}.

XPS Data Analysis

The general XPS spectrum of the L–B film is shown in Figure 2. It reveals the presence of the characteristic elements of the protein, *i.e.* carbon, oxygen, and nitrogen, as well as silicon coming from the glass substrate. As can be seen, the C_{1s} core-level spectrum is composed of various multi-components attributed to carbon in various environments. From the lowest to the highest binding energies, the main ones are respectively due to aliphatic carbon (–CH$_2$–), central carbon* (C(=O)–C*–NH), and amide carbon (C(=O)–N) of the protein backbone.

The intensities and the binding energies of the photoelectron peaks are reported in Table 1. From the integrated intensities, and with Scofield's relative cross-sections,[8] a quantitative evaluation of the XPS data was performed. The experimental ratios given in Table 1 are close to the ones calculated from the amino-acid composition of the protein.[9] These results show that the XPS technique can provide an accurate measure of protein content. However, we can note a slight discrepancy in the atomic ratio O_{1s}/C_{1s} which may be explained by the contribution of hydroxyl OH groups coming from the silanol groups of the glass substrate which increase the O_{1s} content.

Using the attenuation in the silicon 2p signal intensity due to the protein monolayer, and assuming the inelastic mean free path λ of a photoelectron in a polymer at about 1380 eV to be 28 $\mathring{A}$,[10] the thickness d of the film can be estimated as a function of pressure deposition from the relation

$$d = -\lambda \sin \theta \ln (I_{Si}/I_{Si_0}) \tag{1}$$

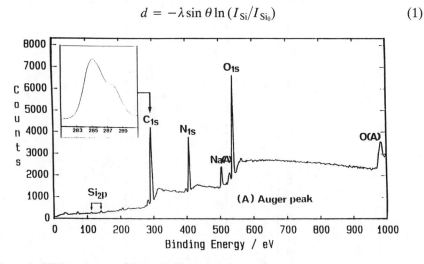

Figure 2 *XPS spectrum of legumin film on a glass substrate*

Table 1 *Binding energies and intensities of the photoelectronic peaks* Si$_{2p}$, O$_{1s}$, C$_{1s}$, *and* N$_{1s}$ *and quantitative analysis*

Core levels	Binding energies (eV)	Glass support without protein	Intensities (kCeV s^{-1}) of the photoelectronic peaks (θ:90) Surface pressure π			Quantitative analysis ($\theta = 90°$, $\pi = 30$ mN m^{-1}) Atomic ratios	
			25 mN m^{-1}	30 mN m^{-1}	35 mN m^{-1}	experiment	theory
Si$_{2p}$	103.1	34.7	2.6	1.0	0.7	0.36	0.33
O$_{1s}$	531.4	286.3	148.2	110.8	104.4	1.0	1.0
C$_{1s}$	285.3		93.3	108.0	108.4	0.27	0.31
N$_{1s}$	400.0		45.7	51.7	50.8		

where I_{Si_0} is the intensity of the signal Si_{2p} obtained without protein, I_{Si} is the signal obtained in presence of protein, and θ is the angle between the sample surface and the analyser (equal to 90° in this experiment).

The protein film thicknesses are 73 Å, 99 Å, and 110 Å for surface pressures of 25 mN m^{-1}, 30 mN m^{-1}, and 35 mN m^{-1}, respectively. The increase in film thickness with the deposition surface pressure, and the decrease in molecular area of legumin observed in the compression isotherm (Figure 1), indicates that the protein is oriented perpendicular to the glass substrate.

The variation in atomic ratios as a function of θ enables depth profiling of the protein layer. As can be seen in Table 2, whatever the deposition surface pressure, the O_{1s}/N_{1s} ratio goes down with the angle θ, whereas the C_{1s}/N_{1s} and C_{1s}/O_{1s} ratios go up. These data indicate a preferential orientation of N and O atoms toward the glass substrate. These results are confirmed by analysis of the ESCA C_{1s} spectrum (not shown). The decrease in aliphatic carbon ($-CH_2-$) content when θ increases implies an orientation of this type of carbon toward the external surface of the protein film. Conversely, the increase in the other carbon signals such as those for C–N, C=O, or C–O with θ suggests an orientation toward the glass substrate.

The XPS data obtained suggest a structural organization of the protein monolayer in which the hydrophobic aliphatic amino acids are at the external surface whereas the hydrophilic amino acids are oriented toward solid substrate. According to these results, the forces involved in the interactions between the glass and the protein film seem to be hydrogen bonding between solid silanol groups and the carbonyl, amine, and amide groups of the protein.

Table 2 *Variation of core level intensity ratios Si_{2p}/C_{1s}, C_{1s}/O_{1s}, C_{1s}/N_{1s}, and O_{1s}/N_{1s} as a function of angle θ and surface pressure deposition*

Angle θ	Atomic Ratios	Surface Pressure 25 mN m^{-1}	30 mN m^{-1}	35 mN m^{-1}
90° ($\lambda_c \sin \theta = 23$ Å)	Si_{2p}/C_{1s}	0.03	0.01	0.01
	C_{1s}/O_{1s}	1.78	2.78	2.94
	C_{1s}/N_{1s}	3.57	3.69	3.78
	O_{1s}/N_{1s}	2.00	1.33	1.31
45° ($\lambda_c \sin \theta = 16$ Å)	Si_{2p}/C_{1s}	0.02	–	–
	C_{1s}/O_{1s}	2.27	3.33	3.33
	C_{1s}/N_{1s}	3.76	3.84	3.84
	O_{1s}/N_{1s}	1.65	1.15	1.15
10° ($\lambda_c \sin \theta = 4$ Å)	Si_{2p}/C_{1s}	–	–	–
	C_{1s}/O_{1s}	3.4	4.18	4.17
	C_{1s}/N_{1s}	4.18	4.30	4.50
	O_{1s}/N_{1s}	1.23	1.03	1.07

4 Conclusion

ESCA is a suitable method for the investigation of a protein monolayer deposited onto a solid substrate. This technique provides information on the chemical elements present in the film, their relative binding energies, and their chemical environment as well as their amount. The angular-dependent XPS studies allow us to obtain the protein film thickness and the structural organization within the protein monolayer.

References

1. C. G. Goelander, V. Hlady, K. Caldwell, and J. D. Andrade, *Colloids Surf.*, 1990, **50**, 113.
2. R. P. Vasquez and R. Margalit, *Thin Solid Films*, 1990, **192**, 173.
3. H. Fitzpatrick, P. F. Luckham, S. Ericksen, and K. Hammond, *J. Colloid Interface Sci.*, 1992, **149**, 1.
4. R. W. Paynter and B. B. Ratner, in 'Surface and Interfacial Aspects of Biomedical Polymers', ed. J. D. Andrade, 1985, Vol. 2, p. 189.
5. C. Dagorn-Scaviner, J. Gueguen, and J. Lefebvre, *Die Nahrung*, 1986, **30**, 337.
6. M. Subirade, J. Gueguen, and K. D. Schwenke, *J. Colloid Interface Sci.*, 1992, **152**, 442.
7. J. Gueguen, A. T. Vu, and F. Schaeffer, *J. Sci. Food Agric.*, 1984, **35**, 1024.
8. J. H. Scofield, *J. Electron Spectrosc.*, 1976, **8**, 129.
9. R. Casey, *Biochem. J.*, 1979, **177**, 509.
10. D. T. Clark and H. R. Thomas, *J. Polym. Sci., Polym. Chem. Ed.*, 1977, **15**, 2843.

Equilibrium and Dynamic Surface Tension of Dough During Mixing

By H. M. W. J. C. Uijen and J. J. Plijter

GIST-BROCADES, PO BOX 1, 2600 MA DELFT, THE NETHERLANDS

1 Introduction

Gas retention capacity and hence good baking quality of a dough may be related to the behaviour of the gluten protein. During mixing, more liquid–air interface is produced, and subsequently the surface-active components have to redistribute themselves. The aim of this study is to obtain more physicochemical information about the surface-active components in dough during mixing.

2 Materials and Methods

The overflowing cylinder technique[1,2] was used to determine the dynamic surface tension behaviour of suspensions of lyophilized dough. The equilibrium and dynamic surface tensions were measured by the Wilhelmy plate method.

3 Results

The equilibrium surface tension of a dough suspension decreases with increasing concentration at constant mixing time as shown in Table 1. A

Table 1 *Equilibrium surface tension γ of dough suspensions as a function of mixing time and dough concentration*

	$\gamma/\text{mN m}^{-1}$		
Mixing Time min	Dough Concentration 0.10 wt %	0.36 wt %	5.0 wt %
2	46.0	41.8	39.0
4	45.0	41.5	40.0
6	42.0	41.5	38.5

decrease in surface tension with increasing mixing time is only seen at the lowest concentration.

The percentage change in surface tension with the flow speed for three samples differing in mixing times is shown in Figure 1. For a 0.36 wt % dough suspension, the dynamic surface tension behaviour of a 2 minutes mixing time sample is typical of that for high-molecular-weight proteins. This is in contrast to the 6 minutes curve which is more typical of lower-molecular-weight surface-active molecules. This phenomenon is in agreement with biochemical results[3,4] relating to S–S/S–H interchange during mixing.

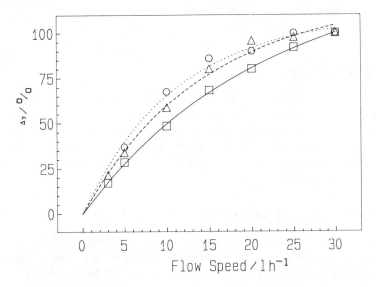

Figure 1 *Change in surface tension,* $\Delta\gamma = \gamma_{dyn} - \gamma_{eq}$, *expressed as a percentage is plotted against the flow speed for three mixing times in a* 0.36 wt % *dough suspension:* □, 2 min; △, 4 min; ○, 6 min

4 Conclusions

The equilibrium surface tension decreases with increasing dough concentration as well as with increasing mixing time at the lowest dough concentration. At the start of the mixing period the dynamic surface tension is determined by the high-molecular-weight gluten protein, whereas towards the end of the period the low-molecular-weight protein fraction is mainly responsible for the surface activity of the dough.

Acknowledgement

This study is carried out as a part of the ECLAIR project on wheat (AGRE 0052).

References

1. G. F. Padday, 'Proceedings of International Congress on Surface Activity', London, 1957, Vol. 1, p. 1.
2. D. J. M. Bergink-Martens and A. Prins, *Materialen*, 1990, **8**, 2.
3. A. Graveland, *Getreide, Mehl Brot*, 1983, **3**, 70.
4. A. Graveland, P. Bosveld, W. J. Lichtendonk, and J. H. E. Moonen, *Biochem. Biophys. Res. Commun.*, 1980, **4**, 1189.

Interfacial Segment Density Profile of Adsorbed β-Casein from Neutron Reflectivity Measurements

By Eric Dickinson, David S. Horne[1], J. S. Phipps[2], and R. M. Richardson[2]

PROCTER DEPARTMENT OF FOOD SCIENCE, UNIVERSITY OF LEEDS, LEEDS
LS2 9JT, UK
[1]HANNAH RESEARCH INSTITUTE, AYR KA6 5HL, UK
[2]SCHOOL OF CHEMISTRY, UNIVERSITY OF BRISTOL, CANTOCK'S CLOSE,
BRISTOL BS8 1TS, UK

1 Introduction

Just as with light, the specular reflection of neutrons from an interface depends quantitatively on the refractive index profile normal to the interface. Neutrons, however, possess a range of advantages over light. Firstly, the wavelengths of neutrons are some two to three orders of magnitude shorter than light, so that the interfacial layers are probed on the scale of small molecule dimensions. A second advantage is the simple relationship between neutron refractive index n and layer composition. This may be expressed as

$$n = 1 - \left(\frac{\lambda^2}{2\pi}\right)\rho \tag{1}$$

where λ is the neutron wavelength and ρ is the scattering length density. The latter quantity is given by

$$\rho = \Sigma Nb \tag{2}$$

where N is the number density of the appropriate nucleus and b its scattering length. Neutrons also possess the distinct advantage that isotopes of the same element may manifest different scattering lengths. The most important two isotopes in interfacial studies are H and D, since they differ very substantially in scattering length. This is exploited by contrast-matching the solvent to air or some other liquid, so that only the solute profile generates the reflectivity function observed. In addition, in favourable circumstances, isotopic substitution may be employed to follow competitive adsorption in systems containing a mixture of adsorbing species.

The advent of dedicated neutron spectrometers has been accompanied by a rapid expansion in applications of the reflection technique to a wide range of chemical systems. Our recent investigation, briefly described herein, represents the first reported study of the protein β-casein adsorbed at both the air–water and oil–water interfaces.

2 Experimental

The essence of the reflection experiment is to measure the neutron reflectivity as a function of the momentum transfer vector κ perpendicular to the reflecting surface. All our experiments were carried out on the CRISP reflectometer at the Rutherford–Appleton Laboratory. Wavelength is varied in this instrument by using a pulsed broad band beam of neutrons and scanning their time of flight so as to vary wavelength λ at the fixed scattering angle θ. A full description of the apparatus and methodology, and its adaptation for implementation at the liquid–liquid interface have appeared previously.[1,2]

The β-casein was prepared in the laboratory from bovine acid casein precipitated from fresh skim-milk. Separation from the other caseins was achieved by ion-exchange chromatography on a Sepharose Q column. Hydrogenous hexane (Aldrich 'Gold Label') was partially deuterated to provide an air-contrast-matched oil for the oil–water interface studies. D_2O was used as supplied by MSD Isotopes Ltd.

Experiments were carried out in a teflon trough specially constructed for the purpose.

3 Theory

The specular component of the reflectivity, $R(\kappa)$, is given by

$$R(\kappa) = \frac{16\pi^2}{\kappa^2} |\hat{\rho}(\kappa)|^2 \tag{3}$$

where κ is the wave vector normal to the interface, defined by $\kappa = (2\pi/\lambda) \sin \theta$, and $\hat{\rho}(\kappa)$ is the one-dimensional Fourier transform of $\rho(z)$, the average scattering length density profile in the direction normal to the interface. In principle, it should be possible to transform the reflectivity data to obtain the adsorbed protein segment density profile. However, as is often the case in attempting to implement a mathematical inversion procedure, the lack of high quality data in ranges of κ generally inaccessible because of experimental constraints precludes this approach. Preliminary analysis of our reflectivity data has therefore been carried out using a Guinier type plot, and a more detailed analysis has been accomplished by

fitting a two-layer model, calculated by the matrix method of Abeles.[1] The
layer boundaries were rendered diffuse by introducing a Gaussian rough-
ness factor, as described by Cowley and Ryan.[3]

4 Results and Discussion

For this series of experiments, a single β-casein concentration of
5×10^{-3} wt % was chosen. This is in the plateau region of the adsorption
isotherm at both the air–water and oil–water interfaces.[4] Two levels of
contrast were used for the unbuffered aqueous phase: (*a*) D_2O only, and
(*b*) air-contrast-matched water. The two sets of data obtained in experiments at the air–water interface are shown in Figure 1, plotted according
to the Guinier approximation. From the extrapolated intercept of the
air-contrast-matched water plot, a value of 3.8 mg m^{-2} is obtained for the
surface concentration of adsorbed β-casein. The slope of these plots gives
the second moment, $\langle \sigma^2 \rangle$, of the adsorbate density distribution function.
The lines being almost parallel, the same value of 2 nm for σ is obtained
for both the contrast variations. This suggests that, as expected, the protein
lies predominantly on the aqueous side of the interface.

The preceding observations are completely model independent. Using a
non-linear least-squares fitting routine, each set of data was fitted independently to a two-layer model. The protein volume fraction distributions
derived as best fits are shown in Figure 2. Though the two profiles are not
too dissimilar, comparison of a reflectivity curve calculated using the
parameters derived for the other profile demonstrates a marked and

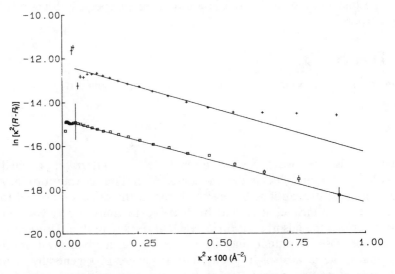

Figure 1 *Guinier-type plot for observed reflectivity behaviour of β-casein at the
air–water interface. Respective subphases are: +, D_2O; and □, air-contrast-
matched water*

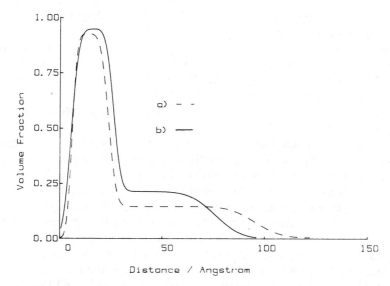

Distance / Angstrom

Figure 2 *Volume fraction profiles of adsorbed β-casein obtained by fitting neutron reflectivity data measured at the air–water interface to a two-layer model: (a) D_2O, (b) from air-contrast-matched water*

systematic deviation from the observed data points, illustrating the sensitivity to structure of the reflectivity technique. The best fit profile parameters are listed in Table 1.

The same two subphase contrasts were used in experiments with β-casein at the n-hexane–water interface, in both cases with air-contrast-matched n-hexane as the upper phase. The presence of this attenuating hexane layer prevented the assigning of a unique value to the adsorbed amount of protein. However, the slopes of the Guinier plots from the two data sets again provide second moment values. The good agreement between the two values suggests that the β-casein does not protrude far into the n-hexane phase. Model fitting again demands a two-layer structure, though

Table 1 *Fitted parameters for adsorbed β-casein segment density profiles*

Interface	Subphase	D_1/nm	ϕ_1	D_2/nm	ϕ_2	σ_f/nm	σ_g/nm
Air–Water	D_2O	1.8 ± 0.2	0.93 ± 0.05	7.2 ± 0.3	0.14 ± 0.02	2.5	2.0
	CM water	2.0 ± 0.2	0.95 ± 0.02	5.0 ± 0.3	0.21 ± 0.03	1.9	2.0
Oil–Water	D_2O	2.0 ± 0.2	0.96 ± 0.03	5.4 ± 0.3	0.15 ± 0.03	1.9	1.7
	CM water	2.0 ± 0.2	0.96 ± 0.03	5.4 ± 0.3	0.15 ± 0.03	1.9	1.7

D_1 and D_2 are the thicknesses of the inner and outer layers; ϕ_1 and ϕ_2 are the corresponding protein volume fractions; σ denotes the square root of the second moment of the adsorbate profile. σ_f is calculated from the model fit; σ_g is obtained from the Guinier analysis on this occasion the same parameters (Table 1) provide fits satisfying both oil–water interface data sets.

The reflectivity behaviour strongly suggests that adsorbed β-casein adopts a similar conformation at air–water and oil–water interfaces. The inner layer, directly at the interface, has a thickness of 2.0 ± 0.2 nm and a high protein volume fraction of 0.96 ± 0.03. The outer layer, extending beyond this into the aqueous phase, is 5–7 nm thick, and has a much lower volume fraction of 0.15 ± 0.03.

Agreement between these neutron reflectivity experiments and existing β-casein adsorption studies is generally very good. The value of 3.8 mg m^{-2} for the surface concentration is within the range of previously published results.[4-6] Bearing in mind the sensitivity of photon correlation spectroscopy to segments present at the very periphery of the adsorbed layer, the layer thicknesses derived in this study show reasonable consistency with the dynamic light scattering data. However, the most important results to be derived from the neutron reflectivity experiments are the estimates of the segment density profile of the adsorbed β-casein normal to the interface. Our predictions here of a dense, thin, inner layer and a diffuse, thicker, outer layer are consistent with the structures inferred previously by Leaver and Dalgleish[7,8] from observations of the accessibility of the adsorbed β-casein to tryptic digestion, and from measurements of the decrease in hydrodynamic thickness of the adsorbed layer during enzyme attack.

The behaviour demonstrated here by the random coil protein, β-casein, contrasts with that observed in a recent neutron reflectivity study of bovine serum albumin adsorbed at the air–water interface.[9] Compared with β-casein, the globular protein was found to adsorb in a much thinner configuration, though again a two-layer model was found to provide the best fit: the dense inner layer was only 1.1 nm thick, with the outer layer extending only a further 2 nm. It is noteworthy, however, that the inner layer of BSA has the same inferred high volume fraction (0.93) as estimated here for β-casein.

A fuller account of this work, including a more extensive comparison with previous adsorption studies, is published elsewhere.[10]

Acknowledgements

The authors wish to thank Dr J. Penfold and the instrument scientists at the RAL for their assistance and SERC for financial support. This research was partly funded by the Scottish Office Agriculture and Fisheries Department.

References

1. J. Penfold and R. K. Thomas, *J. Phys: Condens. Matter*, 1990, **2**, 1369.
2. T. Cosgrove, J. S. Phipps, and R. M. Richardson, *Colloids Surf.*, 1992, **62**, 199.
3. R. A. Cowley and T. W. Ryan, *J. Phys. D: Appl. Phys.*, 1987, **20**, 61.

4. J. Benjamins, J. A. de Feijter, M. J. A. Evans, D. E. Graham, and M. C. Phillips, *Faraday Discuss. Chem. Soc.*, 1975, **59**, 218.
5. J. R. Hunter, P. K. Kilpatrick, and R. G. Carbonell, *J. Colloid Interface Sci.*, 1991, **142**, 429.
6. D. G. Graham and M. C. Phillips, *J. Colloid Interface Sci.*, 1979, **70**, 427.
7. J. Leaver and D. G. Dalgleish, *Biochim. Biophys. Acta*, 1990, **1041**, 217.
8. D. G. Dalgleish and J. Leaver, *J. Colloid Interface Sci.*, 1991, **141**, 228.
9. A. Eaglesham, T. M. Herrington, and J. Penfold, *Colloids Surf.*, 1992, **65**, 9.
10. E. Dickinson, D. S. Horne, J. S. Phipps, and R. M. Richardson, *Langmuir*, 1993, in press.

Dynamic Interfacial Tensiometry as a Tool for the Characterization of Milk Proteins and Peptides

By George A. van Aken and Marjolein T. E. Merks

DEPARTMENT OF BIOPHYSICAL CHEMISTRY, NIZO, PO BOX 20, 6710 BA EDE, THE NETHERLANDS

1 Introduction

One of the research themes at the Netherlands Dairy Research Institute (NIZO) focuses on the preparation of mixtures of peptides by enzymatic hydrolysis of milk proteins. In the past, attention has been primarily directed towards the taste and allergenic properties of these peptide mixtures. The present aim is to extend our knowledge of their physico-chemical properties.

An important physico-chemical property of proteins is their ability to stabilize foams and emulsions. This stabilizing action is mainly due to the formation of an adsorption layer. The protein lowers the interfacial tension and confers upon the interface certain visco-elastic properties. However, once a protein solution is brought in contact with the interface, it usually takes a considerable period of time before the interfacial properties reach their final values.[1] This time-dependence of the interface is due to relaxation processes which can be classified into two general types— exchange of molecules between the bulk and the interface, and dynamic processes within the adsorption layer (such as re-orientations, and progressive unfolding of protein molecules).

Here we report on an explorative study of the dynamic surface tension of bovine serum albumin (BSA) chosen as a model protein. We show that the relaxation behaviour at constant area of a layer of BSA spread at the air–water interface is related to the dilational elastic modulus of that layer.

2 Theory

The variation of the interfacial tension σ and interfacial pressure $\pi = \sigma_0 - \sigma$ resulting from an imposed variation of the area A of the interface is quantified by the interfacial dilational modulus E, defined by

$$E = \frac{d\,\sigma}{d\,\ln A} = -\frac{d\,\pi}{d\,\ln A} \tag{1}$$

For sinusoidal variation of A with an angular frequency ω and a sufficiently small amplitude ΔA, where

$$A = A_0 + \Delta A \cdot \cos(\omega t) \tag{2}$$

it is found that π also varies sinusoidally, but phase-shifted with a phase angle ϕ, i.e.

$$\pi = \pi_0 + \Delta\pi \cdot \cos(\omega t + \phi) \tag{3}$$

This phase difference is due to relaxation processes. By combining equations (1) to (3), we find for these oscillatory variations of A the relation

$$E = E' + E'' \cot(\omega t) \tag{4}$$

with in-phase and out-phase contributions $E' = |E|\cos\phi$ and $E'' = |E|\sin\phi$, respectively, and with $|E| = -\Delta\pi/(\Delta A/A_0)$.

These relaxation processes can also be studied by imposing a step-wise variation of the area. One observes an instantaneous change of the interfacial pressure, followed by a decay of this change towards a new interfacial pressure. We note that, for an insoluble adsorbed layer, in general we have $\pi(t \to \infty) \neq \pi(t = 0)$. A sketch of a decay curve is given in Figure 1.

According to de Loglio et al.[2,3] one can calculate the dilational moduli from the decay curve by Fourier transformation, leading to the expression

$$E' + i \cdot E'' = -\frac{i\omega}{\Delta A/A_0} \int_{-\infty}^{\infty} [\pi(t) - \pi(t = 0)] \exp(-i\,\omega t)\,dt \tag{5}$$

Separation of the real and imaginary parts yields

$$E' = -\frac{\pi(t \to \infty) - \pi(t = 0)}{\Delta A/A_0} - \frac{\omega}{\Delta A/A_0} \int_0^{\infty} [\pi(t) - \pi(t \to \infty)] \sin(\omega t)\,dt \tag{6}$$

and

$$E'' = -\frac{1}{\Delta A/A_0} \cdot \omega \int_0^{\infty} [\pi(t) - \pi(t \to \infty)] \cos(\omega t)\,dt \tag{7}$$

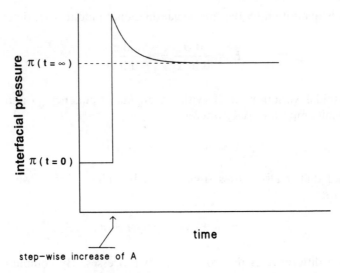

Figure 1 *Sketch of the variation of the interfacial pressure for a step-wise variation of the area*

3 Experimental Results

We have measured E' and E'' of BSA monolayers spread on a 0.1 M phosphate buffer solution (pH 7). The experiments were carried out in a ring trough,[4] which ensures leak-free barriers. For both the oscillatory measurements and the decay measurements, the quantity $|\Delta A / A_0|$ amounted to *ca.* 0.072; and the stepwise increase of the area necessary for

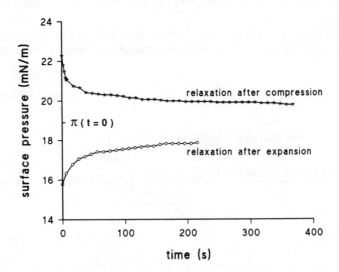

Figure 2 *Measured decay curve for a layer of BSA spread at the air–water interface* ($\pi(t = 0) = 18.9\ \mathrm{mN\,m^{-1}}$, $\Delta A / A_0 = 0.072$)

the decay experiments was achieved in *ca.* 1 second. Figure 2 shows typical decay curves (after expansion and after compression) for a BSA monolayer with an initial surface pressure of $18.9 \, \text{mN m}^{-1}$. Figure 3 shows the dependence of E' and E'' on π. In Figures 4 and 5 the values for E' and

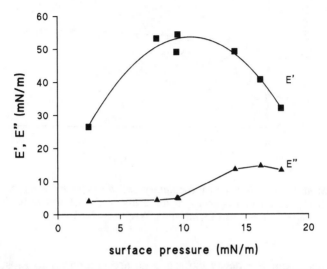

Figure 3 *Measured values of* E' *and* E'' *at an angular frequency of* $0.0628 \, \text{rad s}^{-1}$ *as a function of the surface pressure for a layer of BSA spread at the air–water interface. Values were obtained by oscillatory variation of the area* $(\Delta A / A_0 = 0.072)$

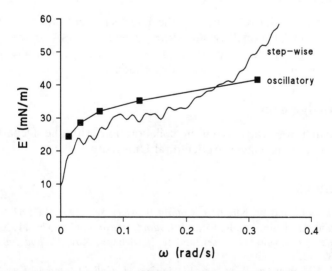

Figure 4 *Modulus* E' *obtained by oscillatory and by step-wise variation of the area as a function of frequency* ω *for a layer of BSA spread at the air–water interface* $(\pi(t = 0) = 18.9 \, \text{mN m}^{-1}, \Delta A / A_0 = 0.072)$

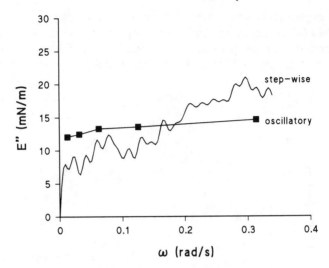

Figure 5 *Modulus* E" *obtained by oscillatory and by step-wise variation of the area as a function of frequency* ω *for a layer of BSA spread at the air–water interface* ($\pi(t = 0) = 18.9$ mN m^{-1}, $\Delta A/A_0 = 0.072$)

E'' obtained from the decay experiments are compared to those obtained from the oscillatory measurements. We see that the two sets of values agree fairly well.

4 Conclusions

We have studied the relaxation of the interfacial tension at constant area and the interfacial moduli of an adsorbed layer of BSA at the air–water interface. Experimental support is given for the theoretically predicted relation between these two types of experiments.

Acknowledgement

This research was carried out in collaboration with the Department of Food Science, Wageningen Agricultural University.

References

1. R. Maksymiw and W. Nitsch, *J. Colloid Interface Sci.*, 1991, **147**, 67.
2. G. Loglio, U. Tesei, and R. Cini, *J. Colloid Interface Sci.*, 1979, **71**, 316.
3. R. Miller, G. Loglio, U. Tesei, and K.-H. Schano, *Adv. Colloid Interface Sci.*, 1991, **37**, 73.
4. A. J. J. Kokelaar, A. Prins, and M. de Gee, *J. Colloid Interface Sci.*, 1991, **146**, 507.

Contribution of Surface and Bulk Rheological Properties to Gas Bubble Stability in Dough

By A. J. J. Kokelaar, T. van Vliet, and A. Prins

DEPARTMENT OF FOOD SCIENCE, WAGENINGEN AGRICULTURAL
UNIVERSITY, BOMENWEG 2, 6703 HD WAGENINGEN, THE NETHERLANDS

1 Introduction

Bread quality is determined by a high bread volume and a fine crumb structure. This quality can be realized if, during mixing, many small gas bubbles are occluded and held in the dough during fermentation and baking. Dough is a foam. Important processes which may cause instability of the foam structure are disproportionation and coalescence. Surface rheological properties as well as bulk rheological properties may contribute to the loss of the gas cells by these processes.

Concerning the surface properties, two parameters are important: the Laplace pressure and the surface dilational modulus. The Laplace pressure Δp is defined as

$$\Delta p = \frac{2\gamma}{r} \tag{1}$$

where γ is the surface tension and r is the radius of the gas bubble. A difference in Laplace pressure between two gas bubbles causes a difference in gas stability, leading to gas diffusion from smaller to larger bubbles. The surface dilational modulus E is defined as:

$$E = \frac{A\,\mathrm{d}\gamma}{\mathrm{d}A} = \frac{\mathrm{d}\gamma}{\mathrm{d}\ln A} \tag{2}$$

where A is the surface area of the deformed surface. The shrinking of a gas cell can be slowed down or even stopped when the surface dilational modulus E exceeds half the value of the surface tension.[1]

The most important bulk rheological property relevant to a growing bubble is the increasing resistance to biaxial extension due to strain hardening.[2]

2 Materials and Methods

Surface Rheological Properties

To imitate a dough system, we prepared 0.5 wt % suspensions of ground freeze-dried dough of the variety *Spring*. The surface tension was measured using the Wilhelmy plate method. The surface dilational modulus was determined as a function of radial frequency using a special Langmuir trough.[3] Experiments were carried out at room temperature.

Bulk Rheological Properties

The resistance of dough (variety *Spring*, water content 35 wt %) towards biaxial extension was determined at 20 °C by 'lubricated' uniaxial compression measurements[4] using a Zwick material testing instrument.

3 Results

The equilibrium surface tension of the 0.5 wt % dough suspension was 42 mN m^{-1}. The surface dilational modulus was *ca.* 35 mN m^{-1} at a frequency of 0.01 rad s^{-1} increasing to about 110 mN m^{-1} at 1 rad s^{-1}.

The measured stress during biaxial extension increased from 0 N m^{-2} at a biaxial strain of 0 to about 2×10^3 N m^{-2} at a biaxial strain of 1.15 (at relevant timescales) due to the strain hardening of the dough.

4 Discussion

Stability towards Disproportionation

To prevent disproportionation, the surface and the surroundings of the gas bubble must be able to exert a certain resistance with respect to deformation. Using relevant values for the different parameters, the surface and bulk stress values given in Table 1 were calculated. It can be concluded that, directly after mixing, the difference in Laplace pressure (*ca.* 1.6×10^4 N m^{-2} for bubbles with a radius of 5 μm and 10 times less for

Table 1 *Comparison of 'surface stress' Δp (Laplace pressure) and 'bulk stress' σ around the gas bubbles in a bread dough during bread-making*

Stage in bread-making process	Gas cell radius $r(\mu m)$	Biaxial strain ε_b	Surface Stress Laplace Pressure Δp (N m^{-2})	Bulk Stress σ (N m^{-2})
mixing	5–50	0	16000–1600	–
first proof	100–135	0–0.3	800–600	0–700
	150–200	0.45–0.75	530–400	700–1000
tin proof	230–300	0.85–1.15	350–270	1000–2000

bubbles with a radius of 50 μm) will cause gas diffusion. The smaller bubbles will shrink and the larger ones will grow. The surface dilational modulus at the relevant timescales is *ca.* 20 to 40 mN m^{-1} which is equal to or higher than half the value of the surface tension (which is 42 mN m^{-1}). This indicates that the shrinking of the small gas cells can be slowed down due to the 'stiffness' of the surface. The modulus will probably increase further as a result of the addition of emulsifiers. Also, it can be seen from Table 1 that, during first proof, the growing of the large cells can be retarded or even stopped altogether because the bulk stress starts to exceed the surface stress due to strain hardening of the dough. These two effects together will result in a rather monodisperse gas cell-size distribution in dough at the start of the fermentation.

Stability towards Coalescence

To prevent the coalescence of two adjacent gas bubbles, the thin dough film between the bubbles must have a certain resistance to extension. The contributions to this resistance are obtained from the film elasticity E_f, which is approximately twice the value of the surface dilational modulus E, and the bulk biaxial stress σ due to strain hardening. The contribution of surface properties to the resistance to biaxial extension of a film with thickness h will be more important than bulk properties if $2E_f > h\sigma$. Using relevant values for the parameters ($E_f = 0.04$–0.2 N m^{-1}; σ (during oven proof) $= 5 \times 10^3$–1×10^4 N m^{-2}), this condition can be satisfied by dough films with a thickness of about 8–80 μm depending on the wheat variety and various other conditions like the water content of the dough. Such films may be present during the last stage of the bread-making process.

5 Conclusions

Directly after mixing and in the early stages of the fermentation process, disproportionation occurs but the process is slowed down due to the 'stiffness' of the surface. Later on, strain hardening of the dough around growing gas cells may also counteract disproportionation.

The stability of dough films to rupture is determined by bulk properties for films thicker than *ca.* 8–80 μm. For thinner films the surface properties may play a dominating role.

References

1. J. Lucassen, in 'Anionic Surfactants', ed. E. H. Lucassen-Reynders, Dekker, New York, 1981, p. 217.
2. T. van Vliet, A. J. J. Kokelaar, and A. M. Janssen, this volume, p. 272.
3. A. J. J. Kokelaar, A. Prins, and M. de Gee, *J. Colloid Interface Sci.*, 1991, **146**, 507.
4. S. H. Chatraei, C. W. Macosko, and H. H. Winter, *J. Rheol.*, 1981, **25**, 433.

Formation and Stabilization of Gas Cell Walls in Bread Dough

By Sarabjit S. Sahi

FLOUR MILLING AND BAKING RESEARCH ASSOCIATION, CHORLEYWOOD,
HERTS WD3 5SH, UK

1 Introduction

Gas cell wall formation and stabilization are crucial events in the processing of bread doughs, but little is known about the molecular aspects of gas cell wall formation and about the composition and properties of the gas–dough liquor interface.

Dough liquor, the liquid phase separated from dough by ultracentrifugation, has been postulated to contribute to loaf volume and crumb texture.[1] In the absence of a dough liquor phase, bread doughs lose their capacity for gas retention.[2] A role for the dough liquor phase in doughs has been suggested in a study which found evidence for small holes in the starch–gluten matrix during proof.[3] It was hypothesized[3] that the foam structure of the dough was kept intact by the presence of thin liquid films stretching over the holes, the liquid films being stabilized by surface-active materials at the gas–dough liquor interface.

Indications are that events at the gas–dough liquor interface of the gas cells of bread doughs influence product quality. These events involve water-soluble proteins and lipids. This report describes some surface properties of dough liquor materials.

2 Materials and Methods

Wheat varieties were selected to represent a range of breadmaking properties. Brock and CWRS1 were chosen to represent poor and good breadmaking varieties, respectively. Avalon was chosen to represent intermediate breadmaking properties.

Dough Liquor Preparation from Bread Doughs

Flour (60 g) and doubly-distilled deionized water (38 ml) were used to prepare doughs in a Brabender Do-corder (DCE 30) at a work input of

11 W h per kilogram of dough; products were centrifuged at 10^5 g for 75 minutes at 30 °C in an ultracentrifuge (MSE Europe 55M). Immediately after centrifugation, the extracted dough liquor was drained off by inverting the tubes for a period of 60 seconds. No attempt was made to scrape off the remaining material from the tube sides. Collected material was freeze-dried and used for further study.

Surface Tension and Surface Shear Rheology Measurements

Solutions of the dough liquor material (1 wt %) were prepared by mixing appropriate amounts of dough liquor solids and doubly distilled deionized water and stirring for 10 minutes at a set speed. Surface tension measurements were made at 30 °C using a Kruss Interfacial Tensiometer K8 employing a du Nouy ring. Surface rheological measurements were carried out at 30 °C using an oscillating ring surface shear rheometer.[4] This instrument works on the principle of the normalized resonance technique. A feedback system is used to shift the resonant frequency and restore the system to resonance at any arbitary frequency within the shifted range.[4] The pre-set frequency in this work was 3.3 Hz and the maximum strain employed was 1%.

Estimation of Lipid Content of Dough Liquor Solids

(a) Total Lipids. The method of Welch[5] involves transmethylation of total lipid, without prior solvent extraction, using concentrated sulfuric acid in methanol. Resultant fatty acid methyl esters (FAME) are then separated and analysed by gas chromatography.

(b) Free Lipids. Dough liquor solids were first extracted with petroleum ether (b.pt. 40–60 °C) and then treated according to Welch's method as described in *(a)*.

Determination of Total Nitrogen

Nitrogen contents of freeze-dried dough liquor solids were determined by the Kjeldahl method.

3 Results

The surface tension measurements in Table 1 show that dough liquor material from Brock, a poor breadmaking flour, produces a much greater reduction in surface tension than does CWRS1, a good breadmaking flour. The same material from Avalon, with intermediate baking properties, produces an intermediate reduction.

Table 1 *Surface tension and lipid and protein levels of dough liquor solids*

Wheat variety	Protein content (wt %)	Total lipids (wt %)	Free lipids (wt %)	Surface tension/mN m^{-1}	
				Dough liquor	Defatted dough liquor
CWRS1	20.0	0.23 ± 0.02	0.07 ± 0.01	48.9	50.0
Avalon	13.2	0.98 ± 0.01	0.57 ± 0.01	43.8	48.3
Brock	15.6	1.13 ± 0.02	0.69 ± 0.01	35.0	46.9

Protein and lipid levels are expressed as wt % of dough liquor solids and are based on at least duplicate runs. Surface tension of water at 30 °C was measured as 71.7 mN m^{-1}. Surface tensions values are averages from triplicate measurements and have accuracy of ± 0.3 mN m^{-1}.

All solvent-extracted dough liquor solids produce higher surface tension values compared with the untreated solids. Extraction with petroleum ether leads to removal of free lipids, suggesting that these materials in the dough liquor are primarily responsible for the surface tension differences observed.

The build up of the surface shear modulus (Figure 1) and the surface shear viscosity (Figure 2) of the three flours was found to be in order of their breadmaking properties. These two factors are a measure of the overall visco-elasticity of the interfacial films. Surface elasticity allows films

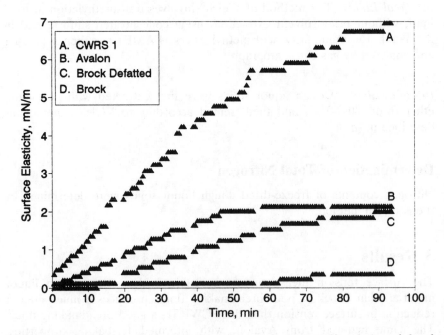

Figure 1 *Plot of surface shear elasticity against time of a* 1 wt % *solution of dough liquor solids*

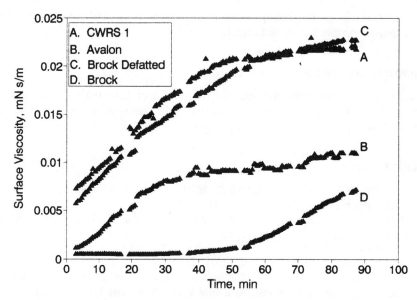

Figure 2 *Plot of surface shear viscosity against time of* 1 wt % *solution of dough liquor solids*

around gas cells to resist rupture by counteracting the stretching of the films. Surface viscosity on the other hand helps to retard the diffusion of molecules close to the interface, thus reducing drainage from films. High values of these two parameters result in more stable films leading to increased foam stability.[6] Mechanically stronger films may contribute to the breadmaking properties of flour by allowing greater bubble expansion before the rupture of the starch–gluten membrane finally occurs to produce a sponge-like structure.

The influence of the lipid material on the rheology of interfacial films was investigated using solvent-extracted dough liquor solids from Brock. Removal of free lipids results in an enhancement of both the surface elasticity and viscosity (curve C, Figures 1 and 2). The destabilizing influence of surface-active lipid-like material on protein-stabilized interfacial films has recently been demonstrated.[7]

4 Conclusions

Surface tension measurements show that dough liquor material from poor breadmaking flours possesses greater surface activity than that from good breadmaking flour. Surface rheological measurements indicate that a good breadmaking flour produces more visco-elastic films than those formed by a poor breadmaking flour. Both surface tension and surface rheology appear to be influenced by the level of lipid present in the dough liquor, higher lipid level producing greater surface activity, but moderating the viscoelastic properties of the interfacial films.

Acknowledgement

I acknowledge the financial support of the Chief Scientist's Group of the Ministry of Agriculture, Fisheries, and Food. The results are the property of the Ministry and are Crown copyright.

References

1. J. C. Baker, H. K. Parker, and M. D. Mize, *Cereal Chem.*, 1946, **23**, 16.
2. F. MacRitchie, *Cereal Chem.*, 1976, **53**, 318.
3. Z. Gan, R. E. Angold, M. R. Williams, P. R. Ellis, J. G. Vaughan, and T. Galliard, *J. Cereal Sci.*, 1990, **12**, 15.
4. M. Sherrif and B. Warburton, *Polymer*, 1974, **15**, 253.
5. W. R. Welch, *J. Sci. Food Agric.*, 1977, **28**, 635.
6. D. E. Graham and M. C. Phillips, in 'Theory and Practice of Emulsion Technology', ed. A. C. Smith, Academic Press, New York/London, 1976, p. 75.
7. M. Coke, P. Wilde, E. Russell, and D. Clark, *J. Colloid Interface Sci.*, 1990, **138**, 489.

Destabilization of β-Casein Foams by Competitive Adsorption with the Emulsifier Tween 20

By D. R. Wilson, P. J. Wilde, and D. C. Clark

AFRC INSTITUTE OF FOOD RESEARCH, NORWICH LABORATORY, NORWICH RESEARCH PARK, COLNEY, NORWICH NR4 7UA, UK

1 Introduction

Food dispersions typically contain two classes of surface-active agents: high-molecular-weight macromolecules (*e.g.* proteins) and low-molecular-weight detergent-like molecules (*e.g.* lipids or emulsifiers). Individually, these two types of surface-active molecule stabilize air–water or oil–water interfaces by distinct and mutually incompatible mechanisms. Briefly, proteins adsorb and interact to form an immobile, gel-like visco-elastic film at the interface, whereas detergent-like molecules adsorb but do not interact and are therefore relatively free to diffuse laterally in the plane of the interface. Several factors can influence which of these two mechanisms controls stability in model food dispersions. In several detailed investigations[1-3] of model systems containing a globular protein and a food emulsifier we have shown that stability is dependent upon: (*i*) the surface activity of the two components, (*ii*) their relative concentration in bulk solution, (*iii*) interactions between the two components, and (*iv*) the strength of protein–protein interactions in the adsorbed film.

The structural properties of the protein component could also influence which mechanism dominates stability. To investigate this possibility, we have examined the foaming and interfacial properties of β-casein solutions in the absence and presence of the emulsifier, Tween 20. The casein fraction of bovine milk represents 80% of the total protein in milk. β-Casein constitutes approximately 30% of the casein fraction;[4] β-casein is a flexible protein which contains very little conventional secondary structure and is classically referred to as a random protein. Previous studies have shown that β-casein can be displaced from the oil–water interface by other proteins[5,6] and surfactants.[7] In addition, the surface films formed by this protein at the oil–water interface have low shear viscosity compared to films formed from globular milk proteins such as β-lactoglobulin and α-lactalbumin.[8]

In this paper we extend some of these investigations to the air–water interface and describe measurements of the stability of foams formed from β-casein alone in the presence of the polysorbate emulsifier, Tween 20.

2 Materials

Surface chemically pure water (surface tension > 72.8 mN m^{-1} at 20 °C) was used throughout this study. β-Casein (C-6905) and fluorescein isothiocyanate (FITC; F-7250) were from Sigma Chemical Co. All other chemicals were of 'AnalaR' grade from BDH Chemical Co. and used without further purification. All experiments were performed in 10 mM sodium phosphate buffer (pH 7.0). The protein concentration was kept constant at 0.2 mg ml^{-1} (8.33 μM) in all experiments. The value R refers to the molar ratio of Tween 20 to β-casein.

3 Methods

Foam stability studies were performed at 20 °C using a conductimetric method as described previously.[1] Air-suspended thin liquid films of the solutions under investigation were formed in a ground glass annulus as described previously.[1] The drainage of the films was monitored visually by viewing through a Nikon Diaphot inverted microscope. The final thickness of the films was measured interferometrically.[1] Lateral diffusion of surface adsorbed FITC-labelled β-casein in air-suspended thin liquid films was measured at different Tween 20 concentrations by the fluorescence recovery after photobleaching (FRAP) technique.[1,3] FITC-β-casein was prepared using the reaction scheme described previously for β-lactoglobulin.[1] Unreacted FITC was removed from the labelled protein by FPLC using a fast desalt column equilibrated in 10 mM sodium phosphate buffer, pH 7.0. The extent of fluorescent labelling was approximately 0.2 to 0.3 moles of FITC per mole of β-casein. Differences in the surface behaviour of FITC-β-casein and unlabelled β-casein were not detected.

4 Results

The stability of foams formed from solutions of β-casein in the absence and presence of Tween 20 was examined by a micro-conductimetric method. A selection of typical foam conductivity decay curves for this system are plotted in Figure 1. In the absence of Tween 20, β-casein formed relatively stable foams. However, addition of very low concentrations of Tween 20 (<1 μM) resulted in measurable reductions in foam stability. The Tween-induced decrease in foam stability is summarized in Figure 2, where the foam conductivity remaining after 5 minutes drainage is plotted as a function of the molar ratio R of Tween 20:β-casein and concentration of added Tween 20. The plot shows a steep fall in foam stability between R values of 0 to 5. Only a gradual decrease was seen between R values of 5

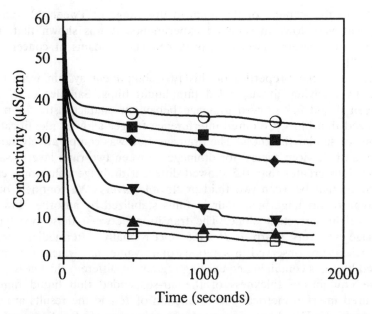

Figure 1 *Foam stability measured by conductivity as a function of time for β-casein in the absence and presence of Tween 20. The molar ratio R of Tween 20 to β-casein is:* ○, 0; ■, 0.1; ◆, 1.0; ▼, 5.0; ▲, 15.0; *and* □, 20.0. *The symbols are not individual data points but are simply to assist in curve identification*

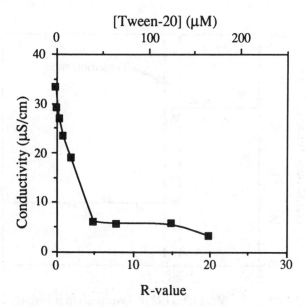

Figure 2 *A plot of foam stability, as determined by foam conductivity remaining after 5 minutes drainage, as a function of molar ratio R and total Tween 20 concentration*

to 20, but the stability of the foam in this range of Tween 20 concentrations was very low. In separate experiments, it was shown that, in the absence of β-casein, Tween 20 produced stable foams at concentrations above 40 μM.

The interfacial properties of this two component system were investigated by studying air-suspended thin liquid films. Samples of R values between 0 and 0.5 showed drainage behaviour characteristic for protein-stabilized thin films. Coloured interference fringes allowed identification of regions of similar thickness. The drainage rate was comparatively slow and exhibited behaviour typical for drainage between two rigid layers. Samples of R values greater than 0.5 showed different drainage behaviour, characteristic of that between two fluid or flexible layers. We routinely observe this type of drainage behaviour in films stabilized by adsorbed low-molecular-weight surfactant alone. The transition between these two types of drainage was very sharp at $R = 0.5$. Occasionally, intermediate drainage properties were observed in solutions of R value 0.5, which were characterized by films containing co-existing regions of different thickness.

The equilibrium thickness of the air-suspended thin liquid films was measured interferometrically as a function of R and the results are shown in Figure 3. The films formed from solutions of $R = 0$ to 0.5 were characterized by an equilibrium thickness of approximately 70 nm. The films formed from solutions of $R > 0.5$ had a characteristic equilibrium thickness of approximately 25 nm. There was a sharp transition in equilibrium thickness at $R = 0.5$.

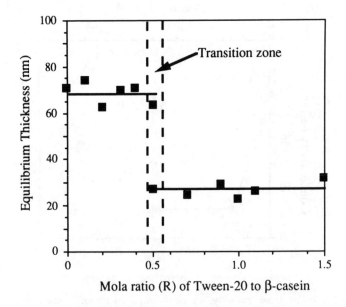

Figure 3 *The effect of molar ratio R on the equilibrium thickness of air-suspended aqueous thin films formed from solutions of β-casein and Tween 20*

The surface diffusion of adsorbed FITC-β-casein was investigated by FRAP. Measurements were attempted in both 25 nm and 70 nm thin films. However, fluorescence was only observed in the thicker of the two films. The lack of recovery of the signal to the prebleach level signifies that lateral diffusion of adsorbed protein did not occur. The absence of fluorescence from thinner films (25 nm) indicates that the surface concentration of β-casein in these films was low and therefore they must have been stabilized by Tween 20. Some films that contained co-existing thin and thick regions were investigated. These films showed consistent behaviour with fluorescence only present in the 70 nm regions and no detectable lateral diffusion of FITC-β-casein.

5 Discussion

The presence of low concentrations of Tween 20 in solutions of β-casein was found to cause significant destabilization of foams. The extent of destabilization was dependent upon the amount of Tween 20 present in the solution. This could be explained by competitive adsorption of Tween 20 at the air–water interface disrupting the weak casein–casein interactions in the adsorbed protein film. We postulate that the protein–protein interactions in adsorbed films of β-casein must be very weak compared to those found with other proteins (*e.g.* β-lactoglobulin[1,2]) since Tween 20 reduces stability of β-casein foam at much lower concentrations. Although β-casein does self-aggregate to form micelles in solution, the absence of 3-dimensional structure compared with other globular proteins may limit its ability to form protein–protein interactions at interfaces. The presence of weak protein–protein interactions of β-casein films at the oil–water interface probably accounts for displacement of this protein by other disordered and globular proteins.[5,6]

We had assumed that instability would continue in the foams until sufficient Tween 20 was added to displace a significant amount of the β-casein from the interface. We anticipated that stability would return to the foams containing $>40\ \mu M$ Tween 20, since Tween alone at this concentration produced stable foams. However, this was not observed (Figure 2). The decrease in foam stability was less pronounced at concentrations of Tween 20 $>40\ \mu M$, but foam stability did not return even in mixtures containing $166\ \mu M$ Tween 20 ($R = 20$). It is difficult to account for this observation. Interaction between Tween 20 and β-casein could in part explain this anomalous behaviour but evidence for such an interaction was not detected in preliminary studies. Indeed, the film drainage and FRAP studies suggest that phase separation occurs in the films of solutions of $R = 0.5$ and greater. Also interfacial tension measurements with β-casein and another non-ionic surfactant $C_{12}E_8$ showed no crossover which indicates the absence of an interaction.[7] Although we did not undertake direct measurements of displacement of β-casein from the air–water interface, the absence of fluorescence from adsorbed FITC-β-casein indicated that most β-casein was displaced from the air–water interface at $R = 0.5$. This is more than 30 times less Tween 20 than required to displace β-casein from the oil–water interface when added after emulsification.[7]

If the surfactant is present before emulsification (*i.e.* more comparable with our foaming studies), complete displacement of β-casein from the emulsion droplets was not observed. It is also notable that major differences are observed in the behaviour of β-lactoglobulin at oil–water and air–water interfaces in the presence of Tween 20.[9]

A mechanism involving abrupt displacement of β-casein from the air–water interface by non-interacting Tween 20 is not sufficient to explain the unusual thin film and foaming properties of this system. An additional step is required to explain the poor stability of the foams when the Tween 20 concentration is $>40 \mu M$. One possibility is that the unusual structure of β-casein allows small amounts of the protein to remain partially adsorbed at the interface. This could impede stabilization of the thin films by the Marangoni mechanism by impeding lateral diffusion of Tween 20 in response to mechanical or thermal disturbances in the films.

References

1. M. Coke, P. J. Wilde, E. J. Russell, and D. C. Clark, *J. Colloid Interface Sci.*, 1990, **138**, 489.
2. D. C. Clark, M. Coke, P. J. Wilde, and D. R. Wilson, in 'Food Polymers, Gels and Colloids', ed. E. Dickinson, Special Publication No. 82, Royal Society of Chemistry, Cambridge, 1991, p. 272.
3. D. C. Clark, P. J. Wilde, and D. R. Wilson, *Colloids Surf.*, 1991, **59**, 209.
4. P. F. Fox, in 'Developments in Dairy Chemistry 4', ed. P. F. Fox, Elsevier Applied Science, London, 1989, p. 1.
5. E. Dickinson, S. E. Rolfe, and D. G. Dalgleish, *Food Hydrocolloids*, 1988, **2**, 397.
6. E. Dickinson, S. E. Rolfe, and D. G. Dalgleish, *Int. J. Biol. Macromol.*, 1990, **12**, 189.
7. J.-L. Courthaudon, E. Dickinson, and D. G. Dalgleish, *J. Colloid Interface Sci.*, 1991, **145**, 390.
8. J. Castle, E. Dickinson, B. S. Murray, and G. Stainsby, in 'Proteins at Interfaces. Physicochemical and Biochemical Studies', eds. J. L Brash and T. A Horbett, ACS Symposium Series No. 343, Washington DC, 1987, p. 118.
9. P. J. Wilde and D. C. Clark, *J. Colloid Interface Sci.*, 1992, in press.

Subject Index